The Woodburners Encyclopedia

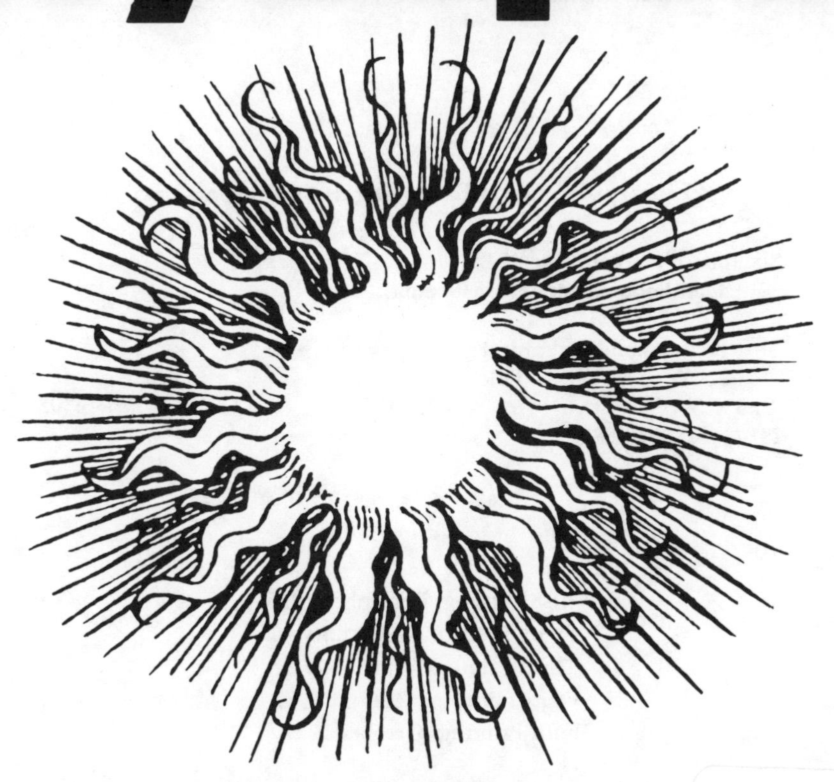

An Information Source of Theory, Practice and Equipment Relating to Wood As Energy.

SECTION ONE

by
Jay Shelton

SECTIONS TWO and THREE

Andrew B. Shapiro
President of The Wood Energy Institute,
President of Vermont Energy Resources

Illustrations by
Vance Smith

Vermont Crossroads Press
Waitsfield, Vermont

International Standard Book Number 0-915248-08-5

Library of Congress Cataloging in Publication Data

Shelton, John Winthrop, 1942-
The Woodburner's Encyclopedia.

Includes bibliographical references and index.
1. Wood as fuel. I. Title.
TP324.S5 697'.04 76-46302
ISBN 0-915248-08-5

First Printing, December 1976
Second Printing, February 1977
Third Printing, April 1977
Fourth Printing, August 1977
Fifth Printing, October 1977

Design & Typography by Jim Dodds

Vermont Crossroads Press, Inc.
Box 333
Waitsfield, Vermont 05673

Table of Contents

Acknowledgements

I am indebted to Larry Gay, Jules Alcorn, Andrew Williams, Richard Hill, Raymond Chany, Richard L. Stone, and Andy Shapiro for their help and suggestions concerning the content of the text; to Vivi Mannuzza for editorial assistance; to Leon Peters for his patient work on the technical illustrations; and to Karen Brownsword for her patient typing of seemingly endless drafts. All contributed significantly to this book, and none is responsible for its shortcomings.

Jay Shelton

The publishers would like to express their sincere thanks to Avery Hall, Consultant, Buildings and Energy Systems, for his technical contributions to the Encyclopedia which appear in Appendix 5.

Introduction

We consume energy whether it be the food we eat or the fuel used to heat houses or power automobiles. The United States with about 6% of the world's population consumes almost 35% of the world's energy. Fossil fuels (coal, oil, natural gas) provide most of our energy. The world supply of these fuels is limited. We need to reduce our energy consumption patterns, and we need to find alternates to fossil fuels.

Wood is a form of solar energy. It is a renewable resource if managed well. Wood will never completely replace other forms of energy, but wood is a good alternative for some energy purposes.

This book is intended to help people understand how to use wood most effectively for heating purposes.

The Publishers

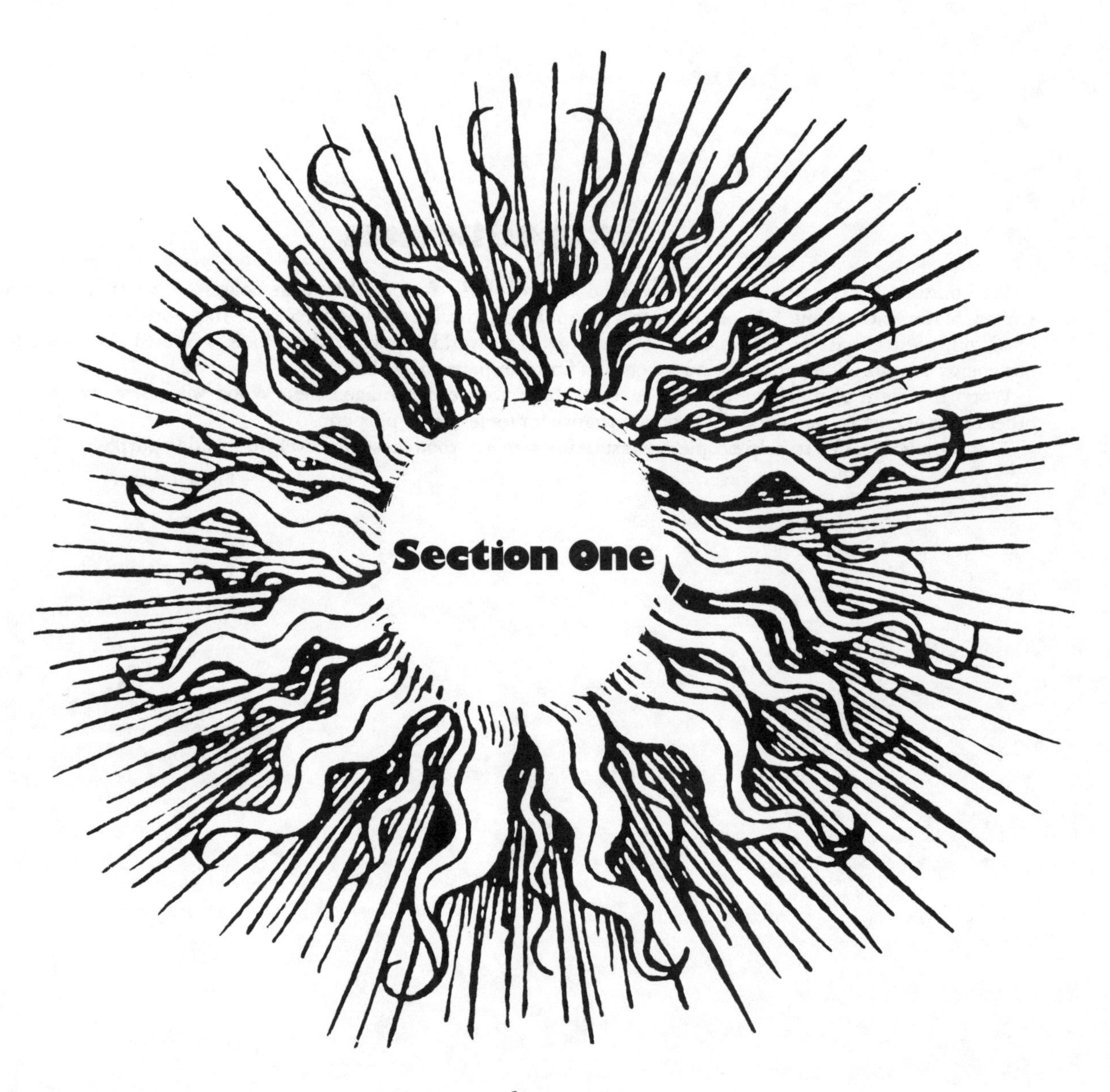

by
Jay W. Shelton

Chapter One
Wood As An Energy Resource

Wood is a beautifully versatile resource. It has always been a fundamental building material, and until the last few centuries, it has been the dominant fuel for heating and cooking in most of the world. Today, 1/3 of the world's population still depends on wood for heating and cooking.[1] Though wood's importance as an energy resource in industrialized countries has fallen, its use in paper making has increased dramatically. In the future, wood will become increasingly important as a raw material for chemicals and plastics, and as a fuel for electrical generation and perhaps even transportation.

In 1970 wood accounted for about 1 percent of all energy used in the United States.[2] The use of wood for heating is no doubt increasing mainly because of the rising cost of fossil-fuel and nuclear energy. Wood heating in America today bears little resemblance to that of 200 years ago in colonial New England, when 30 cords of wood a year might be burned in open fireplaces. Even then much of the house was not very warm. Tighter construction and use of insulation in contemporary houses results in 5 to 10 times less heat loss than from colonial houses. In addition, contemporary closed metal stoves and wood furnaces are probably 2 to 3 times more energy efficient than the open colonial fireplace. Today, the same size house in the same climate can be heated more uniformly using only 3 to 8 cords, which is the sustained annual yield from 5 to 15 acres of woods. Chain saws and trucks have eased the human labor involved in acquiring wood, and controlled burning processes in stoves and furnaces decrease the stoking and reloading effort in the home.

Most wood grown in the U.S. is not used but decays on the forest floor. Forests are not managed for maximum yields. A considerable expansion of wood use is possible. In relatively rural and wooded areas, many more people could heat with wood without serious pressure on wood resources. However, use of wood fuel far from its source remains impractical because of high transportation costs.

Wood is a form of solar energy; the sun's energy made it possible for the trees to grow. Used for house heating, wood has an advantage over heating with solar collector panels because nature has already solved the energy storage problem. The process of photosynthesis stores the energy of the sun in the trees. Compared with oil and coal, wood as a fuel is virtually sulfur-free and thus does not aggravate sulfur-oxide air pollution. Wood also does not contribute to possible carbon-dioxide pollution, which could lead to adverse changes in climate.

A possible problem with wood heating is the incompletely burned gases and particles contained in smoke. Wood (and also coal) heaters emit more smoke than do gas and oil heaters. Some of the chemicals in wood smoke are very unhealthy if ingested or breathed in large quantities. Not enough is known about how much of each chemical is emitted, and how it is dispersed and altered in the environment to be able to assess whether or not the air-pollution problem is serious.

CURRENT AND FUTURE USES OF WOOD

Serious consideration is being given to fueling large central electrical generating plants with wood.[3] One suggestion for such a scheme is to have a 1000 megawatt (electric) plant (typical of large modern nuclear and fossil-fuel plants) in the center of a roughly 30 × 30 mile forested area. This is enough land to supply a generating plant indefinitely. New growth would replenish the harvested trees. Entire trees, including branches, twigs and leaves or needles would be chipped and dried. The dried chips would then be burned to produce heat for a steam turbine which would generate electricity: the major difference compared to existing oil, coal or nuclear plants would be the fuel used to heat water transforming it to steam.

The wood ashes contain most of the nutrients which would have been released if the tree had rotted on the forest floor, the major exception being nitrogen. A significant part of a tree's nitrogen is in its leaves and twigs, and this is one reason for attempting to leave them in the forest. Thus, spreading the ashes back into the forest almost completes the nutrient cycle. The net effect of this proposed scheme is the conversion of solar energy into electricity (and heat); the trees capture and store the solar energy, which is released when they are burned.

Conversion of plant material into liquid and gaseous fuels is another wood-energy scheme which is attractive because these fuels could then be used in many existing furnaces and engines requiring relatively minor modifications. There are four promising techniques: fermentation, pyrolysis, chemical reduction, and enzymatic reduction.

In a moist, warm, and anaerobic environment, some organic materials may be fermented to a gaseous fuel (mostly methane) and carbon dioxide. 60 to 80 percent of the original stored energy is converted in this process. After removing the carbon dioxide and small amounts of sulfur-containing compounds, the gas is essentially identical to natural gas, and thus is readily usable in existing gas-burning furnaces and appliances.

Pyrolysis involves heating the organic matter in the absence of oxygen. This has long been done with wood to make turpentine, methanol, acetic acid, creosote and charcoal. In general, liquids, gases and solids are all formed in this process and all can be used as fuels.

Chemical reduction involves heating under pressure in the presence of water, carbon monoxide and a catalyst. The principal product is an oil with a heating content of about 120,000 Btu per gallon (heating oil has an energy content of about 140,000 Btu per gallon). Enzymatic reduction is not yet as developed as the other three processes. It would involve the use of enzymes (substances produced in living cells) to assist simple chemical transformations of wood to simpler substances.

Wood is also of interest as a source for industrial chemicals. Chemicals derived from wood were more common in the past than they are today. Demand for wood for this purpose is likely to rise, since fossil fuels are now the principal source for the chemicals which could be made from wood. Rayon and acetate are now derived from wood, and it is chemically feasible to make most plastics, synthetic fibers and synthetic rubber from wood.[4] The economic feasibility should constantly improve as research continues and fossil fuel prices rise.

Sixty million tons of wood per year would supply most

of these materials in the U.S. today. That is about the same as the U.S. pulpwood production used primarily for paper products.[5] Since structural and physical aspects of wood matter little if it is to be used to make chemicals, much wood which is unsuitable as pulpwood or timber, could be utilized.

Manufacturing plants in suitable areas may switch to wood as the fuel for "process heat" — heat needed in manufacturing processes. A number of wood-product plants now derive all their space heating and process-heat from their own wood wastes, and increasing interest in wood heat is being shown by all kinds of plants located near available wood supplies. Sawdust, slabs, ends, branches and tops all make excellent fuel.

Wood is a natural but frequently overlooked source of energy. Although today wood constitutes only about 1 percent of the energy consumption in the U.S., in 1900 the amount of wood consumed as fuel was 2½ times what it is today, and constituted 20 percent of the total energy budget. Wood use peaked in about 1870 when it was about 73 percent of the total, and the actual consumption was 3.7 times what it is today[6] (Figure 1-1).

A century ago, wood use was very inefficient; much of it was burned in open fireplaces for heating. Because our energy usage was inefficient years ago the average energy use per person in this country has increased only by a factor of 4 over the last hundred years.[6]

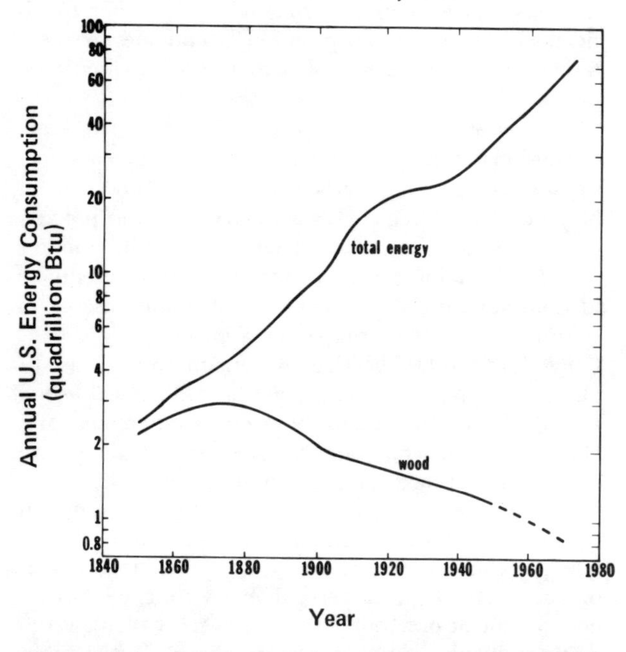

Figure 1-1. Annual total energy and wood energy consumption in the U.S. Wood is assumed to have an energy content of 20 million Btu per cord. [Data from H.C. Hottel and J.B. Howard, *New Energy Technology - Some Facts and Assessments*, (Cambridge, MIT Press, 1971)], Figure 1-1.

ENVIRONMENTAL STRESSES OF SUPPLY AND DEMAND

In many countries there is a very serious firewood crisis far exceeding just a short supply.[7] The Indian subcontinent and central Africa have been hardest hit, but parts of Latin America are also suffering.

Many towns in these areas are dependent on wood for both heating and cooking. A few decades ago the task of collecting a family's daily wood took one person a few hours. But as populations have grown, the cutting has been so intense that a treeless landscape extends out as far as 30 miles from the towns, making a fuel-gathering trip all but impossible to complete in a day. Reforestation has not generally been attempted, and even where it has, the need for fuel is so great that the new seedlings have sometimes been surreptitiously pulled up to be burned.

Serious side effects accompany such intense pressure on wood resources. The exposed soil erodes, loses nutrients and can hold less water, making it more difficult to reforest. The larger and quicker runoff of precipitation intensifies flooding downstream. Perhaps most serious, as wood becomes too hard and expensive to obtain, people turn to dung as fuel — dung which previously was spread on the land as fertilizer for crops. In areas where chemical-fertilizer replacement is not feasible, food production declines and people starve. Thus, in large portions of the world, the current energy crisis is a wood crisis, and it is directly related to soil erosion, flooding and starvation.

These serious problems are unlikely to arise in America for a number of reasons. 1) Because of our economic strength and natural resource base, we have many sources of energy to choose from — oil, natural gas, coal, uranium, geothermal, tidal, and the many forms of solar energy (wood, and other energy crops, wind, hydro,[8] direct solar heating, ocean thermal, photo-voltaic, ocean waves, etc). We will never be wholly dependent on only one energy source. This flexibility permits shifting our energy mix in response to changing supplies, environmental problems, and economics. The day will probably never come when everyone depends on wood, or any other single source, for heating. Wood heating will always be a regional or local phenomenon. In rural forested areas, wood-heating is likely to become more common than it already is. In New York City, wood is never likely to be the most reasonable source of heating energy. (Perhaps some of the city's electrical energy may be supplied by wood-burning power plants.) Wood heating may not be viable in parts of the Southwest due to the scarcity of trees which is related to the scarcity of water. 2) Some of this country's forested areas are very stable ecologically. In New England, soil and climate conditions are such that trees are like weeds. If the land is not managed, it becomes a forest. Pastures revert to woodlands in a few decades. 3) Environmental concern and legislation will probably become stronger in this country and destructive forestry practices will be curbed.

The present demand for wood is high and will increase. Wood will always be an important resource for lumber, paper, chemicals, as well as fuel. But there is a great deal of wood which is not used for any of these purposes. For instance, it is estimated that in Vermont about one fourth of the wood growing has no commerical value (as timber or pulpwood). The harvesting rate of commercially valuable trees is about half of their growth rate. In addition, only about half of the above-ground mass of each tree is taken out of the forest using conventional harvesting techniques — the tops and branches are left behind. Thus, the actual annual sustained growth of all wood in Vermont is roughly 5 times the current annual harvest, leaving considerable room for expansion of the current harvest.[9]

Current growth rates in most natural forests range from ¼ to ¾ cord per acre, per year. This sustained yield could probably be doubled by forest management — including thinning, cutting of less productive individual trees, selecting for (or planting of) more productive species, and harvesting trees at the optimum time. Some species can send out new sprouts from their cut stumps; the already established root system provides such abundant supplies of nutrients that the sprouts grow much more quickly than new seedlings would. The result can be a higher yield of wood per acre.

In the South, tree *farming* is common. Particularly high yield strains are planted in rows and nurtured as carefully as many food crops.

Of course, more intensive use of forests will have environmental effects. Nutrient loss from the soil is a potential problem, especially if the whole above-ground tree is harvested since the nutrient-rich small branches, twigs, and leaves would not be left in the forest. If tree roots are also harvested, as has been proposed, the soil would be loosened and soil erosion could be a problem, especially on slopes. New access roads also frequently cause increased erosion. Nutrient losses can be replaced through fertilization. Erosion is more serious, and as a result tree harvesting will have to be limited or even prohibited in some areas.

There is no denying the environmental impact of greatly increased wood use. The questions are how, and how much, to limit the impact and how the impact compares to that of the alternatives (oil, coal, nuclear power, etc.). With proper care, wood can make a significant contribution to our energy consumption.

WOOD VERSUS FOSSIL FUELS

Wood and other proposed plant fuels offer some outstanding advantages compared to other sources of energy. Foremost is their renewability. Wood energy is renewable in the time it takes to grow a tree. The sun is the ultimate source of the energy. As long as the sun shines and the earth is a healthy place for life wood will be available if it is harvested reasonably.

All forms of solar energy including wood, wind, falling water and ocean temperature differences, as well as direct radiation, have this property of being continuously available or "renewable" (as opposed to fossil fuels which can be used up), but wood is outstanding, compared to other solar options due to the conveniently stored form of its energy. House heating with direct solar energy is made more complex by the intermittency of sunshine, which creates the need to store heat for use at night and on cloudy days. In trees and other plants, nature has provided us with collectors, the leaves with their chlorophyll collect, and the plant stores. A pound of wood has more than 50 times the amount of stored energy than does a pound of hot water in a solar storage tank, and wood doesn't need to be insulated to hold its energy.

The air pollution created by wood combustion is different than the pollution created by fossil fuels. Wood contains very little sulfur — .01 to .05 percent. Oil and coal typically contain 1 to 3 percent sulfur. The emission of sulfur oxides from wood burners is very small. Oxides of nitrogen are also virtually absent from wood smoke. However, the emission of particles and unburned compounds can be substantial. In industrial size burners these pollutants can be controlled — combustion can be complete enough to consume most burnable compounds, and conventional technology can control particles. However, most domestic wood stoves and fireplaces emit a considerable amount of smoke. Unfortunately there is apparently no detailed information about these emissions. A vast number of different chemicals (Table A3-1) have been identified as ingredients in wood smoke generated in particular circumstances and many of these are known to be potentially hazardous to health; many are carcinogenic. The amount of each chemical emitted from typical wood fires, and the way each is dispersed and modified in the environment are unknown. If many people heat with wood in a region of high population density and where air can stagnate, visual pollution can certainly result, but further research is needed to determine if there is a serious health hazard as well.

The most interesting and perhaps significant comparison between wood and fossil fuel emissions concerns carbon dioxide — a compound not normally considered a pollutant at all. If the amount of carbon dioxide in the atmosphere changes significantly, the climates of the earth may change. Carbon dioxide is an absorber of the infrared ("heat") radiation coming from the earth's surface. The loss of energy from the earth in the form of infrared radiation prevents the planet from overheating. Increasing the amount of carbon dioxide in the atmosphere would further inhibit the loss of infrared radiation from the earth, which could lead to global warming. This is often called the "greenhouse effect," since glass also absorbs infrared radiation.[10]

The consequences for climates around the world are perhaps impossible to predict because of the incredible complexities of the atmosphere and its interactions with land and oceans. A warming trend is the most common guess, but this could be accompanied by cooling over parts of the earth, due to shifts in global circulation patterns. Precipitation would be likely to increase in some regions and decrease in others. The consequences of an increase in carbon dioxide in the atmosphere are unpredictable, even if carbon dioxide were the only atmospheric pollutant, but other pollutants, such as particles, may also affect climates.

This lack of predictability is no consolation. Although the changes could be benign in the long run, they are likely to be disruptive particularly for agriculture. As food supplies tighten with increasing population and rising material living standards, more and more marginal agricultural areas will be brought into production. In such areas, a slight change in climate can result in large decreases in yields. It is estimated that a decrease in the mean annual temperature of 1 °C in Iceland would decrease the yield of forage by 27 percent, and the observed 60 percent decrease in carrying capacity for cattle (and formerly bison) in the U.S. western high plains is apparently attributable to a 20 percent decrease in annual precipitation.[11] Of course, in some areas, climate changes would undoubtedly increase production. However, a general warming trend could ultimately result in the flooding of much prime agricultural land because of a rising sea level.

Adding carbon dioxide to the atmosphere may or

may not contribute to mass starvation. But it is certain that we will not know for many decades what the effects will be, by which time it may be too late to do anything about it. Thus the carbon dioxide "problem" like so many environmental problems, involves the *possibility* of a grave situation in the future. Almost no reliable predictions now exist about its probability or seriousness. Yet, if preventative action is to be taken, it must be started now.

Shifting from fossil fuels to wood where feasible would help but not because wood emissions are free of carbon dioxide. Both fossil fuels and wood contain carbon, and when burned, most of the carbon is incorporated in carbon dioxide molecules. The more complete the combustion, the more carbon dioxide is produced. Separating, and storing or chemically altering the carbon dioxide is unfeasible.

The critical difference between wood and the fossil fuels is whether the carbon dioxide would have been generated and released even without our burning the fuel. In the case of wood, it would. If plant matter is left on the ground and allowed to decay, the same amount of carbon dioxide is released as would have been if the plant had been burned. Burning wood releases the carbon dioxide a few years sooner, but there is no long-term effect on the amount of carbon dioxide in the atmosphere.

All of the carbon in plants is obtained from carbon dioxide taken out of the atmosphere by growing plants. Thus, in using wood on a sustained-yield basis (harvesting and burning it at no more than the rate at which new growth is occurring), as much carbon dioxide is taken out of the air by growing plants as is released in burning them.

There are two other quantities whose natural balance is not upset when wood is used for fuel on a sustained-yield basis, oxygen and heat. When any fuel is burned, oxygen from the air is consumed. Significant oxygen depletion is not a serious possibility. However, whereas the burning of fossil fuels results in a net disappearance of oxygen from the atmosphere,[12] the use of wood as a fuel does not. While growing, a tree releases the same amount of oxygen as it will consume when burned. If left to decay in the forest, a tree consumes the same amount of oxygen in the process of rotting.

A potentially more serious effect of combustion of any fuel is the heat released. The climate of large urban areas (e.g., parts of the Northeast) are warmer by a few degrees because of this heat. Wood again is unique compared to fossil fuels. Whether wood is burned or left on the ground to rot has no net effect on the amount of heat released. As wood decays, heat is very slowly released, but the total amount is the same as if the wood were burned. The combustion of fossil fuels releases heat which would otherwise have been essentially locked up forever.

CONCLUSION

Nature has been growing and decaying trees for hundreds of millions of years. The solar energy locked into chemical form in the plants through photosynthesis has always been released mostly as heat during decay. By taking the wood and burning it in a stove, the heat is released in a home rather than on the forest floor. The stored solar energy ends up in the same form in either case; burning the wood merely reroutes the energy through a house on its way back into the atmosphere and eventually back into space. The amounts of carbon dioxide and oxygen in the atmosphere are not significantly affected, whether wood is burned or left to decay. The net effects are the changes in the forests and their watersheds due to harvesting, and the emissions of chemicals and particles from incomplete combustion of wood.

Chapter Two
Energy, Temperature and Heat

ENERGY

The concept of energy is elusive because energy is not a substance that can be seen and touched. The concept "apple" is much easier to learn — a person can be shown one, and he can touch, smell and eat one. Energy is not a material substance, but a property of objects and phenomena.

Most of the energy used to toast bread in American homes came from the sun. It originates there as nuclear energy of hydrogen. When hydrogen nucleii combine deep in the sun (in nuclear reactions similar to those in hydrogen bombs) the nuclear energy is released mostly as heat. This makes the sun hot, which in turn causes it to radiate energy (sunlight) out in all directions in the solar system. The earth intercepts a minute fraction of this radiant energy. Some of it is absorbed in green plants and converted by them into the chemical energy contained in their own molecules. This organic matter sometimes accumulates on the ground or bottom of a lake and ultimately gets buried under mud and sand which become rock in geological time. This happened long ago, and today we mine the organic material, since transformed into oil or coal. Burned at a power plant, the chemical energy is released as heat again, the same form it had once on the sun. The heat is transferred to make steam. When passed through a turbine generator, some of the steam's energy is converted into electricity. An electric toaster contains small wires, which, when electric current is passed through them, convert the electric energy into heat. This makes the wires glow red hot, and some of their heat is then transferred to the bread to toast it. Ultimately, all the heat generated by the toaster finds its way out of the house and into the outdoor air. Its final fate is to travel as infrared radiation back into space. Since space is relatively empty, the energy is unlikely to encounter any matter or undergo any more transformation.

Of the many forms of energy, the most fundamental is motional or "kinetic" energy. The higher an object's speed and the heavier it is, the more kinetic energy it has.[1] Any moving object has kinetic energy — a traveling car, Mars (due to its orbital motion about the sun), and even individual molecules in this book or in the air.

Individual atoms and molecules are in constant random, jostling motion. In solids, each atom is constrained to vibrate around a fixed location, but its speed can still be very high. In liquids the molecules can migrate, but because they are so close together (essentially touching) the motion is not very free; most of the motion is like a shaking or oscillation, as in solids. In a gas like the atmosphere, molecules are far enough apart (about 10 times the diameter of a molecule) that each molecule can travel a distance of many molecular diameters before colliding with another and bouncing off in a new direction. This molecular motion is essentially perpetual.

Kinetic energy is energy by virtue of motion. There are three other fundamental forms of energy which are different kinds of "potential" energy: gravitational, electrical and nuclear. They are potential in the sense that they are potentially kinetic energy: potential energy can be converted into kinetic energy. A rock held above the ground has gravitational potential energy; if it is let go, that potential energy is converted into kinetic energy. Fuel represents a form of electrical potential energy called chemical energy; when the fuel is burned, atoms in the fuel and oxygen from the air rearrange themselves into new molecules. This process is a chemical reaction. During the rearrangement, strong electrical forces act releasing stored potential energy. The forces accelerate the atoms which increases their kinetic energy. In experimental nuclear fusion reactors, nucleii of atoms are made to come close enough together that the very strong nuclear attractive force can act, releasing stored nuclear potential energy.

One of the most fundamental facts about energy transformations is that energy is never lost or created. The total amount of energy (of all kinds) is always the same before and after a reaction or transformation. Thus when a pound of wood is burned, all of its approximately 8600 Btu of energy will be accounted for in some form, including light, infrared radiation, chemical energy (combustion is never perfectly complete), latent heat (p. 14) and ordinary (or "sensible") heat.

HEAT AND TEMPERATURE

Heat is a form of energy. In fact, it is the kinetic energy of molecules due to their random jostling motions.[2] The amount of heat in any object is the sum of the kinetic energies of all its molecules. Temperature is the intensity of heat on a molecular scale. The temperature of a substance is directly proportional to the average kinetic energy per molecule. The faster the average speed of molecules in their random motions, the higher is their temperature.

Heat and temperature are different concepts. A group of molecules with a high average speed (kinetic energy, technically) is hot — they have a high temperature; but the quantity of heat they represent depends on how many of them there are. If there are only a few, there is very little heat (their total kinetic energy is small). A group of molecules whose average speed is low has a lower temperature, but if there are many of them, the amount of heat they represent can be large. Thus, for instance, a given amount of heat can be stored either in a small tank of hot water or in a large tank of warm water. Aluminum foil in a hot oven is as hot as the oven; but because it is so thin, it doesn't contain much heat. When fuel is burned, the chemical forces causing the atomic rearrangements accelerate the atoms, which increases their speed. Since the average speed of each atom is increased during the reaction, the temperature of the reactants increases (flames are hot!). Since the total kinetic energy of all the molecules has increased, there is more heat (flames generate heat).

Once released, energy in the form of heat does not easily change into other forms of energy. The heat in a candle flame spreads over its surroundings (e.g., the air), and the temperature of the heat decreases as it does so. On a molecular level, the high kinetic energy per molecule in the flame is shared with the vastly larger number of molecules in the candle's environment, increasing their energies very slightly. The total amount of kinetic energy (heat) is not changed, but being shared among a much larger number of molecules, each has less; thus the temperature of the heat is less as it spreads out.

The heat used to warm houses is constantly moving out of the house and into the air. Its total quantity is unchanged in the process, but its temperature decreases. Trying to heat a poorly insulated house is frustrating, because, among other things, it is almost literally an attempt to heat the outdoors.

All objects contain heat because their molecules have some kinetic energy. The colder matter becomes, the less kinetic energy it has per molecule, and hence the less heat it has. [There is a special temperature, absolute zero (-460°F), at which molecules have virtually zero kinetic energy.] It may seem strange to speak of a cold (e.g., 32°F) object containing heat, but a colder object, say at 0°F, could be warmed by touching the 32°F object.

CONDUCTION AND CONVECTION

There are a number of ways heat moves from one place to another. Heat is conducted up a silver spoon from a hot cup of tea to your fingers. Most of the heat transferred through solid air-tight walls in a house is via conduction. Heat is always conducted from hotter to cooler regions. The mechanism is molecular collisions and interactions.[3] Whenever an energetic (hot) molecule collides with, or vibrates against, a less energetic one, their energies tend to be equalized in the collision. The lower end of the silver spoon is the hottest; thus throughout the length of the spoon, energy tends to be transferred upwards in all the little collisions and interactions. There is no net motion of matter in heat conduction. Kinetic energy of molecules, or heat, is transported from one end of the spoon to the other, but no silver atom moves more than a very small distance from its location in the spoon — just far enough to touch its nearest neighbors and pass along the kinetic energy.

Materials differ in their inherent ability to conduct heat — in their "conductivity." Conductivity is a measure of the rate at which heat is conducted through a slab of material whose two sides are maintained at a constant temperature difference. Conductivities of some materials are given in the middle column of Table 2-1. The range of values is enormous, from 245 for silver to .02 for glass-wool insulation. Materials, such as glass-wool, which have low conductivities are good insulators. Cast iron and steel have virtually identical conductivities. Wood is a moderately good insulator, and charcoal and ashes are even better.

In contrast to conduction, convection *does* involve transport of matter along with heat. In convection, heat is transported by moving hot matter. Hot air heating systems distribute their heat via convection — by circulating the hot air. The air molecules and heat move together. If a blower or fan (or pump in the case of liquids) is involved, the convection is said to be "forced." If the hot air moves only under the influence of its own natural buoyancy, the heat transport is called "natural" (or "gravity") convection. Natural convection tends to move the warmest air to the ceiling of a room. Forced convection from a fan in the room can force the heat to be more evenly distributed.

RADIATION

Heat can also be transported via radiation. There is really no such thing as radiant *heat*. Radiant *energy* is infrared electromagnetic waves. Light, radar, X-rays, microwaves and radio broadcast waves are all forms of electromagnetic waves. Infrared radiation itself cannot be seen, for it is beyond the deepest red of the visible spectrum, and it is not heat, since it has nothing to do with motion of molecules; in fact, infrared radiation travels most easily in a vacuum. However, if some of it is absorbed on a surface, such as one's skin, its energy becomes heat, which one can then feel. Infrared radiation travels at the same speed as light

	SPECIFIC HEAT (Btu/lb°F)	THERMAL CONDUCTIVITY (Btu/hr sq. ft°F/ft)	DENSITY (lb/cu. ft)
Wood	.5-.7	.06-.14	25-70
Charcoal (Wood)	.20	.03	15
Ashes (Wood)	.20	.04	40
Glass (Soda-lime)	.18	.59	154
Glass Wool Insulation	.16	.02	3.3
Cast iron	.12	28	450
Steel (mild)	.12	26	490
Silver	.056	245	654
Building Brick	.2	.4	120
Firebrick	.2	.6	110
Insulating Firebrick	.2	.08-.2	20-50
Water	1.0	.35†	62.3
Air	.24	.014†	.074*

† This is conductivity only, without any convection.
* at 70°F, standard atmospheric pressure, and 50% relative humidity.

TABLE 2-1. Properties of materials. Specific heats are related to the amount of heat which can be stored in a given mass of material. Thermal conductivities are related to the rate at which heat can be conducted through materials.

and radio waves, which is so fast that in practical circumstances it arrives at essentially the same instant it is emitted. (Contrastingly, conduction and convection of heat usually take perceptible amounts of time.) Infrared radiation cannot penetrate most solids; it is usually mostly absorbed on their surfaces, where it becomes heat. Some may be reflected, but usually only a small fraction, unless the surface is a shiny light metal, in which case most of the radiation is reflected. A few materials allow some of the radiation to pass through them — certain plastics and minerals are examples. Glass usually does not let much infrared radiation through, but absorbs most of it.

All matter is constantly both emitting and absorbing infrared radiation. The emission is related to the constant motion of the molecules and electrons in matter. The intensity of the emitted radiation depends on the surface temperature of the object and its emissivity (discussed on P. 52). The hotter a surface is, the more energy it radiates. Increasing the temperature of a stove's surface from 200° to 400° F more than quadruples the net radiation from it. (See Table 2-3) The temperature dependence of emitted infrared radiation is the principal reason infrared photography works in the dark; warmer objects emit more radiation and hence come out brighter in infrared photographs.

The intensity of radiation emitted by most room-temperature surfaces is about half that of direct sunlight! Thus, our environment is always full of high levels of infrared radiation. Why are we not more aware of it? Since we cannot see the radiation, we can sense its presence only if it results in a change in the temperature of our skin. Normally there is a near balance between the radiation absorbed on one's skin, which was emitted by the surroundings, and the radiation emitted by one's skin. If the same amount of energy leaves as enters, there is no net change in energy; one's skin temperature is unaffected and neither the arriving or leaving radiation is noticed.[4] But in the vicinity of hot objects, such as an operating wood stove, one's skin is warmed because it absorbs more radiation than it emits, resulting in a net energy gain.

The radiation exchange between two surfaces is always a net energy flow from the warmer surface to the cooler, just as in the case of conduction.

When surfaces are sufficiently hot, detectable radiation becomes evident in the visible region of the electromagnetic spectrum. This first occurs at about 900°F, at which temperature most objects appear deep red. As the temperature increases, the brightness of the glow or emission increases, and the color shifts to orange, yellow, and finally white (at about 2800°F). This phenomenon can be used to make a rough estimate of temperatures between 900° and 2900°F (Table 2-2). Most of the radiation, however, remains in the (invisible) infrared part of the spectrum. Both visible and infrared radiation are part of a single phenomenon — the emission of electromagnetic waves from all objects.

The amount of radiation *received* by a surface, say one's face, a certain distance from a hot stove depends on geometrical considerations (in addition to temperature). Radiation is emitted in all outward directions from every point on most surfaces. Thus, anywhere in front of a flat radiating surface, some radiation is received, even towards the sides. For any given emitting-surface temperature, the intensity of the received radiation at any point is approximately proportional to the apparent size of the radiating surface as seen from that point (technically, the solid angle subtended or angular area). Thus, the radiation is most intense directly in front of, and up close to, a hot surface, for then the surface practically fills the whole field of vision. The intensity is less as one moves away from the hot surface, and is in proportion to its smaller angular size. The radiation from a red hot coal will easily ignite a piece of paper held a half an inch away, but will have virtually no affect on paper a few feet away. These geometrical considerations are especially important for safety of installations; radiant intensities on surrounding combustible surfaces must be kept sufficiently low to avoid ignition.

ENERGY UNITS

Energy can be measured in many different units. The basic unit usually used in the heating and air-conditioning field is the British thermal unit, or Btu. A Btu is defined as the amount of energy it takes to increase the temperature of one pound (a pint) of water by one degree Fahrenheit. Thus to heat a pound of water from 60°F up to the boiling point, 212°F, requires 212 - 60 = 152 Btu.

A Btu is a relatively small amount of energy on a human scale. It takes about 75 Btu to heat up a cup of water to boiling (starting at about 60°F). About 370,000 Btu are needed to boil down the 30 to 40 gallons of sap to make 1 gallon of maple syrup. The 3000 calories of food energy that a typical adult eats each day is 12,000 Btu. A two minute shave with an electric

APPEARANCE	TEMPERATURE °F	TEMPERATURE °C
No emission detectable	Less than 885	Less than 475
Dark red	885 - 1200	475 - 650
Dark red to cherry red	1200 - 1380	650 - 750
Cherry red to bright cherry red	1380 - 1500	750 - 815
Bright cherry red to orange	1500 - 1650	815 - 900
Orange to yellow	1650 - 2000	900 - 1090
Yellow to light yellow	2000 - 2400	1090 - 1315
Light yellow to white	2400 - 2800	1315 - 1540
Brighter white	higher than 2800	higher than 1540

TABLE 2-2. The approximate color of glowing hot, solid objects. [Adapted from D. Rhodes, *Kilns*, (Chilton Book Co., Philadelphia, 1968), p. 170.]

shaver uses about 1 Btu of electrical energy. Portable electric space heaters generate a maximum of about 4,000 Btu of heat for each hour they are on.

POWER

A piece of wood contains a given amount of chemical energy. If it is burned completely, a given amount of heat will be released. If burned quickly, the heat will be produced at a high rate for a short time; if burned slowly, the same amount of heat will be produced at a slower rate for a longer time. The rate at which energy is converted from one form to another, or transported from one place to another, is called power. Power has units of energy divided by time. For instance, conventional furnaces are usually rated in Btu's per hour. A 100 watt light bulb converts electrical energy into light and heat at a rate of 341 Btu per hour. If oven-dry wood (with an energy content of 8600 Btu per pound) is burned at a steady rate of 1 pound per hour, the rate of burning, in terms of energy, is about 8600 Btu per hour. If 50 percent of the wood's energy becomes useful heat in the house (i.e., if the stove's energy efficiency is 50 percent), then the heating power of the stove in this circumstance is half of 8600 Btu per hour, or 4300 Btu/hr.

The human body is constantly converting chemical energy from food into other forms, mostly heat. For a sitting person, this conversion rate is about 500 Btu per hour. If 500 Btu of food energy are "burned" each hour for a day, the total energy used is 12,000 Btu (500 Btu/hr x 24 hr), which is the same as 3,000 calories, a typical daily food-energy consumption. The heat output of the human body can be a significant contribution to house-heating needs, particularly in a crowded or snug house.

HEAT STORAGE

The capacity of a stove to store heat affects its performance. The more massive a stove is, the more heat it can store, and the more heat it takes to warm it. An object's heat capacity is the amount of heat it takes to raise its temperature by one degree. The ability of different materials to store heat is usually described using the concept of specific heat, which is the heat capacity per pound. E.g., the specific heat of iron (and steel) is about 0.12 Btu per pound per degree Fahrenheit. Thus, it takes .12 Btu to warm 1 pound of iron 1°F. The heat stored in a 100 lb. iron stove with an average temperature of 470°F (referenced to room temperature, 70°F) is 4800 Btu.[5]

LATENT HEAT

There is another form of energy which, like infrared radiation, is often associated with heat because of the ease with which it is converted into heat, and vice versa. Latent heat is a form of potential energy which is contained in water vapor. When water is at the boiling point, energy in the form of heat must be added to make it boil away. Its temperature does not change in the process, but its physical state changes from liquid to vapor. The water vapor contains the energy as latent heat; if it condenses back to liquid water, the energy it took to boil it away is released again as heat.

The term "heat" is often also called "sensible" heat, to unambiguously distinguish it from latent heat and radiant energy. Only sensible heat is related directly to the motional energy of molecules and atoms. To feel sensible heat requires physically touching an object or substance; it can only be sensed by direct contact (on a molecular level). Radiant energy (or sometimes "radiant heat") becomes sensible heat when absorbed.

Evaporation of water requires roughly the same amount of heat energy as does boiling it away. In the case of evaporation, the necessary sensible heat is "stolen" from the surroundings, mostly the remaining liquid water. This is why evaporation is a cooling process. In either case, boiling or evaporation, the amount of energy required to effect the change from liquid to vapor is about 1000 Btu per pound of water[6], and whenever water condenses, the same amount of energy is released. Water vapor is said to contain latent (concealed) heat because sensible heat is released when it condenses.

STOVE HEAT

In a fire itself, chemical energy is converted into radiation and heat. In a closed stove, the radiation is absorbed by the inner surface of the stove walls. Some

TEMPERATURE of Surface (°F)	TOTAL ENERGY TRANSFERRED Btu per hour per square foot	NATURAL CONVECTION (percent of total)	RADIATION (percent of total)
80°	15	28	72
100°	51	35	65
150°	168	39	61
200°	315	40	60
400°	1230	35	65
600°	2850	29	71
800°	5430	23	72
1000°	9370	18	82
1200°	15100	15	85
1400°	23100	12	88

TABLE 2-3. The amount and type of energy transferred from an exposed hot surface, such as a radiant stove, to its surroundings.[7] The total heating rate per square foot of stove surface area increases dramatically as the surface becomes hotter. At all temperatures, more than half of the energy is in the form of infrared radiation. (In circulating stoves, air is heated as it circulates through interior passages; in this case, the convected hot air carries more heat into the room than does radiation.)

of the sensible heat is convected and conducted to the stove wall; the rest convects up the chimney. The heat at the inner surface of the stove walls then generally conducts to the outer surfaces of the stove. There some of it radiates away, and some is convected away in the air.

For most stoves (all non-convective types), more energy is transferred by radiation than convection. The percentage depends on the temperatures of the stove's outer surfaces, and ranges from 60 to 70 percent (Table 2-3), for surface temperatures up to about 800°F, the approximate maximum surface temperature of most stoves.

Chapter Three
Fuelwood

MEASURES

The most common unit of measure for fuelwood is the standard cord, which is a pile of 4 foot long pieces of wood 4 feet high and 8 feet long, with a volume of 128 cubic feet (Figure 3-1).

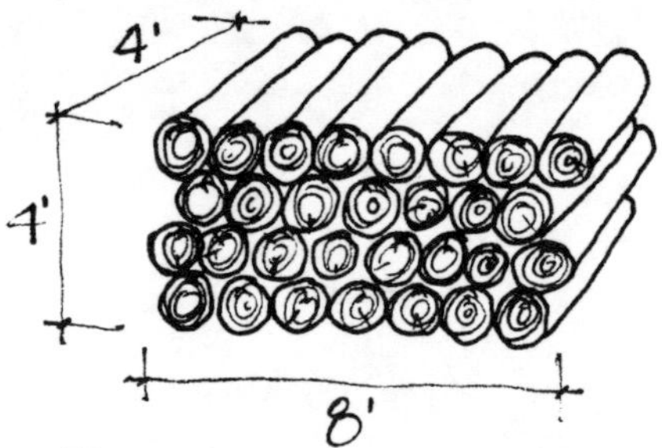

Figure 3-1. A cord of wood.

The 128 cubic foot volume includes some air between the pieces of wood. The actual volume of solid wood can range from 60 to 100 cubic feet depending on both the shape of pieces and how they are stacked. If the pieces of wood are crooked and bumpy, the air spaces will be relatively large and a cord will contain less solid wood. If straight pieces with a variety of diameters and shapes are used, much more wood can be piled into a cord if care is taken to use the smaller pieces to fill in the spaces between the larger, as in a carefully constructed stone wall (Figure 3-2).

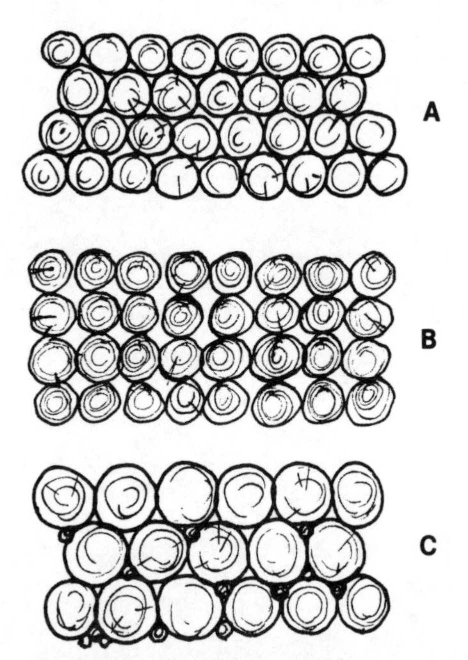

Figure 3-2. Some effects of stacking density on the amount of actual solid wood in a cord. Pattern A represents the densest packing of straight, uniform-diameter logs. Stacked as in pattern B, there are 13 percent fewer logs in a cord compared to A. With small logs filling some of the spaces in pattern A, as indicated in C, a cord contains 2.3 percent more solid wood compared to A.

A standard cord is defined to be 128 cubic foot pile of wood in 4 foot long pieces. If the pieces are cut into shorter lengths and restacked, the volume will be less for two reasons: 1) some wood is lost as sawdust. 2) more importantly, shorter pieces generally fit closer together, by an amount depending on how crooked or irregular the 4 foot pieces were. A 25 percent volume decrease is possible. After splitting the wood, the volume of the pile may also change.

A face cord (or run cord, or a rick), is the amount of wood in a pile 4 feet high and 8 feet long, period (Figure 3-3). The width of the pile (or length of the pieces) can be anything. A face cord of 2 foot long logs (a 2 foot face cord) would be a pile 4 feet x 8 feet x 2 feet. It would probably be a little more than half a standard cord because of denser packing of shorter pieces. A 4 foot face cord is exactly a standard cord.

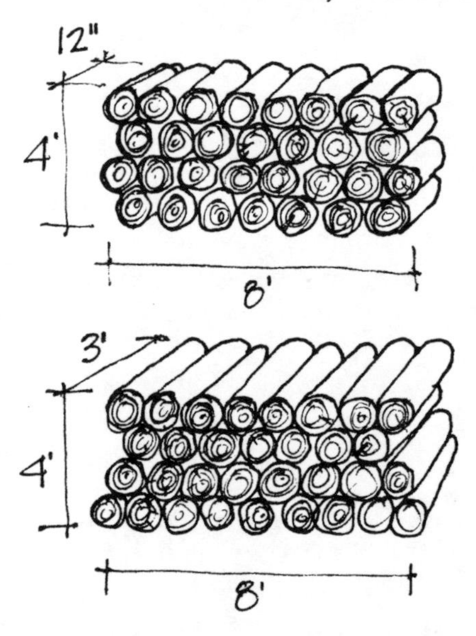

Figure 3-3. Two face cords.

SOURCES

Fuelwood can be obtained from a great variety of sources at a great range of prices. The most expensive and most convenient way to get wood is to buy it, cut to the buyer's specified length, split and "seasoned" or air dried. Prices usually fall between $25 and $175 per standard cord. Green wood (wood recently cut from live trees) is sometimes cheaper. A year is ample for drying (or seasoning), so if one has storage space and can buy a year in advance, one can take advantage of the lower prices some sellers offer on green wood.

Sawmill slabs and edgings are the wastes from making round, tapered trees into square, straight lumber. Some mills, particularly smaller ones, sell (and in some cases deliver) slabs and edgings at a very reasonable cost. Some people prefer these mill wastes because they do not need splitting, and they are inexpensive. However, they usually need cutting because they tend to come in long lengths (typically 8 to 16 feet). Also hardwood wastes, which some people prefer, are less common than softwood wastes. Prices, of course, may rise and availability decline, as demand grows, especially the demand for chips. Increasingly, mills sell or use all their wood wastes. Many mills burn their wastes to heat their buildings, generate electricity, and dry their lumber. The older practice of burning wastes just to make disposal easier is declining due to air pollution laws and the rising cost of energy.

If one is equipped and willing to haul wood, and to cut and split, local dumps, landfills, and construction sites can be sources of already felled and trimmed wood. Landscapers and arborists may be happy to dump their wastes at a residence. Neighbors who do their own landscaping may give away felled trees. Both electric utility and telephone companies clear trees and large branches from their overhead lines, and often the usable fuelwood is left in piles under the lines for the public. Demolition sites can be a source of very dry, and sometimes dirty, lumber. Road construction and improvement projects sometimes have felled and limbed trees which need disposal, and many state and national forests have programs permitting cutting of selected trees by private citizens.

DEFINITION OF MOISTURE CONTENT

The standard method for determining the moisture content of wood is to measure the wood's loss in weight when dried. Essentially all moisture can be driven off if the wood is placed in an oven at somewhat above 212°F for sufficient time (hours to weeks depending on the size of the pieces used). When the wood stops losing weight perceptibly, it is called ovendry and is said to have a moisture content of zero percent. Thus if a piece of wood weighs 10 pounds before such oven drying, and 8 pounds after, 2 pounds of water were driven off, and the moisture was 20 percent of its original weight. Its "moisture content," however, is 25 percent. The most generally accepted definition of moisture content is the weight of moisture lost during oven drying, divided by the wood's ovendry weight, not its original weight (including the moisture). Thus wood with a moisture content of 100 percent is not pure water, but half water and half dry wood. (See Appendix 2.) The moisture content of green woods is typically between 50 and 150 percent. Seasoned firewood has a moisture content of roughly 20 percent.

ENERGY CONTENT

Through experience one develops impressions of differences among the heating performances of various woods. Some people become quite particular about the wood they insist on burning. Others burn practically any wood without apparent difficulty.

Is there a difference in the amount of heat one wood will give compared to another? Different woods certainly burn with different characteristics. Wet wood is more difficult to burn. Dense woods such as apple burn for a longer time than light woods such as pine. Some wood types are easier to ignite than others, and are thus especially useful for kindling. Some produce smoke with an especially pleasing fragrance.

One of the most important characteristics of a wood is its energy content. A reading of some of the available lists can be confusing and frustrating — confusing because large discrepancies (as big as 42 percent) exist between lists concerning the energy content of a cord of a given type of wood, and frustrating because woods one may be particularly interested in may not be listed.

The confusion arises because there are a number of different concepts relating to the energy in wood. Three common measures of energy content are called high heat value (or gross heat value), and low heat value (or net heat value), and available heat. The first two of these measures have unambiguous meanings and are commonly used by heating engineers. The third, available heat, is not a precisely defined concept and different authors have given it different meanings. The units in all cases are usually Btu's (British thermal units) per cord or per pound.

The differences between these energy measures are mostly in how water vapor is handled. All wood has some moisture in it; this water is evaporated or boiled off in a fire. Perfectly dry wood also generates water vapor as it burns; the hydrogen atoms in the wood combine with oxygen to become water molecules. Water vapor contains considerable potential energy (often called latent heat) which is released as (sensible) heat if the vapor condenses to liquid water. The amount of energy is the same as must be given to liquid water to make it boil away or evaporate and is about 1050 Btu per pound of water. In practice, when wood is burned in a stove or fireplace, not much of this latent heat becomes available for use because not much of the water vapor condenses.

The high heat value of wood is its total chemical energy per pound. It represents the amount of chemical energy released when 1 pound of wood is completely burned. Only if all the water vapor (which was evaporated or boiled off or manufactured during the combustion) condenses, does all the chemical energy become usable sensible heat.

The low heat value of wood is its total chemical energy per pound minus the latent heat in the water vapor which would result from the complete combustion of the wood. The low heat value is the amount of sensible heat produced when a pound of wood is burned completely and no resulting water vapor condenses.

The term "available heat" as used in other publications can have any of a number of meanings. In most cases [1] it is an estimate of the actual amount of useful heat given off by a stove; it thus includes arbitrary assumptions about the stove's energy efficiency. All of these concepts are discussed in more detail below.

Actual measurements of energy contents (or caloric values) of woods are made by completely burning very small samples in what is called a bomb calorimeter. The combustion occurs in a closed container (the "bomb") and essentially all the heat generated is measured through measurement of the resulting rise in temperature of the bomb and its surroundings. The bomb is pressurized with oxygen before the wood sample is ignited to help ensure complete combustion.

There are often discrepancies for the same kind of wood. [2] These may be due to regional, seasonal or individual tree differences, or the investigators may not have used identical species. Wood from different parts of the same tree sometimes yield slightly different values. Heartwood and sapwood have slightly different chemical compositions. Sapwood is the lighter-colored outer portion of a log or stem and heartwood is the (usually) darker-colored central portion. In a live tree, the wood cells in the sapwood are alive, and those in the heartwood are not. Earlywood (spring growth) and latewood (summer growth) also have slightly different compositions. Wood with different proportions of these components might be expected to have different overall energy contents. Thus variations of a few percent in measured energy content for a given type of wood are probably to be expected.

Despite apparent discrepancies and problems, it is remarkable how nearly equal are the energy contents of all of the different kinds of wood. *To within a few percent, one pound of ovendry wood of almost any kind has the same energy. The major difference between woods is their weights or densities.* A solid cubic foot of dry black locust weighs about 43 pounds; a cubic foot of dry Douglas fir weighs about 30 pounds. The cubic foot of black locust has more energy in direct proportion to its greater dry weight. The reason all woods have about the same energy content on a weight basis (and with equal moisture content) is that they are all roughly similar in chemical composition.

But there are slight chemical differences among woods, and these probably explain the differences in their energy contents. Species with large amounts of pitch have slightly higher energy contents; an example is pitch pine. Such woods contain more resins, gums, oils and tannins than most. In practice, the very high energy content of resins together with the fairly large range of resin content in different woods probably makes this component the most important factor accounting for the differences in energy content in different woods. For instance, if 8600 Btu per pound is taken to be the energy content of wood excluding the resins, then wood which is 5 percent resins (which contain 17,400 Btu per pound) has an overall energy content of 9040[3] per pound, an increase of 5 percent.

Softwoods tend to have more resin, and this is a reason for their generally slightly higher energy values. Native American trees can be divided into two groups — hardwoods, which have broad leaves, and softwoods which have leaves in the form of needles (e.g., pines) or flat needles (e.g., cedars). Softwoods are also called conifers since all native species have cones. Most softwoods are evergreen; the exceptions are tamarack, larch, and cypress. Most hardwoods are deciduous; they drop their leaves and grow new ones each year. Softwoods are not necessarily softer than hardwoods. For instance, the softwood, Douglas fir, is as hard as yellow poplar, a hardwood, and aspen, a hardwood, is in fact softer than white pine, a softwood. On the average, softwoods contain more resins than do hardwoods.

In summary, the similarities between woods are more striking than the differences with regard to their energy content when reported on a per-pound basis and at equal moisture content.[4] 8600 Btu per pound at zero moisture content is a generally accepted value for all hardwoods and all but very resinous softwoods.

In fact, since good measurements on a large variety of firewoods is lacking, most lists of energy content per cord are computed from the assumption of uniform energy content per unit weight of wood (of the same moisture content), and use solely the difference in weight per cord to estimate the difference in energy per cord. In this approximation, a table of relative densities (Table 3-1) is essentially equivalent to a table of relative energy contents per cord — the denser the wood, the more energy a given volume of it contains.

TABLE 3-1.

SPECIES	RELATIVE DENSITY OR SPECIFIC GRAVITY	MOISTURE CONTENT OF GREEN WOOD (PERCENT)	
		Heartwood	Sapwood
Hardwoods			
Alder, red	.41	——	97
Apple		81	74
Ash:			
Black	.49	95	——
Blue	.58	——	58
Green	.56		
Oregon	.55		
White	.60	46	44
Aspen:		95	113
Bigtooth	.39		
Quaking	.38		
Basswood, American	.37	81	133
Beech, American	.64	55	72
Birch:			
Paper	.55	89	72
Sweet	.65	75	70
Yellow	.62	74	72
Butternut	.38		
Cherry, black	.50	58	——
Chestnut, American	.43	120	——
Cottonwood:			
Balsam poplar	.34		
Black	.35	162	146
Eastern	.40		
Elm:			
American	.50	95	92
Cedar		66	61

SPECIES	RELATIVE DENSITY OR SPECIFIC GRAVITY	MOISTURE CONTENT OF GREEN WOOD (PERCENT)	
		Heartwood	Sapwood
Rock	.63	44	57
Slippery	.53		
Hackberry	.53	61	65
Hickory, pecan:			
Bitternut	.66	80	54
Nutmeg	.60		
Pecan	.66		
Water	.62	97	62
Hickory, true:			
Mockernut	.72	70	52
Pignut	.75	71	49
Red		69	52
Sand		68	50
Shagbark	.72		
Shellbark	.69		
Honeylocust	.66*		
Locust, black	.69		
Magnolia:		80	104
Cucumbertree	.48		
Southern	.50		
Maple:			
Bigleaf	.48		
Black	.57		
Red	.54		
Silver	.47	58	97
Sugar	.63	65	72
Oak, California black		76	75
Oak, red:			
Black	.61		
Cherrybark	.68		
Laurel	.63		
Northern red	.63	80	69
Pin	.63		
Scarlet	.67		
Southern red	.59	83	75
Water	.63	81	81
Willow	.69	82	74
Oak, white:			
Bur	.64		
Chestnut	.66		
Live	.88		
Overcup	.63		
Post	.67		
Swamp chestnut	.67		
Swampy white	.72		
White	.68	64	78
Sassafras	.46		
Sweetgum	.52	79	137
Sycamore, American	.49	114	130
Tanoak	.64*		
Tupelo:			
Black	.50	87	115
Swamp		101	108
Water	.50	150	116
Walnut, black	.55	90	73
Willow, black	.39		
Yellow-poplar	.42	83	106
Softwoods			
Baldcypress	.46	121	171
Cedar:			
Alaska-	.44	32	166
Atlantic white-	.32		
Eastern redcedar	.47	33	—
Incense-	.37	40	213
Northern white-	.31		

Port-Orford-	.43	50	98
Western red cedar	.32	58	249
Douglas-fir:			
Coast	.48	37	115
Interior West	.50		
Interior North	.48		
Interior South	.46		
Fir:			
Balsam	.36		
California red	.38		
Grand	.37	91	136
Noble	.39	34	115
Pacific silver	.43	55	164
Subalpine	.32		
White	.39	98	160
Hemlock:			
Eastern	.40	97	119
Mountain	.45		
Western	.45	85	170
Larch, western	.52	54	110
Pine:			
Eastern white	.35		
Jack	.43		
Loblolly	.51	33	110
Lodgepole	.41	41	120
Longleaf	.59	31	106
Pitch	.52		
Pond	.56		
Ponderosa	.40	40	148
Red	.46	32	134
Sand	.48		
Shortleaf	.51	32	122
Slash	.59		
Spruce	.44		
Sugar	.36	98	219
Virginia	.48		
Western white	.38	62	148
Redwood:			
Old-growth	.40	86	210
Young-growth	.35		
Spruce:			
Black	.40	34	128
Engelman	.35	51	173
Red	.41	34	128
Sitka	.40	41	142
White	.40	34	128
Tamarack	.53	49	––

*Estimates.

TABLE 3-1. Density and moisture content of some American woods. The energy per cord of wood species is approximately proportional to the density; the box at the bottom of the table gives the conversions from density to the more common units of millions of Btu per cord. Densities are given relative to the density of water, or equivalently, in grams per cubic centimeter; to convert to pounds per cubic foot, multiply by 62.3. The densities are averages; 10 percent variations among different samples of the same species are common. The densities are based on ovendry weight and volume at 12 percent moisture content. Moisture contents are based on ovendry weight. Different common names are often used for the same species. The scientific names corresponding to the common names used in this table are in the source below for most species, as well as some alternative common names. Most fruit trees are not included in the above list. Most fruit wood is relatively dense, and is considered to be excellent fuel.

(Data from Forest Products Laboratory, *Wood Handbook*, Tables 3-3 and 4-2, Agricultural Handbook No. 72, U.S. Department of Agriculture, 1974.)

TABLE 3-1 CONTINUED

RELATIVE DENSITY	HIGH HEAT VALUE PER CORD[1] (Million Btu)
.30	12.6
.35	15.0
.40	17.1
.45	19.3
.50	21.4
.55	23.6
.60	25.7
.65	27.9
.70	30.0
.75	32.1
.80	34.3
.85	36.4
.90	38.6

[1]Assuming 80 cubic feet of solid wood per cord, and 8600 Btu per pound of ovendry wood. For very resinous woods, high heat values are a few percent higher. High heat values per cord do not depend on the wood's moisture content except through the shrinkage of wood as it dries.

Some confusion can arise when comparing lists of energies per cord of various woods because the actual amount of solid wood in a cord depends on the straightness and length of the pieces, and how it is piled. Some authors assume 80, others 90, cubic feet of solid wood in a cord (a cord has an overall volume of 128 cubic feet including the spaces between the pieces of wood). The difference between the assumptions of 80 and 90 cubic feet per cord result in a difference of about 12 percent between reported energies per cord. Actual cords may actually contain from 60 to 100 cubic feet of solid wood. Thus, in practice, there is a very large variability in the amount of energy per cord even for a given kind of wood.

The larger (20-40 percent) discrepancies between some lists of woods are due to the different ways of reporting energy contents. All the wood energies given up to this point have been the total chemical energy in wood, as measured in a bomb calorimeter, where the final temperature of the combustion products is essentially room temperature, and where virtually all water vapor generated condenses into liquid water. The heat value measured this way is the high (or gross) heat value. It represents the most heat that could possibly be derived from the burning of wood.

But when wood is burned in a stove or fireplace, the water vapor in the flue gases rarely condenses, especially where the released heat can be used. In fact, such condensation anyplace in the heating system is to be avoided for reasons of safety (related to creosote — see Chapter 12), and chimney life expectancy (due to corrosion). Since the latent heat part of the energy is essentially unavailable for heating, its contribution is frequently subtracted out from the high heat value, yielding the low heat value.

The difference is significant. As mentioned previously, there are two sources of water vapor in the exhaust of a fire, and low heat values take both into account. Wood contains water (its moisture content), and water vapor is also manufactured in the combustion process. When burned completely, each pound of ovendry wood produces about 0.54 pound of water vapor.[5] For wood with a moisture content of 25 percent there is another quarter pound of water vapor going up the chimney for each piece burned whose ovendry weight would be 1 pound. The total amount of water vapor is 0.79 pound, which represents about 830 Btu of potential energy.[6] The assumption behind the concept of low heat value is that this energy is not usable. The low heat value of wood with a 20 percent moisture content is thus about 830 Btu less than its high heat value, or about 7770 Btu per piece whose ovendry weight would be 1 pound, a decrease of about 9 percent.[7] The effect at other moisture contents is illustrated in Figure 3-4.

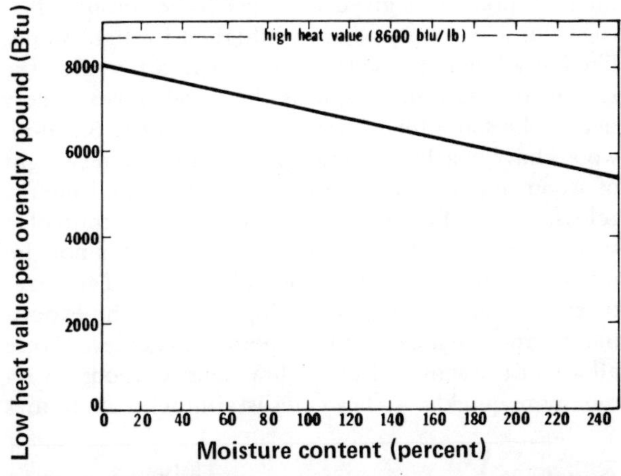

Figure 3-4. Low heat value per piece of wood as a function of moisture content, assuming the piece would weigh one pound if it were ovendry. This represents the maximum amount of useful heat which combustion could yield if no water vapor condenses.

The water vapor in the flue gas, along with everything else, also carries away sensible heat. As long as the gases leave the house at any temperature above room temperature, some of the heat generated in the fire was not recovered as useful heat in the house. Some authors have incorporated an estimate of this loss in their lists of available energy of different woods by making quite arbitrary assumptions about flue-gas temperatures and the amount of combustion air. This is not appropriate. The amount of heat lost up the chimney is not a property of the wood burned, but of the heat-transfer properties of the stove and chimney, and thus belongs rather in a discussion of the energy efficiencies of stoves (Chapter 6), not in a list of wood types.

When assessing the energy content of a cord of firewood, the most important parameter is the ovendry density of that kind of wood, since a pound of dry wood of any kind has nearly the same energy. The densest woods have the most energy per cord (at equal moisture content). Table 3-1 gives the densities of many types of wood. Moisture in wood decreases its useful energy. If all types of wood had the same cost per cord, the better buy would be the denser woods in Table 3-1. If wood were sold by the ton, as is sometimes the case, the best buy in terms of energy would be the dryest wood. No fuelwood dealer I am aware of sells wood by its energy content (e.g., $50 for 20 million Btu). Reasonably accurate Btu assessments would require both weighing and a determination of moisture content.

BURNING QUALITIES

Softwoods are often said to burn faster and hotter than hardwoods and are usually preferred for kindling. Hardwoods are generally felt to last longer in a fire and generate more coals.

Careful experimental measurements support most of these notions, but with the wood's density as the critical variable, not its classification as a softwood or a hardwood.

Figure 3-5 shows that the denser a wood is, the longer it takes for a given size piece to be consumed — the volume consumption rate is highest for light woods like balsa. Denser woods last longer in a fire. But the rate of decrease of weight as the wood burns is very nearly the same for all woods (Figure 3-6). Since all woods have nearly the same energy for the same weight of wood, equal weight loss rates imply equal energy release rates. This means that the heat output of a stove, under steady burning conditions, does not depend critically on the kind of wood used. (Temperatures were also measured during some of the experiments, and the maximum temperatures achieved were all about the same.) But the low density woods do ignite more quickly — they can sustain their own flames

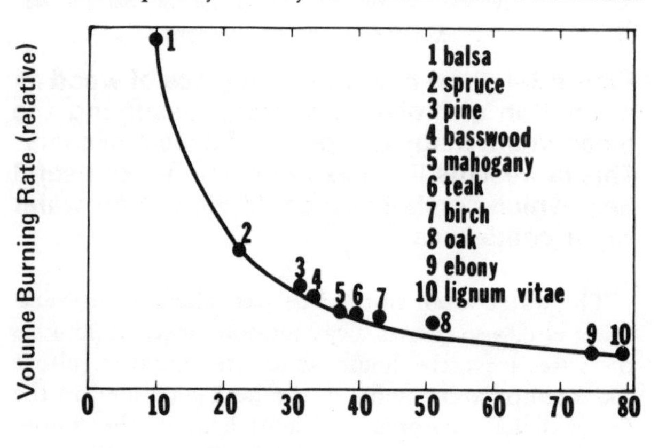

Figure 3-5. Volume burning rate, as a function of density. From L. Metz.[8]

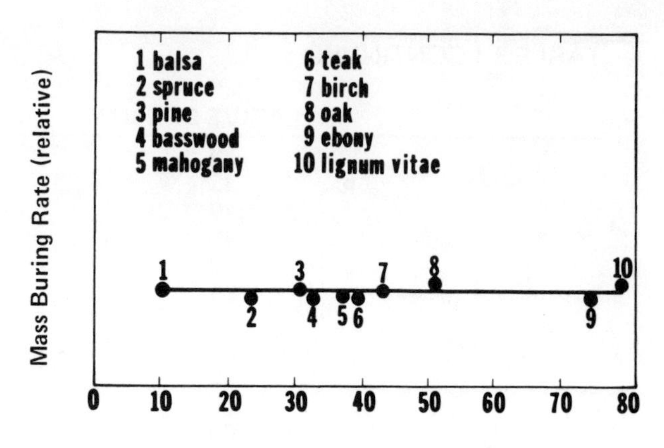

Figure 3-6. Mass burning rate, as a function of density. From L. Metz.[8]

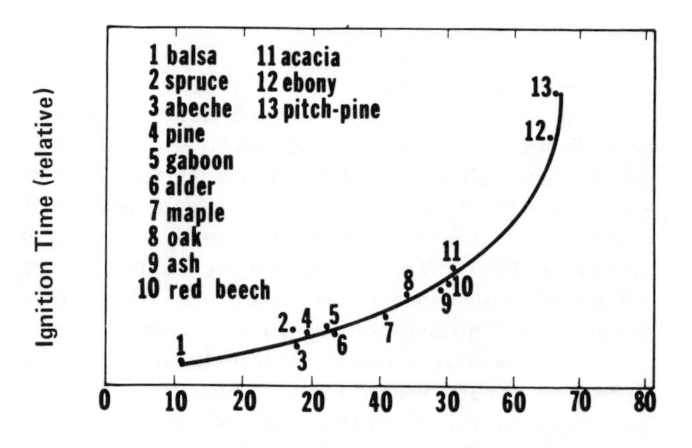

Figure 3-7. Ignition time as a function of wood density. From L. Metz.[8]

sooner (Figure 3-7). This makes light woods preferable as kindling, and it means that a stove will heat up more quickly with low density wood as the fuel. This quickness of ignition may be partly what gives the impression of a hotter fire. Figure 3-8 indicates that when the flames die out, the mass of wood (mostly charcoal) still remaining is highest for dense woods; dense woods produce relatively more coals. In the case of light woods, more of the charcoal is burned while the gases are still burning.[8]

Density is the single most important property of wood that differentiates its burning characteristic. Woods containing relatively large amounts of pitch have slightly higher energy content and probably burn with larger flames. Use of pitchy woods also often results in more creosote. Some woods (such as oak) have relatively concentrated regions in their structure containing pores or open channels which may affect the way they burn (such woods are said to be ring-porous). The mineral content of woods (which is left as ash) may affect burning characteristics.

Some woods burn noisily. Common sounds are hissing and crackling, but some woods can even burn explosively. These sounds are characteristics of woods whose structures are relatively impermeable to gases.

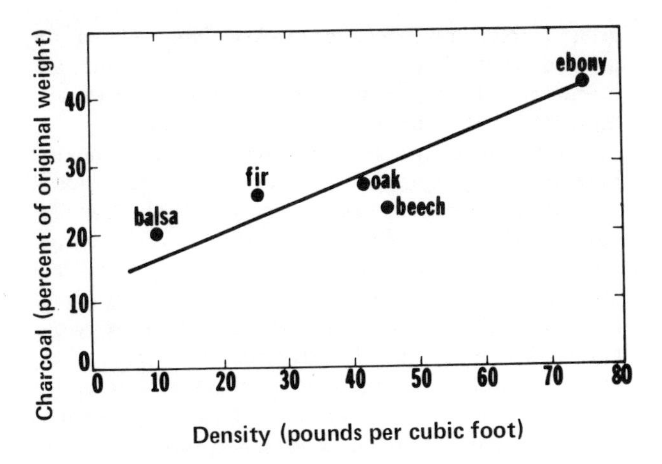

Figure 3-8. Percentage of original weight left as charcoal after large flames have died out. From L. Metz.[8]

When heated, gases inside the wood are generated and try to expand. In woods with a more porous structure, the gases can gradually get out as they are generated and may generate hissing sounds, but nothing more violent. In impermeable wood, particularly those whose structure contains pockets of liquid, the pressure of the gases builds up until the wood structure ruptures, relieving the pressure, letting out the gases suddenly, and making sharp sounds. Resinous woods tend to burn noisily.

ASH

Ash is all the solid residue from the complete combustion of wood. Both the total amount of ash and its composition depend not only on wood species, but also on the amount of bark included, on the ratio of heartwood to sapwood, and whether the wood came from the trunk or the branches of the tree.[9] Ash content typically is between 0.1 and 3 percent.[10] Ash content is rarely considered an important consideration in choosing firewood types.

The composition of ash is indicated in Table 3-2. Wood ash is useful as a fertilizer, particularly because of its potassium content. Potassium barbonate can be extracted from wood ash and used in making soap. Wet ash is caustic, as is lye, and can be used as a degreasing cleaner. Ashes will also decrease the acidity of garden soil.

GREEN WOOD

Green wood has disadvantages. It is always heavier (roughly twice as heavy) and thus a little harder to handle. Green wood is also more troublesome to ignite and keep burning, and contains less available energy. When it is burned in a stove, some of the wood's chemical energy must be used to evaporate or boil the moisture. For this and other reasons (**page 30**), less useful heat is derived from green wood.

Different kinds of green wood are actually quite different from each other in these respects. White ash is relatively easy to burn when green; most softwoods are very difficult. The principle reason is the difference in moisture content. Table 3-1 lists moisture contents of green woods. If one must burn green wood and there is a choice, species with the lower moisture content are the best choice. Heartwood is considerably drier than sapwood; in a desperate situation, heartwood could be separated from sapwood.

Burning moist wood can be advantageous. Because of its moisture content, green wood burns more slowly. Thus, when one is primarily interested in obtaining a steady low heat output over a long period of time, such as overnight, mixing some green wood in with seasoned wood can help. The appropriate proportion of green wood and its size distribution must be learned through experience; the amount depends critically on how moist the green wood is, and on the type of stove it is burned in; if too much is used, the fire will die out.

WOOD STORAGE

After being cut, green wood dries (or seasons), quickly at first and then at a slower and slower rate. Because wood is hygroscopic (it attracts water), it never dries to zero moisture content in any normal storage situation. Rather, its moisture content gradually approaches an equilbrium value which depends on the storage conditions. For covered wood, this final moisture content is given in Table 3-3; the relative humidity of the surrounding air is the main determining factor. In the Southwest in the summer, the moisture content of wood stored outdoors may be as low as 6 percent; in parts of the Northwest in winter, the moisture content of (covered) seasoned wood cannot be much less than 18 percent.

The term "seasoned" has no precise meaning when applied to fuelwood, although it means some degree of dryness compared to green wood. If the term had the

ELEMENT	ASSUMED FORM	TYPICAL QUANTITY (Percent of total ash)
Calcium	CaO	30-60%
Potasium	K_2O	10-30
Sodium	Na_2O	2-15
Magnesium	MgO	5-10
Iron	F_2O_3	1- 2
Silicon	SiO_2	2- 7
Phosphorous	P_2O_5	5-15
Sulphur	SO_3	1- 5

TABLE 3-2. Composition of wood ash. These figures are determined under conditions which ensure complete combustion, and retention of all ash generated. The percentage composition is calculated assuming the elements are in the stated chemical form; the actual chemical forms may be different. [Data interpreted from L.E. Wise, Editor, *Wood Chemistry, (New York, Reinhold Publ. Corp., 1944). p. 435]*

same meaning as it does when applied to paper, seasoned wood would be wood whose "moisture content is uniform and in equilibrium with that of the surrounding air,"[11] i.e., wood which is as dry as it can get, given its surroundings (**Table 3-3**).

The *time* it takes for wood to season depends principally on the size of the pieces, on how freely air can circulate around the piece's surfaces, and on the temperature. The smaller the pieces, the faster they will dry; thus splitting accelerates drying. Air circulation is better in free-standing wood piles than in piles next to buildings. Also, if rows are placed immediately next to each other, air will circulate less freely than if they are spaced apart a few feet. Piling wood with alternating orientations (criss-crossed) also lets more air through. Direct exposure to sunlight helps speed drying by its warming effect, as does indoor storage in a heated room.

Actual drying times will vary significantly depending on the above considerations. Firewood will generally be very close to its equilibrium moisture content (as dry as it can get) after two years of seasoning. Drying is fastest at the beginning; after six months of drying, wood is often called "seasoned," and usually has an acceptibly low moisture content of 25 percent or less.

Covering wood can decrease both the final moisture content, and the drying time, particularly in climates with relatively heavy precipitation. A shed without tight walls, or without walls at all, is excellent, since it protects against precipitation but does not inhibit air flow. However, in very dry climates, a shed may not help because its shading of the wood may offset its protection against the small amounts of precipitation. A plastic sheet which completely covers a wood pile down to the ground is usually worse than no cover since it inhibits circulation and tends to hold moisture in, including moisture coming out of the ground (which can be considerable). One simple way to achieve some protection against precipitation is with a cover made of 2 to 3 foot wide strips of plastic, fastened along their long edges to poles or small boards. The weight of the poles hanging over the sides keeps the plastic strip in place on top of the pile. Ideally, wood piles should be kept a foot or more above the **ground (e.g., on concrete blocks). Not only is the** ground itself often damp in most climates, but the humidity of the air close to the ground is also usually higher than the humidity in the rest of the air.

Storing wood in a heated space, such as a heated basement or garage, will speed up the drying process. The equilibrium moisture content will also be lowered, particularly if the air is not humidified. Wood stored indoors in a heated space in New England will have a moisture content of less than 10 percent after about two years. The reason is that the air is very dry in a heated but unhumidified structure in cold weather. However, indoor storage is not always advisable. Insects which are detrimental to wood buildings, such as termites, may be living in the fuelwood and may spread to the house. State university extension services or the U. S. Department of Agriculture should be able to give advice on the advisability of indoor storage for particular regions of the country.

If stored for too long a time, wood is capable of becoming useless as fuel due to its rotting. Rotting is caused by the growth of fungi; the fungi convert the wood into water, carbon dioxide, and heat, just as does a fire. Thus rotting decreases wood's energy. The fungi are most productive when three conditions are met: the temperature is between 60 and 90 degrees Fahrenheit, the wood's moisture content is above about 30 percent, and ample oxygen is available.

Virtually no rotting takes place in cold winters because the temperature is too low for the fungi. Wood in the interior of a house does not rot because it is too dry. Wood kept continuously underwater seldom rots, partly because of a lack of oxygen, and partly due to the usually cool temperature. On the other hand, wood which is lying on the ground, especially in summer usually provides a productive environment for fungi because the ground keeps the wood sufficiently moist. Rot will appear first at the bottom of a wood pile placed directly on the ground. The most practical way to retard rotting of fuel-wood is to help it dry quickly and to keep it dry, especially during the warm summer months during which most decay occurs.

Woods differ in their natural resistance to decay. The sapwood of virtually all species is highly susceptible to decay, but there is a considerable difference

RELATIVE HUMIDITY (percent)	EQUILIBRIUM MOISTURE CONTENT (percent)
5	1.2-1.4
10	2.3-2.6
20	4.3-4.6
30	5.9-6.3
40	7.4-7.9
50	8.9-9.5
60	10.5-11.3
70	12.6-13.5
80	15.4-16.5
90	19.8-20.0
98	26.0-26.9

TABLE 3-3. Equilibrium moisture content of wood as a function of the surrounding air's relative humidity.[12] The actual moisture content of very thin pieces of wood varies daily and even hourly as the air's relative humidity changes. The moisture content of wood logs varies much more slowly, following monthly and seasonal changes in average relative humidity. [Data from Forest Products Laboratory, *Wood Handbook*, Agricultural Handbook No. 72, p. 3-8, U.S. Department of Agriculture (1974).]

among heartwoods, as indicated in the Table 3-4. Given the choice, the more susceptible species should be burned first so that less of their energy content will be dissipated by the fungi. The most susceptible woods can decay significantly in two years of outdoor, uncovered, exposure in climates with warm, humid summers.

SUMMARY

The two most important aspects of fuelwood are its dryness and its density. Wood containing too much moisture is often very difficult to ignite and keep burning. In addition, stoves and fireplaces have lower energy efficiencies when burning moist wood.

On the basis of ovendry weight, all woods have approximately the same energy content. Thus a pound of (ovendry) oak and a pound of (ovendry) pine will yield about the same amount of heat when burned. But a cord of oak contains considerably more energy than a cord of pine, because it weighs more — oak is a denser wood. Thus, if the price per cord is the same, the best buy is the denser or heavier wood type.

All wood types are satisfactory as fuel. Some types burn more quickly, requiring more frequent refueling, some generate longer-lasting coals, some burn more noisily, some require less time to season, and some ignite more easily and are thus especially suitable as kindling. But no common wood type is truly unsuitable as fuelwood. In practice, people burn whatever is most readily available.

MOST RESISTANT	HIGHLY RESISTANT	MODERATELY RESISTANT	LEAST RESISTANT
Locust, black	Baldcypress	Douglas-fir	Alder
Mulberry, red	Catalpa	Honeylocust	Ashes
Osage, orange	Cedars	Larch, western	Aspens
Yew, Pacific	Cherry, black	Oak, swamp chestnut	Basswood
	Chestnut	Pine, eastern white	Beech
	Cypress, Arizona	Southern Pine:	Birches
	Junipers	Longleaf	Buckeye
	Mesquite	Slash	Butternut
	Oak:		Cottonwood
	Bur	Tamarack	Elms
	Chestnut		Hackberry
	Gambel		Hemlocks
	Oregon white		Hickories
	Post		Magnolia
	White		Maples
			Oak (red and black species)
	Redwood		Pines (other than long leaf, slash, and eastern white)
	Sassafras		Poplars
	Walnut, Black		Spruces
			Sweetgum
			True firs (western and eastern)
			Willows
			Yellow-popular

TABLE 3-4. Relative resistance of the heartwood of various species to decay. The sapwood of all species has little resistance to decay. [From U.S. Forest Products Laboratory, *Wood Handbook*, Agriculture Handbook No. 72, p. 3-17, U.S. Department of Agriculture (1974).]

Chapter Four Combustion

It is curious that blowing on a smoldering wood fire can ignite the gases, while blowing on an already burning candle will snuff out the flame. Why? All fuels need oxygen to burn, so it is reasonable that blowing on a fire increases the combustion rate because more oxygen is provided. But why does blowing on a candle flame have the opposite effect?[1]

Open wood fires normally burn with a yellow-orange flame, but sometimes a little blue color is also present. If the same gases (wood smoke) are burned in a properly adjusted laboratory bunsen burner or a cooking range burner, the flame color is almost all blue. What is the difference? What determines flame color, and what does the color indicate about the efficiency of combustion?

Wood fires often have no flames. At the end of a fire the charcoal can "burn" flamelessly, and even wood can smolder and smoke and generate heat, without flames.

When wood is smoldering, why doesn't the very high temperature of the glowing portions of the wood always ignite the combustible gases? Many stoves have "secondary air" inlets intended to add enough oxygen to gases coming off the main fire to permit complete combustion. Many of them fail to help the gases burn. Why?

FUNDAMENTALS

Combustion involves energy conversion — stored chemical energy in the fuel is converted into heat, light, infrared radiation and other forms of energy. Oxygen is required, and is consumed (incorporated into other molecules) in the process. Many common fuels, including wood, are made mostly of carbon, oxygen and hydrogen. When burned completely, these elements are transformed into carbon dioxide and water vapor.

Combustion is different from many other chemical reactions because it is self-propagating — if one corner of a piece of paper is ignited, it in turn will ignite adjacent portions, and ultimately the whole piece is burned. Explosions are like combustion in this respect, but are sometimes distinguished from combustion by the velocity at which the burning spreads through the fuel; in an explosion or detonation, this velocity exceeds that of sound, so that a shock wave is formed. In combustion or burning, the velocity is less than the speed of sound.

For all ordinary kinds of combustion, there must be fuel, oxygen and high temperatures. Wood in air does not burn unless first ignited — i.e., unless it is raised to a high temperature. Similarly wood, no matter how hot, will not burn if no oxygen is present (e.g., commercial charcoal production); it will char, and gases will be driven out of the wood but there are no flames and no or little heat is evolved. Wood only burns if it is both sufficiently hot and free (gaseous) oxygen is present.

A burning candle is an example of the combustion of a solid fuel which is similar to the combustion of wood but is simpler and hence a better example for illustrating basic ideas.

The candle's fuel is wax, or paraffin. The surrounding air contains oxygen (air is about 21 percent oxygen by volume). The ignition temperature of paraffin is about 1000°F — paraffin cannot burn unless it is at or above this temperature. But as its temperature is increased, paraffin first melts and then vaporizes before it reaches this temperature. Hence, only wax vapor burns.

The energy necessary to melt and vaporize the wax comes from the candle flame. Radiant energy from the flame melts the wax. The liquid wax then climbs up the wick, just as a sponge soaks up water. Radiant energy also evaporates wax from the wick. The wick is in the middle of the flame, but the flame (the glowing gases) does not actually touch the wick (except perhaps at its tip). The wax molecules constantly evaporating from the wick constitute a wax-vapor wind which, in a sense, prevents the flame from touching the wick. What actually happens is that when the wax vapor is a little distance away from the wick, sufficient oxygen has mixed with it making combustion possible.

In order for a wax molecule to start burning, it must have a sufficiently energetic collision with an oxygen molecule. The bonds within each molecule are fairly strong, and these bonds must be weakened enough in the collision that new bonds between oxygen and the fuel are able to form.

This is why heat must be supplied to light a fire or ignite a flame. On a molecular level, high temperature means high speeds of the molecules in their random motions. Molecules in gases (or in vapor phase) are constantly moving around and bumping into each other. When the temperature is raised, they move at higher speeds; thus, when a collision occurs, it is more violent. At sufficiently high temperatures, the collisions are energetic enough to loosen the old bonds and allow new bonds to form.[2] The chemical reactions of combustion can then take place, releasing chemical energy as heat. This heats the gas further, which helps more reactions to occur, and a flame results.

It is this release of large amounts of energy which makes combustion self-sustaining (or self propagating). The energy released, particularly that in the form of heat, can be used to prepare more fuel so that it too can react and release more heat. In a candle flame, the energy released in the outer portions provides not only the necessary radiation to melt and then evaporate the wax, but also to raise the temperature of the wax and oxygen molecules as they move into the combustion region so their collisions will be energetic enough to stimulate the chemical reactions.

In order for a fuel gas to be flammable, it must have some oxygen mixed with it, but not too much. If the mixture is too rich (not enough oxygen), it cannot burn; if one imagines attempting to ignite such a mixture with a source of heat, a few chemical reactions can be induced, releasing some heat, but due to the rarity of oxygen molecules, not enough reactions will occur to generate enough heat to compensate for the heat losses from the reaction region. If energy gets away (by either radiation or conduction) faster than it is replaced by the chemical reactions, the reactions are not self-sustaining and they cease. The same situation can occur if the mixture is too lean — if there are just too few fuel molecules around. In a candle, the bulk of the wick and the adjacent wax vapor do not burn because of insufficient oxygen. Just beyond the luminous regions there is no combustion despite the presence of some fuel (incompletely burned wax) because the fuel is too dilute. In between, the mixture of fuel and air is flammable and combustion occurs. The flammability limits of a gaseous fuel are the concentrations of the gas in air which are the leanest and richest mixtures which will burn.

Table 4-1 gives the limits of flammability of some common fuel gases, many of which are evolved from hot wood. The ranges of flammability are quite large; most of the gases are flammable with concentrations (in air) anywhere from about 10 percent to 60 percent.[3] (I.e., the air-to-fuel-ratio may vary over a six-fold range and the mixture can still burn.)

When burning any fuel for heat, it is desirable to burn it completely. Most fuels are composed principally of carbon, hydrogen, and oxygen. Complete combustion means the conversion of all the carbon to carbon dioxide (CO_2) and all the hydrogen to water (H_2O). In practice, a considerable amount of extra air must be supplied to achieve relatively complete combustion, mostly because there is never perfect mixing of fuel and air; a fuel rich region will run out of oxygen and cannot use the excess oxygen in a distant lean region. It is common in oil and natural gas burners in furnaces to supply 20 to 40 percent excess air to assure moderately complete combustion. However, supplying excess air is no guarantee that combustion will be complete. For instance, a candle flame has virtually no limit to its air supply and yet under some circumstances can smoke prodigiously (indicating unburned carbon particles). The same is true of a smoldering piece of wood in an open fireplace.

As a flammable substance (solid or gas) is gradually heated in air, it will burst into flame at a temperature which is called its spontaneous ignition temperature. Spontaneous ignition temperatures of solids are often determined by heating the surface of the solid with radiation (such as igniting leaves with a magnifying lens in sunlight). Approximate values for some gases are given in Table 4-1. On a molecular level, these are approximately the temperatures at which the collisions between molecules are energetic enough for new bonds with oxygen to form at a sufficient rate for the energy release to exceed the heat losses away from the reaction region. Thus the reaction region warms up still more, enabling more reactions to take place — i.e., a flame has started.

If a flame laps against a solid surface, combustion can be hindered or not, depending on the surface's temperature (Figure 4-1). Contact of burning gases with a cold surface quenches the flames. Gases within the "quenching distance" of the surface are below their ignition temperature despite the hot burning gases next to them, because of their loss of heat to the cool surface on their other side. The quenching distance, the thickness over which no flames exist, depends on the surface's temperature. It is largest for cold surfaces (a few millimeters) and decreases with increasing temperature. For surfaces hotter than the ignition temperature, combustion is not inhibited. In fact, some kinds of hot surfaces seem to enhance combustion.

Another possible effect on flames near large surfaces at any temperature is suppression of combustion due to inadequate oxygen. Wood-fire flames are diffusion flames — the needed oxygen gets to the fuel by diffusing into the gaseous fuel from the air around the flame.[4] When a flame laps against a surface instead of standing freely in air, the normal oxygen supply is restricted. Incomplete combustion can result; soot (unburned carbon) may deposit on the wall or leave the fire as visible smoke, and other incompletely burned compounds may also be formed.

Flammable mixtures of gases have characteristic velocities at which flames can propagate through them. If a long tube is filled with a flammable mixture of a fuel gas and air, and the mixture is ignited at one end, the flame front (combustion zone) propagates to the other end. The speed is called the flame velocity. Gas burners are designed so that gas velocity equals the flame velocity. Then the flame stays fixed in one place while the gas moves through it. Flame velocities in some gases commonly given off by burning wood are given in Table 4-1. A typical value for most common gaseous fuels is about 1.5 feet per second, or about 2 miles per hour.

This is not very fast, and that is why it is so easy to blow out a candle flame. If one blows very gently on a candle flame (not so hard as to blow it out), the wax vapor molecules near the wick are swept along with about the same speed. Air, and especially oxygen, still diffuses into the fuel and the mixture has no trouble burning because the flame front can easily travel fast enough upstream to remain stationary near the wick. But as one blows harder, the velocity of the burning gases exceeds the maximum flame velocity,[5] the flame

		IGNITION TEMPERATURE (°F)	FLAMMABILITY LIMITS (percent gas in mixture with air)	MINIMUM AIR NEEDED FOR COMPLETE COMBUSTION (air volume / gas volume)	FLAME VELOCITY (feet per second)
Hydrogen	H_2	1000	4-75	2.4	.2-10
Carbon Monoxide	CO	1100	12-74	2.4	.3-1.5
Methane	CH_4	1200	5-15	9.5	.1-1.3
Acetic Acid	CH_3COOH	1000			
Formaldehyde	HCHO	800			
Pine Tar		670			

TABLE 4-1. Combustion properties of some compounds in wood smoke. (Data from F.L. Browne, "Theories of the Combustion of Wood and Its Control," Forest Products Laboratory Report No. 2136, 1963; C.G. Segeler, Editor, *Gas Engineers Handbook*, Industrial Press, 1969; R.M. Fristrom and A.A. Westenberg, *Flame Structure*, McGraw-Hill, 1965.)

front gets pushed away from the wick and to the tip of the combustible gases, and there it disappears. The shape of the flame while being blown out can be quite irregular due to turbulence, and nonuniform gas composition, and irregularities in the wick. But the basic idea is simple — blowing can force the air and gaseous fuel away from the fuel source (the wick) faster than the flame can propagate through the gases. For similar reasons an explosive, set off near a gushing and burning oil or gas well, can sometimes blow out the flame.[6]

Figure 4-1. Effect of a cold surface on a flame.

WOOD COMBUSTION

The detailed process of the combustion of wood is more complicated (and hence, interesting) than the combustion of a candle or any common liquid or gaseous fuel. The reasons are: (1) wood is chemically complex; (2) wood is a solid, and in most home applications, fairly large pieces are used, so combustion can be quite inhomogeneous, parts of each piece being reduced to ash while other parts are just beginning to warm; and (3) the heating necessary to get wood to the temperature at which it can burn causes very complex chemical reactions in the wood. In a sense, raw, natural wood itself does not burn — its chemistry is so fragile that substantial changes take place due to the heating necessary to get whatever products of wood which *do* burn, up to their ignition temperatures.

The major chemical components of wood are cellulose and lignin, which are made of carbon, hydrogen and oxygen atoms. When wood is heated, as described in more detail in Appendix 3, many chemical compounds come out of the solid wood in the form of gases and tiny liquid droplets (one ingredient of visible smoke). Most of these compounds were not in the wood originally, but were formed from the wood's chemical structure. In a typical wood fire, these gases and droplets burn when they move out of the solid wood; they are the fuel of wood-fire flames. For laboratory study of these compounds, their burning must be prevented; hence, the wood is placed inside an almost closed container which is heated from the outside. The generated gases and smoke then fill the container and come pouring out, preventing oxygen from getting in and hence, making combustion inside the container impossible. This process of thermal degradation is called pyrolysis. Pyrolysis is the chemical break-up induced by high temperatures in the absence of oxygen. (The terms "carbonization" and "distillation" are also used). The final solid product after all the chemical changes have taken place is charcoal.

Total yields from the pyrolysis of wood depend critically on the conditions under which pyrolysis takes place. Generally, the quicker the heating of the wood, the larger is the yield of gases and tars, and the smaller is the yield of charcoal. Slow heating, taking about 24 hours to complete the pyrolysis reactions, yields about 50 percent charcoal (by weight, on a dry wood basis). Very quick heating, taking 1 minute, of small pieces of wood, yields only about 13 percent charcoal. The remainder, in both cases, is in the form of gases and tars; the tar fraction is largest for quick pyrolysis, and smallest for slow pyrolysis. Taking into account the above yields, and the known energy contents of dry wood and charcoal (8600 Btu per pound, and about 12,500 Btu per pound, respectively), one can roughly estimate that in typical wood fires between 1/3 and 2/3 of the energy content of wood is in the gases and tar droplets, which are normally burned in the long flames. The rest is the charcoal. The fraction will be different in different fires, depending on the fire's intensity and the type of wood, its moisture content and the sizes of the pieces.

COMBUSTION

As wood undergoes pyrolysis, combustion takes place if oxygen is available. When a new piece of wood is added to a fire it is heated by its hot surroundings. Since wood is a moderately good thermal insulator, the heat cannot quickly be conducted to the wood's interior; so only a very thin layer at the surface is affected initially. It dries very quickly and starts to char. The gases, evolved as the temperature rises to about 540°F, are not ignitable because the concentrations of non-combustible carbon dioxide and water are too high. (Most of the water passing out through the surface of the wood at this stage originates in deeper layers where the temperature is warm but less than about 212°F.)

As the surface layer temperature of the new piece of wood is raised to above about 540°F, pyrolysis becomes much more vigorous. The gases (and tar) evolved are ignitable partly because the water vapor contribution from deeper layers is now a smaller part of the total. However, the gases will not ignite and burn unless both adequate oxygen is available and the gases' ignition temperature is reached or exceeded. Both these conditions must be met at the same location and at the same time. The ignition temperature of the gas mixture is roughly 1100°F (600°C); it cannot be ignited by the surface layer of charred wood out of which it has emerged since it is not yet hot enough. Thus, if flames are to appear, they must be lighted either by other flames or by burning charcoal from a nearby piece of already ignited wood.

If the gases do ignite, they burn at a temperature of about 2000°F (1100°C), and there can then be a dramatic increase in the heating rate of the wood from which they come. Flaming requires oxygen, and the constant flow of gases passing out through the surface of the wood prevents oxygen from getting close to the surface. The amount of oxygen is inadequate until the fuel has moved some distance from the wood. Thus, the base of the flame is rarely in contact with the wood. The radiation from a wood-fire flame is substantial — it is estimated that between 10 and 30 percent of the energy released in a "luminous" (yellow) flame is radiant energy. Some of this radiation is absorbed by the wood's surface, and this contributes to its heating.

The charred surface itself does not usually burn for some time. Since charcoal does not vaporize at any temperature achievable in a wood fire, it can only burn when, and to the extent that, oxygen comes to it, and oxygen can get to the surface only when the flow of gases coming out of the wood has subsided. Charcoal starts to glow visibly when its temperature reaches about 900°F (about 500°C). The glow is not necessarily an indication that it is burning, just as the glow of a fire poker or an electric heating element is not evidence that it is burning. However the glowing of charcoal usually indicates that its surface is burning.

Charcoal burns with little or no flame. Charcoal is mostly carbon, but also contains hydrogen, oxygen, and minerals. Oxygen which wanders onto the surface combines with carbon to form carbon dioxide and carbon monoxide. Carbon monoxide is a combustible gas. Small quantities of other gases from the charcoal, particularly hydrogen, may also be emitted. The quantities are small, so if ignited, only small flames result. The flames from charcoal are hard to see because they usually emit only a faint blue light.

As the carbon (and hydrogen and oxygen) disappear from the charcoal, those few elements (some of the minerals) which do not form gaseous compounds are left as ash. The ash layer which is always forming (or being left) on burning charcoal is very light, and is easily blown off by the natural air movement in the fire. Some is carried up the chimney in a typical home heating fire, and some falls and accumulates under or around the fire.

Charcoal is a better thermal insulator than raw wood. Thus the thicker the charcoal layer is on a piece of wood, the more it slows down the penetration of heat to the deeper portions of the wood. This contributes to the steadiness of the burning of wood, especially in larger pieces.

When wood is heated by a flameless source such as radiant energy, it is observed to ignite spontaneously around 600° to 750°F. This is the temperature of the wood's surface when flames first start to appear. It is interesting that this temperature is below the spontaneous ignition temperatures of the gases coming out of the wood (which range from 800° to 1200°F). The ignition temperature of tar from pine is about 670°F, but charcoal's ignition temperature is still lower — perhaps as low as 300°F for fresh charcoal.[7] Probably, spontaneous ignition of wood starts with the spontaneous ignition of charcoal. The burning charcoal's temperature then becomes high enough to ignite the gases.

All of these various stages of pyrolysis and burning are often taking place simultaneously in each piece of wood. At sharp edges of the wood, charcoal can be burning and ash accumulating in less than a minute after adding the piece to the fire; oxygen can get to these areas relatively easily. On the other hand the center of a large log may not even start getting warm for half an hour or more. Also, the transitions between various stages of pyrolysis and burning are not sharp but continuous. For example, small amounts of pyrolysis doubtless continue to about 900°F; and some surface charcoal may be burned even while the gas flow through it is quite strong.

But this description of the basic processes, even if somewhat arbitrarily categorized, is very useful for understanding aspects and problems of practical fire building and control such as starting a fire, keeping it going, controlling its rate, minimizing smoke, and using green (moist) wood as a fuel.

WOOD FIRES

The ignition of wood fuel and maintenance of the resulting fire require both oxygen and adequately high temperatures. Lack of either will cause a fire to subside and sometimes die. For instance, lighting a fire built with closely packed straight pieces of wood is difficult because not enough air can move between the pieces. On the other hand, wood dust (such as is generated from sanding wood) can burn explosively if suspended in air at a concentration where each particle is surrounded by enough oxygen to burn it, and the particles are close enough together for the combustion of a few to ignite others.

The ability of a single isolated piece of wood to burn completely once ignited, depends critically on its size. At one extreme is a thin wood sliver or match stick, which burns readily. On the other hand, a large log will burn as long as an external heat source is used (such as a pilot flame or intense radiant energy), but when isolated, the fire will almost always die out. The difference is in how effectively the combustion of one part of the fuel can heat the other parts to the temperature where they too burn. Hence, for any size piece of wood, the combustion will be more complete if it is ignited at the bottom rather than at the top, for then the flames lap against other parts of the wood rather than the air.

But a larger single piece of wood cannot burn completely no matter where it is lighted because the heat spreads out (is conducted) into a larger amount of wood and hence can not warm the surface to a high enough temperature to help the fire grow. A wood sliver or match stick has much less mass per unit surface area, so the same intensity of surface heating will result in a much larger temperature rise of the wood.[8] It is principally the thinness of a piece of solid fuel which makes it possible for burning started at one location to spread over and through the whole piece.[9] Thus, newspaper and small size kindling burn easily since they can self-sustain a flame; they only require ignition and adequate oxygen to assure that they will burn completely. The critical thickness of a stick of dry wood below which it can sustain combustion if lighted at the bottom, is about 3/4 inch).[10]

Air velocity can also affect the ability of a single piece of wood to support a flame. A gentle breeze is detrimental. The convective cooling of the wood and increased distance of the flame from the wood are stronger effects than the increased oxygen supply. A stronger wind has the additional detrimental effect that it can blow the flame out if the wind's velocity exceeds the flame velocity of the burning gases.

When a two-piece-or-more wood fire is attempted, success depends critically on relative positions, assuming now that the pieces are large enough that they cannot individually sustain combustion. The principles are most clearly illustrated by a somewhat artificial example. Two thick vertical slabs of wood facing each other can support combustion of the inner facing surfaces if the spacing is just right. If brought too close together, the slabs suffocate each other by overly restricting the air supply in the space between them. If the gap is too wide the fire will also die; the further apart the slabs are, the less they keep each other warm.

In practice, the most important principle in laying and sustaining a fire, particularly an open fire, is to position the pieces of wood near enough to each other so that they will keep each other hot, but far enough apart to allow an adequate oxygen supply to move between them. Straight and smooth pieces of wood usually need to be crossed to create adequate air passages among them; crooked or bumpy firewood often need not.

Wood can burn without any flames. This is typically what happens in stoves operated at very low powers with large charges of wood. The low powers are achieved by severely limiting the air supply. Under this circumstance the gases and tars emitted from the wood may never burn, yet combustion of the charred wood can proceed until it is all consumed. Flameless combustion of the surface layer of charcoal is the main source of heat which pyrolyzes the next layer, but radiant energy from nearby pieces of similarly smoldering wood is usually necessary to sustain the process.

A smoldering piece of wood usually does not ignite its own combustible gases, i.e., it does not burst into flame. The glowing charcoal layer of smoldering wood is very hot — about 1200 to 1600°F judging from the color of its glow (Table 2-2). This is above the ignition temperature of the gases and tars. Ample oxygen is also around — typically much less than half of the oxygen in the air passing through a stove is used by a smoldering fire. In open smoldering fires, such as in a fireplace, the oxygen supply is clearly not preventing flaming. And yet, the smoke often does not ignite. The reason is that the oxygen and the high temperatures are not at the same place. The combustible gases are very hot as they pass through the glowing charcoal layer, but have cooled substantially by the time they are a little distance from the surface. Oxygen is abundant some distance from the surfaces, but because of the constant gentle "wind" of gases evolving from the wood, its concentration close to the glowing charcoal is low (though not zero — the charcoal can be burning). Thus, it can happen that at no one place in the fire is there both adequate temperature and adequate oxygen for the gases to ignite.

The higher the air velocity around an ordinary multi piece (more than one piece of wood) fire, the more likely it is that the gases will ignite. This is in contrast to the case of a candle or an isolated single piece of burning wood. In fact, blowing on a smoldering fire is one of the best ways to make it flame. The resulting increased oxygen supply makes the charcoal burn (and therefore glow) more vigorously, which in turn can ignite the smoke. Blowing on a smoldering match does not help reignite the gases because the match has no neighboring glowing wood to contribute heat to the gases. Any combustible gases between pieces of glowing wood in a multi piece fire have a reasonable chance of ignition if the charcoal is hot and oxygen is available, and blowing contributes on both counts. It is possible for the wind speed to be high enough to blow out wood-fire flames, just as a match can be blown out, but only temporarily; when the wind stops, the gases usually reignite.

Green (moist) wood needs more coaxing than dry wood to burn, for a number of reasons, all relating to its water content. The moisture must be evaporated as the wood burns. Since this consumes sensible heat (and converts it to latent heat) the temperature rise of the wood is slowed, which tends to inhibit combustion.[11] The temperature rise at the surface is also slowed because moist wood is a better conductor of heat than dry wood; a larger amount of the energy received at the surface conducts and spreads out into the wood's interior. As the water vapor passes through the hot charcoal on the wood's surface, some of it reacts with the carbon, producing carbon monoxide and hydrogen; this reaction uses heat energy and thus has a cooling effect on the charcoal. And finally, the water vapor which does leave the wood dilutes the combustible gases coming out of the wood and hence, makes them more difficult to burn. All these effects slow the combustion rate; this is why some people like to use some green wood in a fire when a low combustion rate is desired (such as overnight in a stove). But a fire of only very moist wood can be extremely difficult to sustain.

FLAME COLOR

Most, but not all, organic fuel gases are capable of burning either with a mostly bluish flame or an orange-yellow flame. The latter are called "luminous" flames because they are so much brighter. A wood fire can have either. In early and mid stages of combustion, wood flames are long and luminous, while pure charcoal at the end of the combustion cycle may burn partly with a small blue flame.

For those fuels capable of either color flame, the main determining factor is oxygen. With adequate oxygen throughout a flame, the flame color is predominantly blue. The blue light does not come from the ultimate combustion products CO_2, CO, H_2O but originates from intermediary and unstable molecules such as CH, C_2 and OH. Most natural gas and propane flames in appliances (such as ranges, clothes dryers, water heaters and furnaces) are predominantly blue because they are premixed flames — air is thoroughly mixed with the fuel before it burns, ensuring adequate oxygen throughout the flame.

The yellow light in luminous flames comes from very small glowing carbon particles which form in fuel-rich regions.[12] They are glowing simply because they are hot, just as the hot fire poker or any hot solid does.[13] The yellow light is so much brighter than the blue associated with the chemical reactions that the human eye and brain sense only the yellow; in fact, the fainter blue light is still there also. The glowing particles are much too small to be seen by the unaided eye as individual sources of light (they are about a millionth of an inch in size) . One piece of evidence that the particles really exist is that they can be collected on a cool surface inserted into the luminous region of a flame — they form a deposit of soot on the surface.

Almost all diffusion flames are luminous or have luminous regions. Since the oxygen is not premixed with the fuel, there will be regions where the temperature is high enough to induce partial forming reactions but where the oxygen concentration is too low to allow complete combustion. This is the case for the long luminous flames in wood fires. The blue flames from charcoal do not form carbon particles because the gaseous fuel which is burning is mostly carbon monoxide and hydrogen. Hydrogen gas, of course, cannot form carbon particles, and carbon monoxide is one of the few exceptions among carbon-containing fuels in that it cannot burn with a luminous flame under any circumstances.

The yellow color of a flame is not necessarily the sign of an incomplete combustion process. Whenever a flame is yellow, there are carbon particles, but the particles are highly combustible. Thus, in moving out of their regions of formation in the flame, if they pass through a region with sufficient oxygen and at a high enough temperature, they will be burned to carbon dioxide. This is largely the case in a non-smoking candle flame and in wood-fire flames under many circumstances.

SMOKE

The term "smoke" is sometimes used to designate all airborn byproducts of combustion, and sometimes just the visible components, and sometimes just the combustible ingredients. In practice, whenever the concentration of smoke particles is low, so is the concentration of combustible gases.

The total chimney effluent from a wood fire contains at least hundreds of different components. Nitrogen is by far the most abundant ingredient — about 70-78 percent by volume, just a little less than its concentration in air. Its source is the combustion air. Unused oxygen from the air is the next most abundant, constituting about 10 to 20 percent of the volume. Pure air is about 21 percent oxygen. The two products of complete combustion, carbon dioxide and water vapor, come next, carbon dioxide usually being between 2 and 12 percent, and water vapor between 3 and 15 percent of the total, by volume. All other ingredients taken together rarely amount to more than a few percent of the total. They include combustible gases, solid particles and liquid droplets. The gases and liquid droplets include mostly the (mobile) products of pyrolysis mentioned previously, plus derivatives which may be produced from them in the fire and chimney after they have emerged from the wood. The number of identified chemicals is large (Table A3-1), but the actual total number is doubtless still larger. (The liquid droplets are wood tar, and are sometimes called a tar fog. Their detailed chemical composition is not well known.)

There are two kinds of solid particles in wood fire effluents — carbon and ash. Some ash is always swept up the chimney by the moving air and gases. The carbon particles are some of the same ones which made the flames yellow-orange in color, although, often the small particles in the flames become agglomerated together to form larger particles in the flue gases. They are called soot or carbon-black when deposited on surfaces such as fireplace, stove, or chimney walls. Suspended in the gases, they are part of the smoke.

The visible part of wood-fire emissions consists of only the liquid and solid particles. (The gases are all transparent and, hence, invisible.) If the particles are all very small (smaller than can be seen in an ordinary microscope, which means smaller than the wavelengths of light) the smoke looks white or bluish-white.[14] If the particles are big enough to be seen individually, they appear either black (carbon) or brown (tar, and condensed, mixed, organic vapors) . Thus smoke with a sufficient number of larger particles can appear dark (this is rare from small wood fires, but not from candle flames) . The visible smoke from a flameless (smoldering) fire is mostly tar droplets, and they are usually so small that they appear white or bluish. Cigarette smoke is also high in tar content, and is also generated in a smoldering process.

In cold weather, droplets of condensed water can make any flue effluent look quite dense as it emerges out of the chimney. When the water vapor, which is always in flue gases, encounters the cold air outside the chimney, it condenses into droplets forming essentially a cloud. But as this cloud drifts away, it mixes with air, and soon the water droplets evaporate, making the water invisible again.[15]

CONCLUSION

The two necessary conditions for combustion of all fuels is adequate oxygen and sufficiently high temperatures. In wood fires, each log requires the presence of other burning logs or coals nearby in order to stay hot enough to keep burning. If the logs are too close together the supply of oxygen is restricted and combustion can slow or cease. These are the principles of fire building and maintenance.

Flame color is not a reliable indication of completeness of combustion in wood fires. Wood flames are usually yellow. The color comes from small glowing carbon particles, but the particles can be mostly burned by the time they leave the flame. If premixed with air, wood gases can burn with a blue flame; this cannot happen in an open fire, but may be possible in carefully designed stoves and furnaces. Blue flames do indicate relatively complete combustion in the flame, but flame color gives no information about how much combustible gas may be outside of the flaming region. Furthermore, the combustion in many yellow flames may be as complete as in blue flames, particularly in a stove designed to have a very hot combustion chamber. The only reliable way to determine the completeness of combustion is to test the flue gases for unburned gases and particles.

Chapter Five Chimneys

Chimneys are just as important as stoves for successful heating with wood. If a stove lets smoke into the room, an inadequate chimney is the most likely reason. Creosote problems are not exclusively related to chimney design but can be lessened by good chimneys. Heating efficiency can usually be improved by increasing the length of exposed stove pipe in a room, but at some cost to chimney performance. A large fraction of house fires associated with wood heating is due to unsafe chimney installations. The practicality and installation cost of wood heating is significantly affected by whether or not an existing chimney can be used. Chimneys are critical to successful wood heating.

FUNDAMENTALS [1]

The two functions of a chimney for a wood stove, furnace, or fireplace are to carry the undesirable combustion products (smoke, etc.) out of the house and to supply the draft or pressure forces necessary to feed air to the fire. The force necessary for both functions comes from the tendency for hot air to rise, an effect called "buoyancy." The flow up a chimney is restrained by flow resistance from chimney walls, bends, dampers, etc. It is the balancing of these two forces, buoyancy and flow resistance, which determines the quantity of gas flow in a chimney.

The buoyant effect of the hot stack gases effectively applies suction to the stove, which is to say, the air pressure outside the stove is greater than the air pressure inside the stove. Thus if the stove's air inlets are open, or if there are any cracks or holes in the stove, air will be pushed (or drawn) in. The term used to describe and quantify this effect is "draft." The draft at any point in the system is the pressure difference between the point of interest inside (the chimney or stove pipe or stove) and the air just outside (at the same elevation).[2] Draft is a measure of the force making gases flow. At a place where the draft is high, air would be drawn hard into any opening, but if the opening is small, not much air would be let in.

The most common device used to measure draft is beautifully simple — a transparent tube (glass or plastic) in the shape of a "U," partly filled with water. One end is left open and "senses" the pressure of the atmosphere outside the chimney or stove. The other end is connected to a metal tube which is inserted into the chimney or stove (perpendicular to the flow); it senses the pressure inside the chimney. If there is some draft, the pressures are unequal, and this fact will be indicated by a difference in level of water on the two sides of the U-tube. In America, the common unit used to measure draft is "inches of water," referring directly to this height difference. Drafts in domestic chimneys are generally no larger than 0.1 inch. This difference in water level is too small to easily be seen in a U-tube. Thus many draft gauges use different geometries to amplify the visual effect.

The draft is zero at the top of the chimney (the pressure of the flue gases as they emerge essentially equals that of the surrounding air) and is usually maximum at the smoke-pipe collar on the stove. If there is some draft everywhere in the system, no smoke can leak into the house even if there are cracks, since air will be pulled into the cracks rather than smoke being pushed out. This is usually the case.

Three main factors affect draft: the height of the chimney, the average temperature of the gases in it, and the velocity of the gases. Height and gas temperature both affect draft due to buoyancy. The velocity is important because when more gas is let into the chimney (making the velocity high) the draft or pressure difference tends to be relieved somewhat as it is when opening a bottle of champagne or a can of vacuum packed nuts. The important difference is that for the operating chimney the source of the pressure difference (the buoyancy of the hot gases) is always there, and so the pressure difference is never fully relieved. It is simplest to consider this velocity effect separately and to look first at the "static" draft. This is the draft that would exist if there were *no* flow as would happen if the smoke pipe (or chimney) were suddenly sealed off at the connection to the stove (or fireplace). This static draft [3] then depends only on the average flue-gas temperature and the chimney height.[4]

Figure 5-1 gives the static draft at the bottom of a chimney as a function of its height and the average temperature of the gases in it. Since buoyancy depends on air-density *differences*, the figure uses the difference in temperature between the gases inside the chimney and the outdoor air. To use the figure, locate the intersection of the desired temperature-difference and height lines; the static draft is then given (or interpolated from) the number(s) on the nearest curve(s). For instance, the static draft of a 20 foot high chimney with flue gases at an average temperature of 260 degrees Fahrenheit above outdoor temperature is about 0.1 inches of water. The draft increases both with increasing height and with increasingly hot flue gases. Maximum drafts correspond to the upper-right corner of Figure 5-1, and minimum drafts to the lower-left corner.

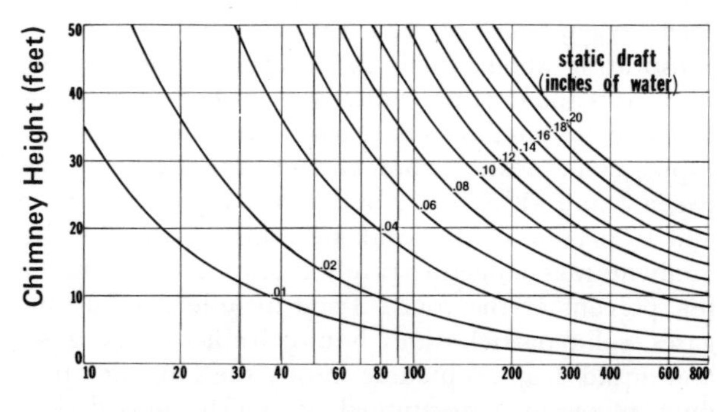

Figure 5-1. Static (or theoretical) draft developed by a chimney, as a function of its height and the temperature difference between the flue gases and the surrounding outdoor air. The flue gas temperature is its average temperature (see text). (An outdoor temperature of 40°F and sea-level pressure of 30 inches of mercury have been assumed.)

The number used for chimney height should be the total vertical rise about the point of interest, e.g., the draft at the stove pipe collar on the stove includes the effect of vertical portions of stove pipe in the house, and the vertical rise of slanted portions, as well as the chimney proper. The temperature used in the figure is the average temperature of the flue gases, a number which is often changing when a stove is in use. The gases leaving a stove may be as hot as 1500° F., but typically are between 300 and 800°F., since they cool as they progress through the venting system, and thus their average temperature is less. The amount of cooling is most for slow moving gases (e.g., from an air-tight stove operated at low power) in a single-wall metal chimney[5], and the least for high velocity gases (e.g., from an open Franklin stove) in an insulated chimney. Typical flue-gas temperatures, averaged over the whole venting system, usually fall between 200 and 600° F.

The ultimately critical question about chimneys is not draft but flow. Draft is the cause — force or push — behind the flow. The flow is the net effect — the number of pounds or cubic feet of flue gases that get through the chimney in a given time. If a chimney has inadequate flow capacity for a particular stove, the stove will not be able to operate at its maximum heat output rate, even with its air-inlet damper wide open; the suction of the chimney will not be able to draw combustion air into the stove at a high enough rate. Also, smoke may leak out of cracks in the stove and/or stove pipe when the air-inlet damper is opened beyond a certain point.

Flow in a venting system is constrained by friction between the moving flue gases and their surroundings. Stovepipe elbows, chimney rain caps, and even straight sections of pipe or chimney all offer resistance to the flowing gases. The relative size of the resistance encountered by gas flowing through various kinds of pipe, chimneys and fittings is given in Table 5-1. These numbers can be used to estimate the overall capacity of a venting system, but are perhaps more useful for estimating the effect of contemplated changes in an existing system, such as the addition of extra stovepipe lengths and elbows, and increasing the height of the chimney.

The total flow of gases through a wood heating system is approximately determined by four parameters: the temperature difference between the flue gases and the outdoor air, chimney height, chimney diameter, and the whole-system resistance coefficient. To illustrate the effects of each of these variables on chimney performance, a particular example will be worked out.

A stove connects to a 2 foot vertical length of 6 inch diameter stovepipe, followed by a rounded elbow, 3 feet of horizontal pipe, a sharp cornered connection to a 13 foot high chimney (also of 6 inch diameter), topped with a rain cap (Figure 5-2). The resistance

	ESTIMATED RANGE
Round elbow, 45 degrees	0.2 - 0.7
Round elbow, 90 degrees	0.5 - 1.5
Tee, or 90 degree sharp elbow (mitered), or breeching (smoke pipe inserted into brick chimney)	1.0 - 3.0
Straight pipe or flue	
4″ diameter	.08 - .12 per foot
6″ diameter	.05 - .08 per foot
8″ diameter	.04 - .06 per foot
Expansions (e.g., 5″ to 6″ stove-pipe adapter)	.05 - .15
Chimney top	
Open	0.0
Spark Screen	0.5
Rain and Wind Caps	0.5 - 3.0
Stove-pipe damper,	
Open	negligible
Closed	5 - 20

Stove air-inlet damper effective resistance (damper fully open)[1]

	Chimney diameter (inches)				
	4	5	6	7	8
small stove	2-8	5-20	10-40	20-80	30-120
large stove	2-4	3-10	5-20	10-40	15-160

[1] With damper closed, the equivalent resistance typically exceeds 10 times the above figures. (If the damper is literally air tight, its resistance is infinite.) A fireplace or open stove has an "inlet" resistance of about 2.

TABLE 5-1. Resistance coefficients of chimney components, and effective resistance coefficients of stove air-inlet dampers. Resistance coefficients are pressure drops (due to resistance, not buoyancy) in units of "velocity heads" ($\frac{1}{2}dv^2$, where d is the mass density of the gas, and v is its velocity in the chimney or fitting.) Effective resistances are the same, except d and v are evaluated in the chimney, not at the air inlet. (Data are from *ASHRAE Handbook and Product Directory, 1975 Equipment Volume,* Table 9 in Chapt. 26, American Society of Heating, Air-conditioning and Refrigerating Engineers, New York, 1975, and from the author's measurements.)

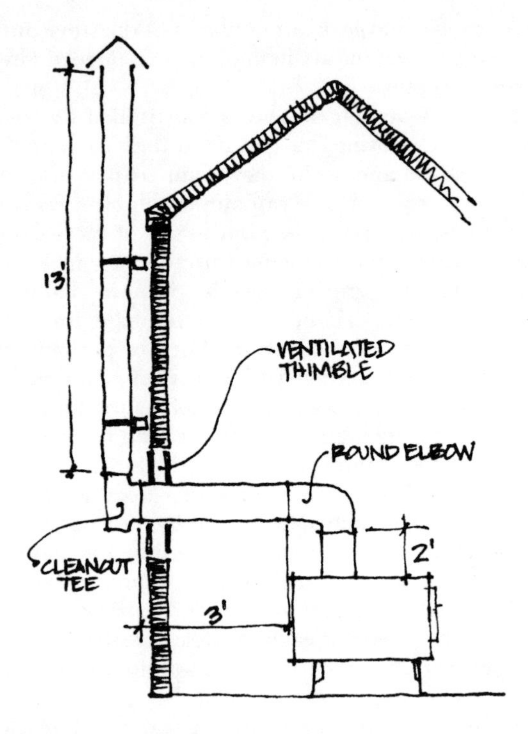

Figure 5-2. A chimney installation used as a quantitative example of a flow-resistance calculation in the text.

coefficient of this venting system (chimney plus chimney connector) is the sum of the appropriate coefficient as given in Table 5-1.

18 feet of 6 inch diameter pipe at .065 per foot	1.2
1 round 90 degree elbow	.7
1 clean-out tee	1.3
1 rain cap	.8
Total	4.0

The large range of numbers in Table 5-1 reflects reality: there are large variations in the resistances of various kinds of caps, pipes and elbows due to differences in surface smootheness, gradualness of curves, shapes of caps, etc. Typical values for the resistance coefficients have been used in this example, but there is clearly a large uncertainty involved. The estimates are nonetheless useful, as will be apparent below.

The stove itself also offers resistance to gas flow, especially at the air-inlet damper. In fact, the resistance of this damper is often so large, even when fully open, that the resistance of the rest of the stove is negligible. If we adopt from Table 5-1 a resistance coefficient of 15 for the stove with its air damper open, we obtain a total resistance for the whole system of 4 + 15 = 19. This is the minimum resistance for this system, since if the air inlet is closed somewhat, or if a stove-pipe damper is used, the resistance can be much more.

Given the system's resistance, the gas flow can then be estimated from Figure 5-3. Curves corresponding to various amounts of flow are drawn on the graph. The use of the graph is similar to the one for determining draft (Figure 5-1): the flow is determined by the location of the intersection of a horizontal resistance-coefficient line and a vertical temperature-difference line. In our example, the resistance is 19. For a typical average flue-gas temperature of 300° F., and outdoor temperature of 40° F., the temperature difference is 260° F. Entering the graph with these two numbers, 19 from the left and 260 from the bottom, we find the intersection lies just above the curve labeled 200 — corresponding to a flow of about 180 pounds per hour. (Typical flue-gas flows from wood heaters are indicated in Table 5-6).

The usefulness of looking at these chimney details quantitatively is not to be able to predict exact performance. This is not only unnecessary (one does not need theory to convince oneself that typical chimneys are adequate), but impossible since there are unpredictable details in each situation which are influential on chimney performance. The usefulness is principally in understanding the *relative* importance of various parameters and changes.

For instance, increases in flue-gas temperature does not always result in better chimney performance. It *is* always true that the higher the temperature of the flue gases, the larger is the static draft. It is also true that for a given system resistance, higher flue-gas temperatures always imply higher flue-gas velocities. But, as the temperature of a gas rises, it expands and becomes less dense, and this effect actually decreases the mass flow at sufficiently high temperatures. As illustrated in Figure 5-3, for a given resistance the mass flow is maximum at about 500°F temperature difference, and actually decreases slightly at higher temperatures. However the effect is small, and in practice the mass flow up a chimney in any particular system is essentially the same for any (average) flue-gas temperature higher than about 350° F.

HEIGHT

Figure 5-3 is based on a chimney height of 15 feet (including the chimney connector, if applicable). Increasing chimney height improves flow capacity, but not in direct proportion to the change in height. If all other things remained unchanged, a doubling of chimney height would result in an increase in capacity of 41 percent; a 10 percent height increase would improve performance by 5 percent.[6] But other things do change. The higher the chimney, the more the flue gases cool before emerging out the top of the chimney.

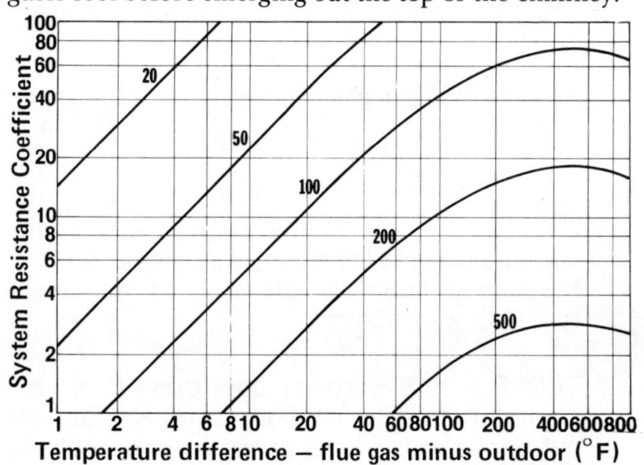

Figure 5-3. Flue-gas flow, in a 15-foot high, 6-inch diameter chimney. See Table 5-1 and text for explanation of system resistance coefficients.[5]

Thus the average temperature of the flue gases decreases, which has a negative effect on chimney capacity.[7] There is always an improvement in chimney capacity with increased height, but the amount of improvement is diminished if heat is conducted out through the chimney walls. Insulated chimneys benefit the most from height increases.

Table 5-2 illustrates the *net* effect of height on chimney capacity. The effect is not very large (compared to the effect of chimney diameter — (Table 5-3). A 50 foot chimney is more than 8 times taller than a 6 foot chimney, but its capacity is only 2¼ times more. A 2 foot addition to an existing 20 foot high chimney will increase its capacity by only about 4 percent.

Thus ever higher chimneys yield diminishing returns. I.e., if a stove lets smoke into the house in calm weather, poor draft due to chimney height is not the most likely cause.[8] In practice, height is determined by the need to terminate the chimney above wind-pressure influences, and by architectural considerations. These usually require a minimum chimney height of about 10 feet.[9] If the chimney diameter is the same as the stove's collar, this height is generally adequate. If it is not, increasing the chimney diameter by only one size (1 or 2 inches) will usually improve performance much more than increasing the chimney's height by 50 percent.

DIAMETER

Chimney diameter is a more decisive variable than chimney height. Assuming the whole venting system (chimney and chimney connector) is of the same diameter, the capacity of the system is approximately proportional to its cross-sectional area. (The proportionality is not exact since heat loss through chimney walls and the resistance to flow also are affected by chimney diameter.) Table 5-3 illustrates the relation between chimney diameter and capacity. An 8 inch chimney has about 90 percent more capacity than a 6 inch chimney, and a 6 inch system is about 55 percent better than a 5 inch system. In other words, diameter *is* critical. In systems with mixed sizes and/or shapes of flue, for instance 6-inch stovepipe connecting to an 8 inch chimney, it is usually reasonable to base the flow estimate on the smaller size portion. In practice, it is usually safe to take the stove manufacturer's recommendation concerning minimum chimney diameter. Lacking explicit instructions, a chimney of the same size as the flue-pipe collar on the stove is usually adequate.

The inside dimensions of an existing masonry chimney can be determined most reliably by direct measurement taken through a hole used for an appliance connection, or at the top of the chimney. Alternatively, one can estimate the internal dimensions from the outside dimensions, but here an assumption concerning chimney construction must be made. Most, but not all, brick chimneys are of single brick thickness, and bricks vary in their dimensions. Bricks are usually *about* 3½ inches wide; thus a good guess is that the inner dimensions are 2 x 3½ inches or about 7 inches less than the outer dimensions.

If a flue liner is present, the inner dimensions are slightly less. Table 5-4 gives estimates of the inner dimensions of the liner based on the outer dimensions of masonry chimneys. These are only estimates. If it

CHIMNEY HEIGHT (feet)	RELATIVE CAPACITY
6	60
8	69
10	76
15	88
20	100
30	115
50	135

TABLE 5-2. The influence of height on chimney capacity. Capacities are relative to a 20 foot high chimney which is arbitrarily given a capacity rating of 100. Heat loss effects are included. The numbers are approximately valid for any type of, and diameter, chimney, but are specifically for a masonry chimney with an internal area of 38 to 50 square inches (6 to 7 inch diameter, if round). (Data is adapted from Table 17 in Chapter 26 of the 1975 Equipment Volume of *The ASHRAE Handbook and Product Directory*).

CHIMNEY DIAMETER (inches)	RELATIVE CAPACITY
3	20
4	38
5	64
6	100
7	139
8	192
10	330
12	506

TABLE 5-3. The influence of diameter on chimney capacity. Capacities are relative to a 6 inch diameter chimney, which is arbitrarily given a capacity rating of 100. The numbers are approximately valid for any type of, and height, chimney. (Data is adapted from Tables 13 and 14 in Chapter 26 of the 1975 Equipment Volume of *The ASHRAE Handbook and Product Directory*).

appears from this kind of guess that a given chimney is marginally adequate for the intended use, an actual measurement should be attempted, or, at least, caution should be exercised.

CHIMNEY CONNECTOR

Is the complexity (number of elbows) and length of a chimney connector critical to chimney performance? Most authorities recommend that stoves be located as close as possible to their chimneys and that the connection be as direct and short as possible. This recommendation is based on conservative safety and performance considerations. But there can be reasons for other kinds of installations. Aesthetic and architectural constraints may be better satisfied by placing the stove some distance from its chimney. Such arrangements generally improve the energy efficiency and the heat distribution of the system.

There are two effects which decrease chimney capacity when chimney connectors are increased in length or complexity from their normal 4 to 6 feet and 1 or 2 elbows. One is increased resistance to gas flow, the other is decreased static draft due to a cooler average temperature of the flue gases. The question is, are the effects significant? Will the stove let smoke into the room or be unable to operate at normal maximum power because of inadequate chimney capacity?

Clearly the answer is, not necessarily. Many people today have long chimney connectors. In the previous century, such installations were very common. The stoves in churches and Shaker buildings were typically placed as far away from their chimneys as possible, and, not uncommonly, 50 feet or more of stove pipe was used to connect them. Obviously it can work.

Table 5-1 can be used to estimate the effect of extra stovepipe on system resistance. In a system with a slightly oversize chimney, the system resistance is typically more than 20; adding 0.065 for each additional foot of 6 inch stovepipe and 0.7 for each of 1 or 2 more elbows will have only a small relative effect on the total resistance. If the system resistance is only 5, then additions of stove pipe and elbows may be significant; this is more often the case for fireplaces and open stoves than for closed stoves, since their air-inlet resistances are much lower. The decrease in static draft due to adding a few extra feet of stovepipe is typically less than 15 percent. This will not be critical in most cases.

Combining the effects of increased resistance and decreased static draft, one can estimate that doubling the length of the chimney connector from 5 to 10 feet and adding an extra elbow or two will not be detrimental to venting if the venting system was slightly (30 percent) oversized (in terms of capacity) to begin with. The most important single clue which suggests excess chimney capacity is a significantly larger chimney diameter than stove-collar diameter.

EXCESS CAPACITY

Excess chimney capacity can be detrimental. As the flue gases rise in a chimney they lose heat through the chimney wall and thus cool. In general, the larger the chimney diameter, the slower the flue gases rise, and the more time they have to cool. Thus they are less buoyant and draft is reduced. This problem is most likely to arise when a large fireplace flue is used for a small stove. The problem can be aggravated by outdoor air descending into the chimney; this cools the gases still further and increases the resistance to their flow. If a stove which is connected to an excessively large chimney smokes, two remedies worth trying are 1) installing an open cone on the top of the chimney (Figure 5-4) to taper the opening down to a smaller size (this should inhibit cold-air down drafts), and 2) using a metal chimney or stovepipe within the larger chimney.

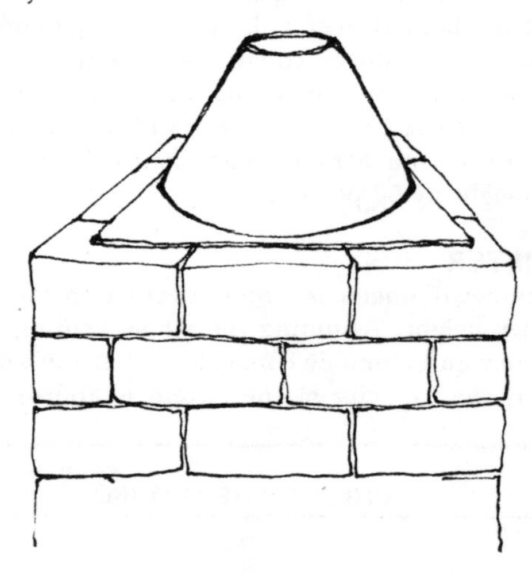

Figure 5-4. A chimney cone. A constriction at the top of an oversized chimney can help prevent downdrafts into the chimney.

Another problem with oversized chimneys is condensation, of both water and creosote. Condensation represents a failure of a chimney system in the sense that the chimney is not venting to the outdoor air all the products and by-products of combustion. Oversized chimneys invite more condensation because they invite more cooling.

CONDENSATION

Minimizing condensation in chimneys is important for many reasons. The condensate is corrosive and thus decreases the life expectancy of chimneys. The

TYPICAL OUTSIDE DIMENSIONS OF CASING	NOMINAL LINER SIZE	INSIDE DIMENSIONS OF LINER	AREA OF LINER	EQUIVALENT ROUND DIAMETER
16 x 16 in.	8 x 8 in.	6¾ x 6¾ in.	46 sq.in.	7.4 in.
16 x 21 in.	8 x 12 in.	6½ x 10½ in.	68 sq.in.	9.0 in.
21 x 21 in.	12 x 12 in.	9¾ x 9¾ in.	95 sq.in.	10.4 in.

TABLE 5-4. Approximate dimensions of rectangular masonry chimneys and liners. (Data from Table 10 in Chapter 26 of the 1975 Equipment Volume of *The ASHRAE Handbook and Product Directory*).

organic condensate (creosote) is flammable and is the fuel for chimney fires. More frequent chimney cleaning is needed when condensation is heavy. If liquid condensate leaks out of the chimney or chimney connector into the living space, it creates a mess — its odor is not pleasant, it can stain, and it can be a fire hazard.

Chimney design can help minimize condensation by keeping the flue gases as warm as possible. There is no one well defined temperature above which no condensation will occur, because flue gases contain a vast array of gaseous compounds, and also in small droplet form. If various types of chimneys are connected to the same stove, condensation is least in insulated chimneys — double-wall factory built chimneys with high-temperature insulation packed between the walls (e.g., Metalbestos). Such chimneys are 3 to 4 times better insulators than single-wall metal chimneys (Table 5-5). Masonry chimneys, even with tile liners, are perhaps surprisingly not much better (only 20 to 30 percent) than single-wall metal chimneys for preventing heat loss from flue gases.[10] Flue gases emerging from an insulated chimney will typically be 50 to 200° F. warmer than from an uninsulated chimney. Thus insulated chimneys help significantly to maintain an adequate draft and minimize condensation. This is particularly significant during the low-power high-smoke firing conditions under which stoves often operate at night.

Masonry chimneys are also especially conducive to condensation whenever a fire is started. Their large mass takes time to warm, which virtually guarantees condensation on the inner surfaces at the beginning of each fire.

CHIMNEY MATERIALS

Almost all new chimneys built today are either prefabricated metal, or masonry. The metal chimneys have 2 or 3 concentric walls with air spaces or insulation in between (Figure 5-5). Masonry chimneys are usually built of brick, but special concrete blocks, adobe and stone are also used. Tile liners are standard (Figure 5-6); they contribute to both the safety and durability of the chimney.

Masonry chimneys with tile liners can be more expensive than prefabricated metal chimneys, partly because there is more labor involved, but they are very durable. Their performance offers both advantages and disadvantages. A disadvantage is their high heat

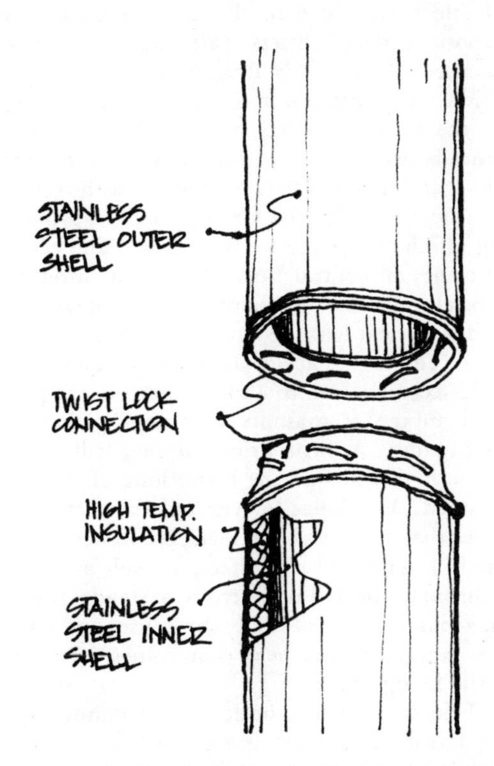

Figure 5-5. A section of a prefabricated insulated metal chimney.

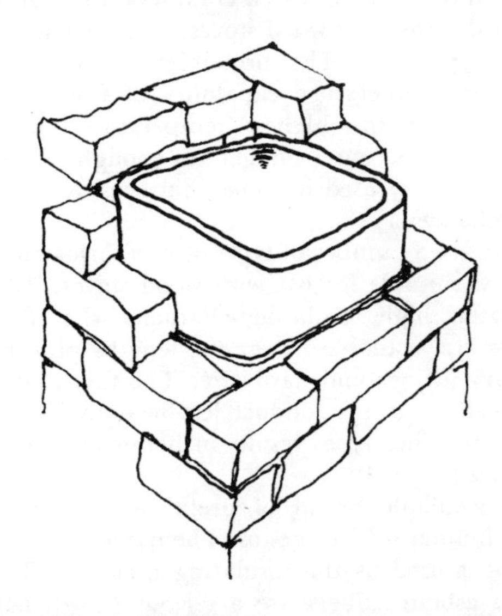

Figure 5-6. A tile liner in a masonry chimney.

TYPE	HEAT LOSS COEFFICIENT[1]
Heavy steel, clay or iron pipe	1.2
Dark or weathered steel stove pipe	1.2
Shiny steel stove pipe	1.0
Masonry, unlined	1.0
Masonry, tile lined	1.0
Double wall metal, air space	0.5
Double wall metal, insulation in between, stainless-steel inner	0.3
Triple wall metal, air ventilated (no insulation)	effectively greater than 1.2

[1] Suggested design value — actual measured values span a considerable range. Units are Btu per hour per square foot of inside area per °F temperature difference (flue gas to exterior).

TABLE 5-5. Heat loss coefficients of some common chimney types. (From Table 6 in Chapter 26 of the 1975 Equipment Volume of *The ASHRAE Handbook and Product Directory*).

loss (Table 5-5), which results in more creosote and soot deposits and slightly less draft, as discussed above. This is especially true for chimneys on the outside of a house. Masonry chimneys in the interior of a house are better; the heat loss is less because the temperature difference between the inside and outside of the chimney is less, and much of the heat loss from the chimney can be a heat gain for the house, which improves the heating efficiency of the whole system. The effective net efficiency of central furnaces vented through interior brick chimneys is about 15 percentage points higher due to heat leakage from the chimney into the house.[11] The use of an insulated metal chimney reduces this contribution to about 5 percentage points.

An advantage of masonry chimneys is their heat storage capacity. A warm brick chimney will continue to let its stored heat into the house long after the fire has subsided. This helps to even out the temperature variations that usually characterize wood heat. To derive the full benefit of this effect, as well as the heat gains through the chimney from a steady fire, the chimney must be located inside the house, and have all its sides exposed so the heat that comes off its sides enters the living space.

Prefabricated (factory-built) metal chimneys (including installation costs) are usually less expensive than masonry chimneys. Installation is not too difficult and is done by some owners themselves. There are many kinds of prefabricated chimneys, most of which are not designed for wood stoves, but for gas- or oil-fueled appliances. The principle difference is the maximum temperature capability of the chimney; stoves require the highest temperature chimneys. (Wood stove flue gases can get hot enough to melt the aluminum alloy used in some chimneys designed for gas appliances.)

There are a number of types of prefabricated metal chimneys suitable for use with wood stoves. All have moderately high insulating characteristics. Straight sections are available in various lengths plus all the necessary fittings and hardware. The fact that these multi-wall chimneys conduct less heat through their walls than other types results in better chimney performance.

One available brand of prefabricated, insulated metal chimney is Metalbestos. The name suggests that asbestos is used as the insulating material. Though inhaled asbestos fibers are a serious health hazard, there is, in fact, no asbestos in Metalbestos chimneys. Many years ago, asbestos was used, but today the insulating material is a mixture of fiberglass and agglomerates of silica particles.

The life expectancy of prefabricated chimneys is not well known because they have not been in use for a long enough time. Some chimneys with all stainless steel metal parts are still in service after 20 years of use. Corrosion of the flue can occur due to chemicals in the exhaust from the stove. Burning of household trash including plastics is very hard on any kind of chimney due to the hydrofluoric and hydrochloric acids generated. The next worst fuel is coal, due to its sulfur content. Wood smoke is also slightly corrosive, but good tile liners (for masonry chimneys) and stainless steel are both fairly resistant. Corrosion is always much worse when chimney surfaces are moist; this is why condensation shortens chimney lifetimes.

There are prefabricated metal chimneys designed for commercial and industrial applications which are more durable than the standard residential types. They feature either a particular stainless steel alloy (Type 316) that is even more corrosion-resistant than most other alloys, or a double coat of acid-resistant porcelain enamel. However there is not yet a demonstrated need for these more corrosion-resistant materials in wood-stove chimneys.

One particular kind of prefabricated metal chimney is unsatisfactory in some wood-stove applications. It is a triple-wall chimney which has no insulation but uses circulating air for cooling (Figure 5-7). The air between the inner two layers is kept warm by the hot exhaust gases in the flue; similarly the air between the outer two pipes is kept relatively cool, there being only one layer of thin metal between it and the outside air. The two air spaces are connected at the bottom of the chimney. Thus the difference in temperature induces a natural-convection air flow. This system can work so well that the outside of the chimney feels cold due to the downward flow of outdoor air inside it.[12] However, the innermost pipe is also cooled, which can result in substantial condensation and reduction in normally expected draft. Cold climates and exterior chimney installations exacerbate these problems.

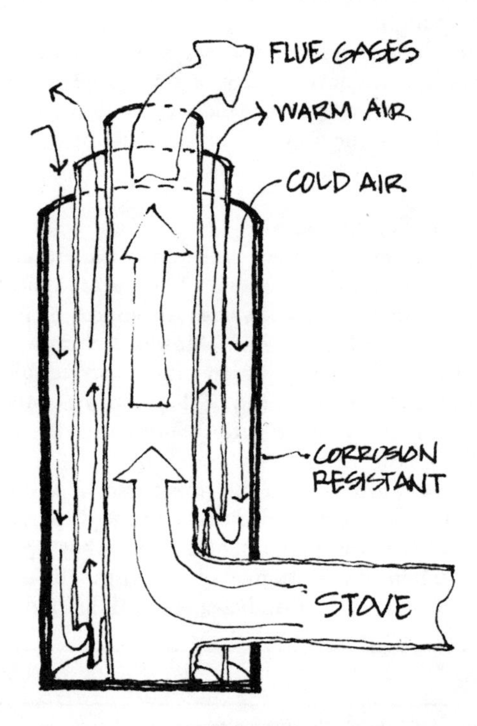

Figure 5-7. A schematic of a triple-wall air-cooled metal chimney.

The cheapest chimney is made of stovepipe. Such chimneys do not satisfy the National Fire Protection Association standards for safety, nor most (probably all) building codes. Much heavier (16 gauge or thicker) single-wall metal chimneys are permitted in commercial and industrial installations, but are not recommended for residential use. Nonetheless, many stove pipe chimneys are in use. Although stovepipe chimneys can be safe when installed (by observing the smoke-pipe chimney-connector safety standards in Chapter 9), safety hazards are likely to crop up later, principally due to unnoticed decay of the pipe. The light gauge (28) stovepipe in common use often lasts only one to three years as an exterior chimney. Unless inspection is unusually frequent and conscientious, the chimney will likely continue to be used after it has become unsafe.

Two other chimney types may have merit but are seldom used in residences. They are heavy steel chimneys made of steel sewer pipe or well pipe, and heavy, clay pipe chimneys. Unless the pipe is of a type specifically listed[13] for use as a chimney for wood stoves, risk is involved.

EFFECTS OF HOUSE ON CHIMNEY PERFORMANCE[14]

Many draft and smoking problems are, interestingly, not related to chimneys but to houses. A building itself, and appliances in it, can generate pressures which may help or impede the normal performance of chimneys. The effects are most noticeable when starting a fire and when firing rates are very low, as is often the case at night, for under these conditions, the normal draft generated by the chimney is small and is thus more likely to be affected by extraneous influences.

A house which is too air-tight can cause wood heaters to smoke, no matter how good the chimney may be, by restricting the supply of air to the heater. Fireplaces and open stoves (e.g., Franklin's) are most often affected, because of their very large air consumption. Listed in Table 5-6 are the typical natural air-ventilation (or exchange) rates of houses (due mostly to very small cracks around windows and doors), and the air needs of both closed stoves, and open stoves and fireplaces. Closed stoves need only about 10 to 20 percent of the amount of air that is naturally entering and leaving the house anyway; thus even a very tight house is unlikely to significantly restrict the needed combustion-air supply. Even if the air supply were restricted by the house, a closed stove is not likely to let smoke into the house, but would be forced to operate on less air, resulting in lower combustion rate and/or more smoke going up the chimney.

Open wood burners usually need air at a rate which is on the order of the natural air-exchange rate of houses. Thus especially tight houses may prevent the needed air from getting to such burners, and smoking can result. Open stoves and fireplaces depend on a certain minimum air velocity into the opening to prevent smoke from coming out into the room. The remedy usually requires increasing the air supply by opening a window, or ducting air under the floor, or through a wall, directly to the wood burner. Decreasing the area of the fireplace opening can also help.

Another problem is chimneys which are not "self starting" but need help from a piece of burning newspaper thrust into the chimney, or chimney connector, to start the flow in the right (upward) direction. This problem is related to the fact that a house itself behaves somewhat like a chimney. When it is cold outdoors, the warm air inside a house is buoyant with respect to the outdoor air. It tends to rise, but is of course, somewhat constrained by the house. In the upper floor(s) and attic, the warm air inside the house pushes its way outside; in the lower portions of a house (foundation and first floor) cold outdoor air is drawn in (Figure 5-8). This is often aptly called the stack effect. In very tall commercial buildings the resulting wind coming into open street-level doors can be very fierce unless special counter measures are taken.

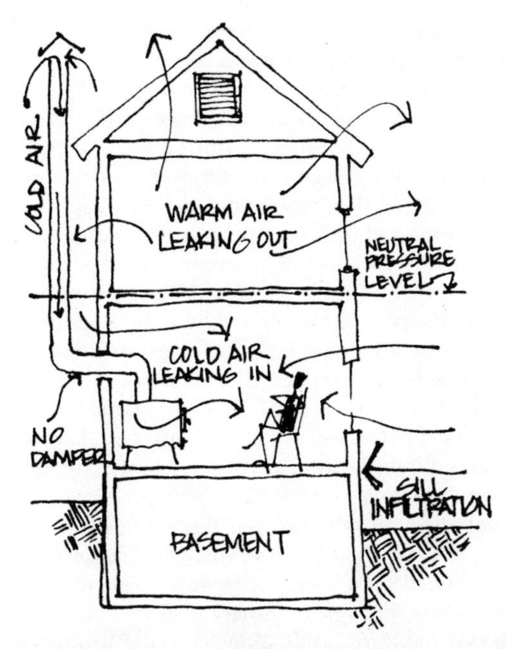

Figure 5-8. Some effects of the stack effect of a house, including flow reversal in a non-operating chimney. The arrows indicate the natural air flow generated by the buoyancy of the air in a warm house on a cold day. If the stove were upstairs, flow would not be reversed. Exterior masonry chimneys are also susceptible to flow reversal.

	POUNDS PER HOUR	CUBIC FEET PER MINUTE (at 70°F)
Small closed stove	100-200	22-44
Large closed stove	150-300	33-67
Open stove (e.g. Franklin)	250-500	56-110
Fireplace (with windows and doors closed)	400-1500	90-330
Typical house ventilation rate	500-2000	125-500

TABLE 5-6. Typical air consumption of wood heaters and air exchange rate of houses. Since most of the mass flow up chimneys is air, these numbers also represent typical flue-gas flows. (Data from various sources, including the author's measurements).

There is always some level in the house at which the inside air pressure equals the pressure outdoors at the same level. Below this level, the outdoor pressure is higher, tending to push air into the house, and above this level, the pressure inside is higher, tending to push air out. This neutral-pressure level is about half way from the ground level to the top of the house, but can be anywhere.

One consequence of these natural pressure differences in and around a house on a cold day is that the air flow in chimneys which are not in use can be backward — down the chimney and into the house. This is most likely in a single-wall metal chimney which runs up the outside of a multi-story house, serving a stove on the ground floor or in the basement. The air inside such an exterior and non-operating chimney will normally be at the same temperature as the outdoor air. Thus it has no tendency to rise (buoyancy). But the warm air inside the house does, and this tends to draw air into the house at its lower levels through all available openings, including the stove and its chimney (Figure 5-8). The bigger the temperature difference between indoors and outdoors, and the taller the building, the stronger is the effect.[15]

Such chimneys are not self starting. A fire lit in a connected stove is likely to burn backwards — fresh air to feed the fire will descend through the chimney, and the smoke will emerge out of the air inlets into the house. To avoid this, newspaper can be put into a vertical section of chimney connector, or the chimney itself, and lit. This usually corrects the flow direction immediately, and if the fire is lit while the paper is still burning, proper operation will have been established. Flow reversal can also occur in the middle of a fire during low firing rates. This can be dangerous, particularly if it happens while the occupants are asleep.

Exhaust fans, such as in kitchens and bathrooms, and forced ventilation systems, can increase the susceptibility of any chimney to these problems of flow reversal and non-self-starting. Particularly in relatively tight houses, use of an exhaust fan can decrease the pressure inside a house, which can result in air being drawn down through chimneys.

The air pressure inside a house is also influenced by the location of open windows or doors. In a closed house, the neutral-pressure level is usually about half way up the house as stated previously; outdoor air tends to be drawn in through any available openings below this level, and house air tends to be pushed out openings above this level. However, if there is an especially large opening anywhere in the house, such as a wide open window, it will neutralize the pressure at that level. Opening a window on the ground floor lets air in and makes the pressure inside the house at the ground floor rise to that outside the house at the same level. This increases the pressure throughout the house, and as a result, all chimneys operate better. With an open window at ground level, and none open elsewhere, flow reversal should not occur (in calm weather) as long as stack temperatures remain above outdoor temperatures, which is always true.

Opening a window near the top of the house equalizes the indoor and outdoor pressures at that level, by letting some of the house air out. This decreases the pressure throughout the house and thus inhibits the proper functioning of all chimneys in the house. The flow will be backwards in any chimney in which the average temperature of the gases falls below the indoor temperature.[16]

It is the location of the neutral pressure zone which is critical. Its location depends on the relative leakiness of the house in its upper and lower portions. The more leaks in the lower portions the lower the neutral zone, and the better will be chimney performance. The leakier it is in the upper portions, the higher will be the neutral zone, and the more problematic chimney performance will be at lower powers and start up.

Interior chimneys (no surface exposed to the outside air until the chimney pierces the roof) are least influenced by the house stack effect. Since the lowest temperature of the flue gases in an interior chimney is approximately the temperature in the house, draft reversal should rarely occur (in calm weather). By the same token, single-wall exterior chimneys are most likely to cause difficulties; a masonry chimney with from 1 or 3 sides exposed should have slightly warmer flue gases than a fully exposed, uninsulated metal chimney, and thus be somewhat better. Insulated chimneys such as Metalbestos chimneys, are the least susceptible to these problems.

Another strange effect caused by the tightness of some houses is smoke coming out of one fireplace while another one is in use. Use of a fireplace or open stove in a tight house can result in decreased pressure through the house due to the chimney draft. The result is that air is pulled into the house through every available opening, which can include other flues. If two fireplaces are served by separate flues in the same chimney, the suction due to the lowered pressure in the house can pull smoke from the active flue down through an inactive one and hence into the house through an unused fireplace or stove. Smoke can also enter the house through any other opening or crack near which it happens to drift or be blown.

WIND

Windy weather can also affect chimney performance, often adversely by reducing or even reversing the flue-gas flow — i.e., by making the fireplace or stove smoke. Properly locating the chimney top relative to the roof can alleviate some of the problem, and special chimney caps can be very effective.

There are two kinds of wind effects. If the wind direction is into an opening through which smoke is trying to come out, the smoke flow is impeded. For example, a low, open (uncapped) chimney on the downwind side of a sharply peaked roof will not perform well if the wind blows too hard[17] (Figure 5-9).

There are two ways to alleviate this problem. If the chimney extends high enough, the wind will be blowing mostly across it, not into it. This is one reason why it is usually recommended that the chimney top be 2 feet higher than any portion of the roof within 10 feet of it and 3 feet higher than the roof area through which it penetrates (Figure 5-10). Use of a chimney cap is the other way to alleviate this wind back-pressure effect. The objective of wind caps is to make the calm-weather smoke discharge omni-directionally. Then, with wind blowing from any direction, some smoke can still always get out on the other side. A flat

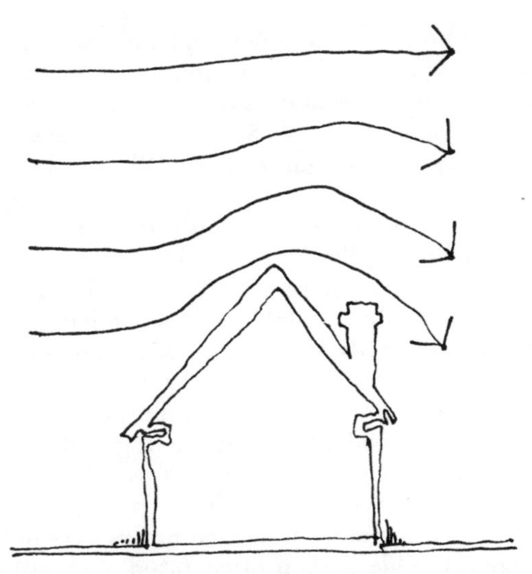

Figure 5-9. Wind interfering with the performance of a chimney without a cap or cover. Smoke can be pushed down a chimney.

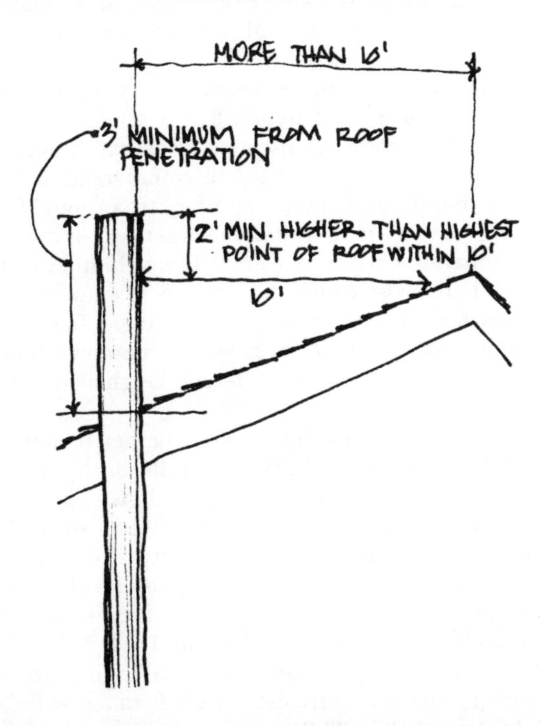

Figure 5-10. Minimum chimney heights above a roof, according to National Fire Protection Association Standard No. 211.

plate on posts is a common and simple wind cap for masonry chimneys, but fancier ceramic and metal caps are also used (Figure 5-11). A conical rain cap on a metal chimney helps, but if it extends too far beyond the outside of the chimney it can catch upward wind gusts and turn them down into the chimney. Chimney caps which rotate or have moving parts are usually unsatisfactory for use with wood fuel; creosote deposits tend to foul up the caps to the point where they become stuck.

The other wind effect always helps chimney performance. The first effect, discussed above, is caused by what is called the "velocity pressure" of the wind, which is the pressure it exerts due to its velocity. Air

Figure 5-11. Some examples of the large variety of chimney caps.

also exerts what is called "static pressure," which, in still air, is the same as atmospheric pressure. But in moving air, the static pressure is less.[18] In principle, a barometer in moving air would read a lower pressure than in nearby still air. Thus if a chimney top is in a region of strong winds, the general (static) pressure level all around it is reduced. This effectively applies suction on the gases in the chimney, and so it draws better – the wind aspirates smoke from the chimney. The stronger the wind, the larger is the effect. This effect is in addition to the velocity pressure effects previously discussed.

When the two effects are combined, and if the chimney has a good cap, winds never cause serious problems. For wind velocities up to the flue gas velocity, there is very little effect. (Typical stove flue-gas velocities are 3-6 miles per hour.) For wind speeds between 1 and 3 times the flue-gas velocity, there is a slight reduction in chimney flow. At wind speeds greater than about 3 times the flue-gas velocity, chimney performance is better than in still air. Thus, with an effective chimney cap, wind speeds between roughly 5 and 20 miles per hour will cause a slight decrease in chimney capacity, slower winds will cause no effect, and faster winds will improve performance.[19]

Wind can also change the pressure inside a house, which then affects chimney performance. If the house has some open windows or doors on the side facing into the wind, the house pressure increases, which helps the chimney; openings on the downwind side decreases the pressure inside the house, hurting chimney performance. Thus a temporary solution to a badly smoking stove or fireplace on a windy day is to open a downstairs window or two on the upwind side of the house.

WEATHER AND ALTITUDE

Wood heating systems can be slightly affected by weather. The two strongest effects have already been discussed; they are due to wind and temperature. Wind pressure can either enhance or detract from a chimney's ability to exhaust the smoke from a fire. Outdoor temperature can affect chimney performance because the draft depends on the difference between the flue-gas temperature and outdoor temperature. The colder the weather, the better the draft. As the temperature changes from 50 to 0° F., chimney capacity can be increased by as much as 10 percent.

Barometric pressure also has a small effect on chimney capacity. When the pressure is high the combustion air and the flue gases are a little denser. An increase in pressure of 1 inch of mercury increases the capacity by about 4 percent.[20] Typical variations in barometric pressure span no more than 2 inches of mercury, which would cause an 8 percent change in chimney capacity.

It is not as certain whether humidity in the air has an effect on stove performance. Even when the relative humidity is 100 percent, water vapor can never constitute more than 1.5 percent of air by weight (at 68° F. or less). Thus changes in humidity can affect the composition of air by only this small amount. The maximum effect on the air's oxygen content, density, thermal conductivity, and specific heat is less than 1.5 percent, which is negligible.

None of these weather effects is very big by itself, but they frequently all act together. When barometric pressure is high in the winter the weather is often clear, cold, dry and calm; when the pressure drops, the weather is often windy, more humid and not as cold. Adding all the effects together suggests that weather can affect chimney capacity by as much as 20 percent, the higher capacity often coinciding with high barometric pressure.

Altitude also affects chimney performance. The thinner air at high altitudes decreases chimney capacity just as does low barometric pressure. For every 1000 feet of elevation above sea level, chimney capacity is decreased by about 4 percent relative to the same chimney at sea level.[20] At 5000 feet a chimney would have to be designed with 20 percent extra capacity compared to a chimney at sea level serving the same appliance. At 10,000 feet the effect is about 40 to 45 percent.

These effects will be most noticeable in systems which are marginal to begin with. In many cases there is enough excess chimney capacity that these weather effects will never be noticed. But in cases where the chimney is actually not quite big enough and/or tall enough and/or well enough insulated, weather effects can be more significant.

This elevation effect is much larger than the barometric pressure effect. Even so, people seldom take it into account for wood heaters. This does not usually lead to any noticeable problems. Chimneys usually have generous excess capacity at sea level, so that if the same chimney design is used at high elevations, it still has adequate capacity. But it would be wise in new installations to use a slightly oversize (in diameter) chimney at elevations above 5000 feet.

MULTIPLE USE OF SINGLE FLUE

Ought one to attach a wood stove to a chimney which is also used by another appliance, such as a furnace, water heater, fireplace, another stove, etc.? The National Fire Protection Association says, "Don't, but if you do, here's how."[21] Similar mixed messages are put out by other organizations. Why is there such equivocation?

Chimneys today are often built to serve more than one appliance by having separate flues or tile-liners within the overall masonry structure. Each flue or liner is essentially an isolated independent channel. Thus it is perfectly safe to attach a number of appliances to the same chimney as long as only one is attached to each flue, and each flue is of appropriate size for the individual appliances. If building a new house with masonry chimneys, it is wise to install enough separate flues to handle each anticipated appliance. In many older houses chimneys have only one passageway. One is then often faced with using the same flue for the furnace and the wood stove.

The principal reason multiple use of a single flue is not advised is possible draft interaction, resulting in some back flow or smoking of the stove or appliance. An additional danger of multiple use of fireplace (or open stove) flues is the possibility of sparks coming out from the fireplace into the room. The possible spark hazard is discussed in Chapter 9.

The hazard due to draft interaction can be significant, but is not necessarily so. In commercial and industrial buildings containing more than one fuel-burning device, it is common practice to connect many separate devices to a single large chimney (with one flue). It is also common in houses to connect both the furnace and water heater to a single flue. It is considered good practice to have the chimney connectors enter the chimney at different heights by a few feet.

The important consideration is whether or not the single flue has the capacity to handle both (or all) appliances under all operating conditions. This question is especially complicated when wood stoves are involved because of their great variety and range of operating characteristics. A rough guide for the maximum safe capacity of a single flue which is used by more than one appliance is given in Table 5-7.

As an example of the use of Table 5-7, consider the following situation. Suppose an oil furnace with an *input* rating of 120,000 Btu per hour is vented through a 20 foot high, tile-lined, masonry chimney which is 16 x 16 inches on the outside. Does this chimney have the capacity to be used also by a wood stove with a rated power *output* of 40,000 Btu per hour? Use of the table requires power input ratings. Since stoves are roughly 50 percent efficient, the input rating is about twice the output, or 80,000 Btu per hour in this case. Thus the combined input ratings of both appliances is 120,000 + 80,000 = 200,000 Btu per hour. The area of a tile-lined flue inside a 16 x 16 inch chimney is about 46 square inches (Table 5-4). From Table 5-6, it can be estimated that the capacity of a 20 foot high chimney with an area of 46 square inches is about 215,000 Btu per hour, enough to handle the furnace and stove. The effective chimney height for the two appliances will not be the same; the conservative course is to use the lesser of the two in making this capacity estimate.

PRACTICAL CONCLUSIONS

1) For a wood heater which is used no more than a few hours a day (e.g., most evenings, or only once or twice a week) I suggest a prefabricated insulated metal chimney. Its low mass and its insulation assure quick warm-up, lessening possible draft and creosote problems. Ideally it should rise inside the house.

2) For a wood heater which is in use most of the time, I recommend a masonry chimney inside the house with most of its surfaces exposed. For continuous wood heating, start-up problems are of little concern, and a massive chimney's heat storage capability contributes to evening out the temperature in the house. The higher heat losses through a masonry chimney's walls contribute more heat to the house compared to insulated metal chimneys.

3) If a stove or fireplace is letting smoke into the house, the best ways to solve the immediate problem are a) to make sure any flue-gas dampers are not shut, and b) open windows and/or doors in the first floor or basement of the house, and close any openings in the upper parts of the house. If the weather is windy, the open windows and/or doors should be on the windward side of the house.

4) If a stove or fireplace tends to smoke only at the beginning of a fire, opening a window or door for a few minutes during light-up can be an easier and safer solution than shoving lighted newspaper up the chimney.

5) If the wood-burning appliance only smokes when the weather is windy, three possible remedies are installing a chimney cap if there was none, installing a better chimney cap if there was one, and increasing the height of the chimney. Changing the height of a chimney is relatively easy with prefabricated metal chimneys.

6) If chronic smoking is a problem even in calm weather, check the venting system for obstructions and clean if necessary. Birds' nests and creosote are possible causes of blockage. Try opening a window or door (in the same room as the stove or fireplace); if smoking stops, direct feeding of outdoor air to the wood heater's vicinity may help.

7) If when the chimney is clean and a window is open the chimney still has inadequate capacity, possible remedies vary from case to case. If the chimney serves a stove, moving the stove closer to the chimney and eliminating elbows in the stovepipe connector may help. If the chimney serves a fireplace, making the fireplace opening smaller may be the only solution. This can be done by raising the hearth or installing a canopy hood extending down from the top of the fireplace opening. Prefabricated chimneys can be made taller by adding another section or two. Weatherstripping or otherwise sealing upstairs windows and attic doors can help.

8) If after all this smoking still occurs, the only remedies are either a smaller stove or fireplace, or a new chimney. Increasing the diameter of a chimney is the surest way to get more capacity.

	AREA OF FLUE (square inches)					
	19	28	38	50	78	113
HEIGHT OF FLUE (feet)	Combined Appliance Input Rating (thousands of Btu per hour)					
6	45	71	102	142	245	NR
8	52	81	118	162	277	405
10	56	89	129	175	300	450
15	66	105	150	210	360	540
20	74	120	170	240	415	640
30	NR	135	195	275	490	740

TABLE 5-7. Input maximum capacity (in thousands of Btu per hour) of chimneys used by more than one appliance. NR means not recommended. Input ratings of wood stoves are about twice their rated heat output (since their efficiencies are about 50 percent). Input ratings of other appliances, such as furnaces and water heaters, are usually indicated on them (The Table is adapted from Table 17 in Chapter 26 of the 1975 Equipment Volume of The ASHRAE Handbook and Product Directory).

Chapter Six The Energy Efficiency of Stoves

ENERGY EFFICIENCY

The stove characteristic which usually receives the most attention is efficiency. Manufacturers devote a considerable amount of space in their promotional literature to efficiency claims, both absolute and comparative, and consumers naturally seek such information.

The term "efficiency" is sometimes misunderstood. In most contexts, the phrase "energy efficiency" would be more precise. The energy efficiency of a stove installation is the fraction (or percentage) of the chemical energy in the wood which is converted into useful heat (heat in the building's living space).

$$\text{energy efficiency} = \frac{\text{useful heat energy output}}{\text{wood energy input}}$$

Other things being equal, more efficient stoves use less wood to do the same heating job.

Many properties of stoves other than their energy efficiencies are important. For instance, a stove's heat output or power, the steadiness and duration of its output between fuel loadings, its tendency towards creosote formation, and its durability are all important, even more so than energy efficiency in many applications. These other stove characteristics are discussed in detail in Chapter 7.

Any appliance with energy as the principal output can be described by an energy efficiency. Incandescent light bulbs have electric energy as their input and light energy as their main functional output; the conversion efficiency is 3 to 8 percent depending on bulb type and size. (Over 90 percent of the electrical energy leaves the bulb as heat, not light!) The top, surface burners of gas cooking ranges average about 50 percent efficiency for cooking or heating water, but may vary from 30 to 60 percent depending on the kind and relative size of the pan;[1] almost all of the chemical energy in the fuel is converted into heat (combustion is very efficient) but only about half of that heat gets into a pan or pot; the rest enters the air in the kitchen.[2]

The efficiencies of non-wood home heating systems are of special interest here. Electric space heating is essentially 100 percent efficient — all of the electric energy entering the house heating system is converted into useful heat in the house. Electricity *generation* is less efficient; in fossil-fuel and nuclear plants, only about 30 to 40 percent of the energy in the fuel is converted into electrical energy. (The "waste" heat is usually dumped into nearby water, or dissipated in the air as sensible and/or latent heat.) The net heating efficiency of electric space heating, including generation, storage (pump storage) and transmission losses is about 30 percent (heat energy in the house divided by the corresponding fuel energy consumed at the electric plant).

Typical efficiencies of gas and oil fired furnaces in homes vary from 40 to 70 percent. The most important factors determining where, in this rather large range, a particular system may lie are: 1) the design of the furnace, 2) maintenance of the furnace, and 3) the sizing of the furnace relative to the building's needs, or equivalently, its duty cycle or fraction of "on" time. (The kind of heat-distribution system [steam, hot water or air] matters some, but usually not as much as the above three factors.) Basic design principles for fossil-fueled furnaces are the same as for wood stoves, and are discussed at length later in this chapter. Furnace maintenance is critical in two areas: 1) keeping the fuel-to-air ratio optimal, and 2) keeping heat-exchange surfaces clean of soot (which is a good thermal insulator and this inhibits heat transfer). The latter is especially important in oil furnaces since oil, being a relatively non-volatile liquid fuel, tends to burn less completely, generating soot (very small carbon particles).

Manufacturers of new furnaces advertise efficiencies as high as 80 to 85 percent. Efficiencies this high are rarely obtained in actual installations because these numbers assume that the furnace is on continuously (a condition making the measurement of efficiency much easier) and no (unused) heat losses from pipes or ducts. In practice, furnace capacities are chosen conservatively, so that they exceed by a comfortable margin the maximum expected heating needs of buildings in which they are installed. As a result, furnaces are constantly cycling on and off, the fraction of "on" time rarely exceeding 75 percent and, of course, approaching 0 percent at the beginning and the end of the heating season.

This cycling is detrimental because furnaces have lower average efficiencies when cycled on and off than when operated continuously. There are many reasons for this, some of which may not apply in particular situations, depending mostly on the type of fuel (gas or oil), the type of heat distribution system, and on whether or not one wants to heat the basement of the house. Some of the possible reasons are the following: gas pilot lights are on at least throughout the heating season, but may not be contributing useful heat to the house except when the main burner is also on; during the first few minutes of "on" time, when the furnace and its flue are not hot, combustion is less complete since the combustion region is not warmed and since normal stack draft has not developed; some of the heat which is stored in the furnace walls and chimney, and in the heat-distribution fluid, ducts or pipes, is lost each time the system cools off — some goes up the chimney, and some may leak out into parts of the house where heat is not desired or needed.

Thus, when all factors are taken into consideration by measuring the actual heating efficiencies of fossil-fuel systems in people's homes, the net result is efficiencies in the range from about 30 to 75 percent.

STOVE EFFICIENCY AND NET EFFICIENCY

There is an important distinction between the energy efficiency of a stove (or fireplace or furnace) itself and the net energy efficiency of its use in a particular installation. The difference arises because 1) the use of a stove *may* result in the house needing more total heat (from whatever source) to maintain the same temperatures, as discussed below, and 2) the heat which conducts through the walls of a chimney may or may not contribute to the house's needed heat depending upon the type of chimney, its location, and its exposure, and 3) when the chimney is not in use, some house heat usually is lost, either in warm indoor air leaking into the chimney and up out of the house, or in heat being conducted through the chimney walls and then convected up the chimney. The back of a fireplace which is built into an exterior wall also can provide a relatively easy path for heat to be conducted directly outdoors, whether or not the fireplace is in use. (A brick wall would have to be many feet thick to have an insulating value equal to that of an ordinary wall with 3½ inches of fiberglass insulation; see Table 2-1).

When a stove is operating it draws air from the house (which supplies the oxygen necessary for combustion). Even a non-operating stove-chimney system usually draws air out of the house unless it is literally air tight. This air originates outside the house and thus has heating energy invested in it by the time it reaches the stove, assuming outdoor air is not ducted directly (unheated) to the stove. This heating of the combustion air up to room temperature may, but does not necessarily, constitute an extra heating load, because whether or not the stove is there and operating, outside air is always moving into houses, and heated indoor air is moving out. This natural air exchange or air infiltration is due to air leakage through small, sometimes invisible, cracks throughout all houses, especially around windows and doors, and due to the use of outside doors and the use of exhaust fans in kitchens and bathrooms.[3] Typical air exchange rates range from the equivalent of ½ to 2 complete air changes in the house per hour.[4] One air change per hour in a house with a volume of 15,000 cubic feet constitutes an air exchange rate of 15,000 cubic feet per hour, or 250 cubic feet per minute.

This natural ventilation rate is largest on windy days, when the wind pushes more air through the cracks on the house's windward side, and on cold days, when the stack or chimney effect of the house itself tends to pull air in through the cracks in the lower half of the house. The net effect is that 15 to 60 percent of a house's need for heat is due to this natural air exchange — the need to warm to room temperature the air which leaks into the house. Any activity or appliance which causes an increase in the air exchange rate increases the amount of heat needed to keep the house warm.

The critical question is whether the needed combustion air is supplied by the house's normal air exchange rate or whether additional air exchange is induced by the operation of the stove or other appliance. Table 5-6 gives the air consumption rate of stoves and fireplaces. Since closed stoves draw so little air compared to the normal air-exchange rate for houses, they do not add to the heating needs of most houses. Most stoves just redirect some of the air, which would have been passing through the house anyway, through the stove. Virtually all wood heaters have a net heating effect in houses; the question here is whether the net heating effect is the same as the stove's actual heat output, or whether some of the heater's output must be used to heat the air it breathes from the outdoor temperature to room temperature.

For open stoves (such as a Franklin) and fireplaces, the situation is rather different. When operating, they may draw 50 to 300 cubic feet per minute. Clearly extra outdoor air must usually be drawn into the house to supply these air guzzling wood burners. This can cause as much as a doubling of the heating needs of the house. The exact increase depends on how much air the burner draws and on how leaky the house is.

This is an important consideration when choosing a wood heating system, but one which is best considered separately from the energy efficiency of the heater. The performance of the stove itself is, of course, insensitive to whether the air it uses is "extra" or "normal" infiltration air. Thus, such effects should not affect a heater's energy efficiency rating. A wood heater operated in a relatively tight house may increase the house's air exchange rate, but the same heater operated at the same power, but in a leaky house may not; the difference is in the house, not the heater. The amount of extra heat needed to warm the extra air flow into the house depends on how cold the outdoor air is, which again is not a property of the heater. The *net* heating effect is what matters, but that has to be derived from the two separate considerations: 1) the efficiency of the stove, and 2) the possible increase in the total heating needs of the house due to operation of the stove.

In practice, only those wood-burning appliances with 8 inch or larger flue-pipe connections are likely to be able to cause significantly increased heating loads. Most Franklin stoves, other open stoves, "free standing" fireplaces and ordinary fireplaces fall into this category. (The effects of ducting outdoor air directly to a wood burner are discussed in Chapter 8).

THE THEORY OF STOVE EFFICIENCY

Wood in an operating stove combines chemically with oxygen in the combustion air, releasing the stored chemical energy in the form of sensible heat, latent heat (uncondensed vapors) and radiation. Some of this released energy then enters the room through the stove walls, stove pipe and chimney.

The maximum possible energy efficiency is achieved when both combustion and heat transfer are complete. Complete combustion is occurring if all the carbon, hydrogen and oxygen atoms in the wood are converted to carbon dioxide (CO_2) and water (H_2O); under these ideal conditions there would be no smoke and no odor, and all possible chemical energy in the wood would have been released. Heat transfer is essentially complete if the flue gases have cooled down to room temperature before leaving the house. If both complete combustion and complete heat transfer were achieved in a heating system, its efficiency would be nearly 100 percent.[5]

Although complete combustion is desirable, complete heat transfer is usually not, for two reasons. In some installations, such relatively cool (room temperature) flue gases would not induce adequate draft or suction on the stove; this would likely result in some smoke entering the room and/or the impossibility of getting very high rates of combustion due to the small air flow. In addition, considerable condensation would occur inside the stove pipe and chimney. Since in practice combustion is never complete, the condensate would contain tar, acids, etc. (ingredients in creosote), as well as water. This can create a mess, a fire hazard after the water has evaporated away, and increased corrosion of chimneys. This conflict between maximizing heat extraction and minimizing creosote deposition is a fundamental problem in stove design and use.

COMBUSTION EFFICIENCY

Combustion efficiency is the percentage of the chemical energy in the wood consumed which is converted to sensible heat, latent heat and radiation inside the stove. Conditions necessary for complete combustion are easy to state. Adequate oxygen in the right places and high temperatures are the two essential ingredients. An excess of combustion air entering the stove is usually necessary to assure an adequate supply to the regions in the stove where combustion is taking place, since some air usually bypasses the fire and the mixing of air with combustible gases coming out of the wood is not ideal.

Thorough mixing of the air supply with the combustible gases is important, for fuel molecules can only burn if oxygen molecules are present. Oxygen naturally wanders or diffuses into fuel-rich regions but not always far enough or fast enough to ensure complete combustion. Better mixing can be achieved by inducing turbulence in the gas mixture by having the air enter the combustion region at relatively high velocity, or by forcing the burning gas mixture to make abrupt turns as it travels through the stove; the intent is to make the air and combustible gases swirl around and hence become mixed.

The importance of high combustion-zone temperatures has been clearly demonstrated experimentally. [6] A small wood-burning furnace was built and equipped with electric heating elements in its walls so that any temperature could be maintained irrespective of other conditions. Completeness of combustion was assessed by measurements of carbon monoxide, total hydrocarbons, particulates, and smoke density. The variables investigated were temperature in the combustion region, total combustion-air flow, the ratio of secondary to primary air (see page 49), moisture content of fuel, type of fuel, fuel size, preheating of secondary air and type of grate. Of these variables, the combustion-region temperature was found to be the single most important factor. As long as the temperature around the fire was about 1100° F. or higher, combustion was very nearly complete. At lower temperatures, the emission of incompletely-burned materials was always significant.

There are two general methods for increasing combustion-region temperatures: by impeding outward heat flow, and by preheating the combustion air. Firebrick and metal liners (Figures 6-1 and 6-2) both can help, although they are often intended more for the protection of the main stove body against excessive temperatures than for their insulating effect. [7] Horizontal baffle plates also help (such as in S-flow-pattern stoves like the Jøtul 602); they are usually hotter than an exterior wall since their "back" sides are heated by the hot flue gases (and, in some cases, flames). Thick metal stove walls are no better at preventing heat from leaving the combustion region than are thin walls.

Figure 6-1. A firebrick liner in a stove (the Fisher Grandpa Bear).

Figure 6-2. Metal liners in a stove (the Lange 6302A).

Thick walls store more heat, but with a steady fire, just as much heat gets through. [8] Cast iron and plate or sheet steel are also identical in this regard (Table 2-1).

Unfortunately, the quantitative relations between the insulating characteristics of the stove walls and completeness of combustion are not very well known. However, theoretical estimates suggest that the insulating value and the amount of area covered in most contemporary stoves would have to be improved considerably to become very effective in helping combustion to be more complete.

There are two reasons why hot surfaces around the combustion zone would help. There is more infrared radiant energy being emitted from them, which, when

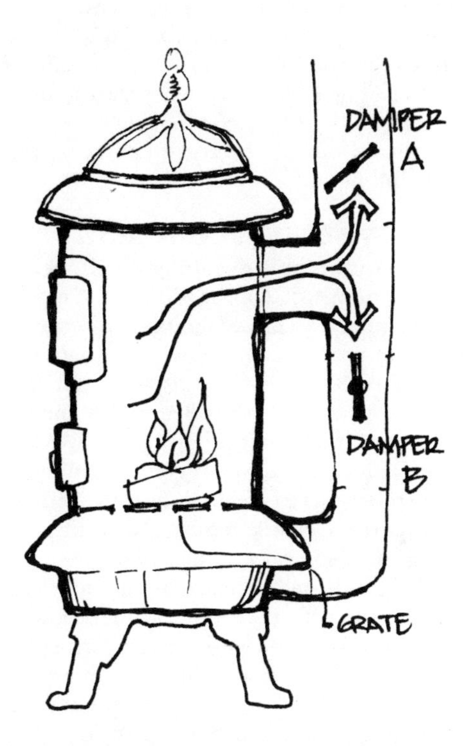

Figure 6-3. A stove design in which there can be some recycling of gases through the combustion zone, contributing to more complete combustion. Recycling is achieved by opening damper B and partially closing damper A.

absorbed in the gases, flames and wood, helps to maintain high temperatures. In addition, if a flame extends out from the wood and touches a surface, the hotter the surface is, the less the flame will be quenched. Combustion reactions of gases cannot take place very close to relatively cool surfaces; soot and other incompletely oxidized products are formed.

In relatively large stoves (or any stove with a relatively small fire in it) the flames may never touch a surface, and thus quenching cannot occur. However, under such circumstances the surfaces will be less hot and hence will not contribute as much to radiant heating of the burning fuel. The net effect on combustion efficiency cannot be safely predicted.

The other way to maintain high combustion-zone temperatures is to preheat the incoming air by ducting it through channels or pipes in or near the fire. The heating must be substantial to be effective. The theoretical minimum amount of air needed to completely burn a pound of typical wood is 6 pounds. Stoves draw in 1.5 to 3 times this amount, or about 9 to 18 pounds of air for each pound of wood burned.[9] Thus, since most of the mass flowing through or leaving the combustion chamber is air, for every 100° F. by which the combustion air is preheated, the average temperature in the combustion region would also increase by roughly 100° F. Since 1100° F. or higher is the desired temperature around the fire, it is likely that preheating of the combustion air must be substantial — hundreds of degrees Fahrenheit — to be significant in improving completeness of combustion. This is consistent with the experiments of Prakash and Murray.[6]

Another way to make combustion more complete is to recirculate some of the flue gases back through the combustion region, giving the combustible gases a second chance to burn as they pass through or near the hot fire. Figure 6-3 illustrates a stove which is purported to accomplish this. Although the basic principle of flue-gas recycling for more complete combustion is valid, I am not aware of any careful measurements verifying the effectiveness of such designs.

HEAT TRANSFER EFFICIENCY

Heat transfer from inside a stove (and stove pipe) out into a room is aided by having 1) high thermal-conductance stove walls, 2) a large surface area, 3) high temperatures and turbulence inside the stove (turbulence and mixing of the gases keeps bringing fresh [hot] gases in contact with the walls), and 4) long residence times, or equivalently, low average flue-gas velocities, to allow as much time as possible for heat to get out of the gases.

Hence if maximum heat transfer were the only objective, plain metal walls would be favored over fire-brick-lined or metal-lined walls. Neither the thickness nor type of metal (cast iron or steel or stainless steel) has a significant effect on how much heat gets through a simple single-layer wall. Heat comes through a thin wall more quickly; thick walls store some heat for later use. On the average, or under steady conditions, the same total amount of heat gets through.

Large surface areas relative to the combustion region or fire size enhance heat transfer efficiency significantly. Examples are double-drum stoves (Figure 6-4), the Sevca and other stoves with an extra, physically separate, "smoke chamber" through which the flue gases pass. An extra-long stove pipe connecting the stove to its chimney serves the same function. Some stoves have arched chambers mounted on top of the

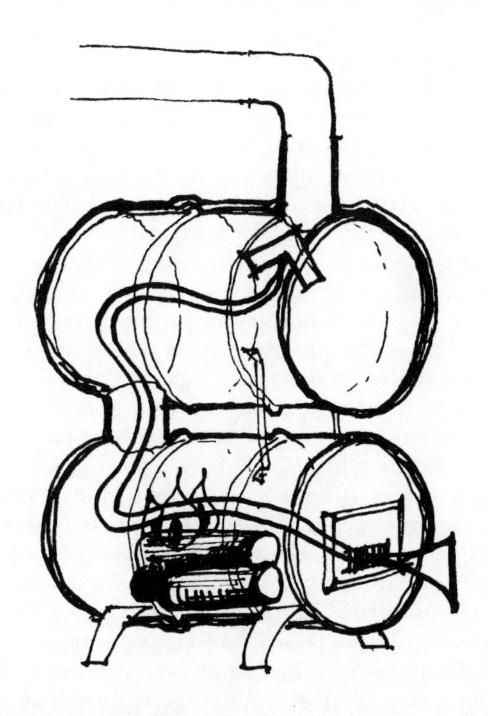

Figure 6-4. A double-drum stove — an example of a stove with a "smoke chamber," resulting in more complete heat transfer and hence improved energy efficiency.

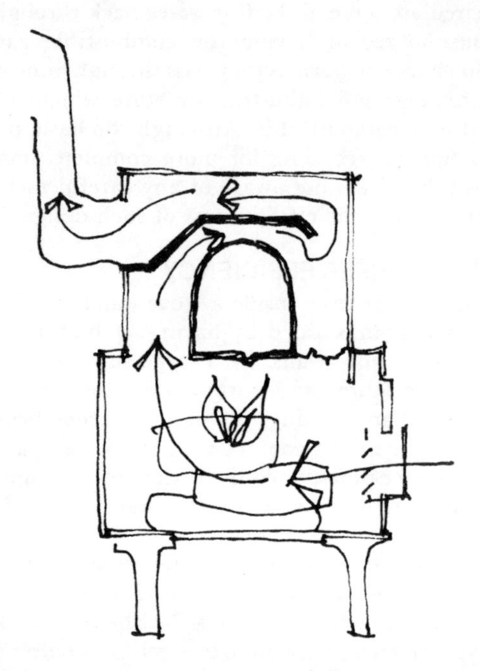

Figure 6-5. The Lange 6303, a stove with an arched smoke chamber which allows additional heat to get out of the hot gases, resulting in higher energy efficiencies, as in the double-drum stove.

basic combustion chamber (Lange 6303 [Figure 6-5], Morsø 2B0). A number of stoves have smaller and more integral smoke chambers built into the basic stove, created through the use of baffle plates (Figure 6-6). Such internal baffling limits the size of the fire which can be built, ensuring more cooling of the flue gases by the time they leave the stove (as compared to an unbaffled stove of the same outside dimensions). A stove with a large surface area, simply because it is a large stove with a large combustion chamber, such as a single-drum stove, will have a high heat output rate or power rating, but the efficiency of heat transfer will not necessarily be enhanced; only a typical fraction of the heat generated inside the stove will pass out into the room. A large surface area *relative* to the size of the fire is what helps heat-transfer efficiency.

Keeping the flue gases stirred up as they move through the stove and stove pipe improves heat-transfer efficiency by prohibiting gas which has lost heat and cooled from remaining against the walls. Baffle plates affect heat transfer because they force hot gases into corners which might otherwise be relatively stagnant (Figure 6-6).

The longer time the hot gases spend in the stove (and in the exposed stove pipe) the more of their heat will be transferred out, other things being equal. If the gases pass through a large volume smoke chamber like the second drum in a double-drum stove, the "residence time" is increased, as well as the stove's outside surface area. A baffled drum would be even better. Using a large-diameter stovepipe has a similar effect.

A stove which uses a minimum amount of air for a given combustion rate has an advantage. The less air is admitted, the less must be exhausted; hence the velocity through the stove and stove pipe will be slower. Minimizing the amount of excess combustion air also increases the *temperature* of the hot gases, for excess air dilutes the combustion products, making the mixture cooler. The higher the temperature of the gases, the larger will be the fraction of their heat which will be transferred out of the stove and stove pipe, other things being equal.

SOME STOVE-DESIGN CONFLICTS

A basic conflict in stove design and operation concerns the amount of combustion air. Complete combustion requires a certain amount of excess air, and usually, the more air supplied, the better. But any amount of excess air both cools and increases the velocity of the flue gases, thus decreasing the heat transfer efficiency. It does not follow that stoves which have the most efficient (complete) combustion are necessarily the most energy efficient overall. Efficient use of air is important. This requires directing the air to where it is needed and mixing it with the combustible materials. Small details such as air-inlet shapes and placements can affect stove performance, and in ways which are not predictable based on theory alone.

Another conflict in stove design concerns the possible use of metal or firebrick liners, or even insulating firebrick around the combustion region. All of these features increase insulating characteristics, which help to keep temperatures high, aiding completeness of combustion. But heat transfer through the insulated walls (despite the higher temperatures) is reduced. An obvious possible solution to this conflict is to physically separate in the stove the locations where combustion takes place and where heat transfer is featured. This could be done by having a highly insulated, hot, low-heat-transfer region where combustion takes place, followed by uninsulated, large surface area chambers and/or passageways where the now completely-burned

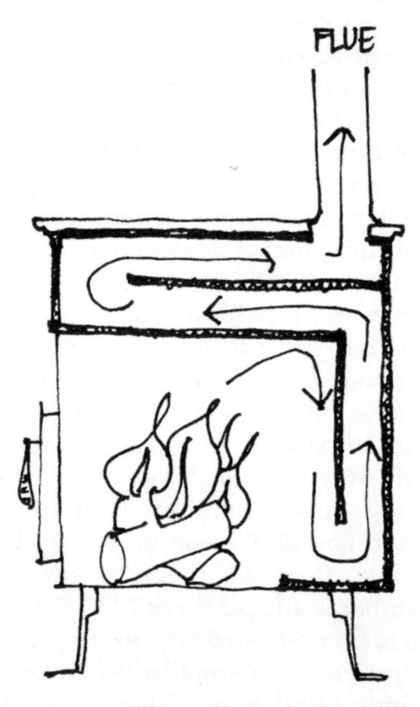

Figure 6-6. A schematic cross-section through a stove with extensive internal baffling, intended to ensure good contact of the hot flue gases with the stove's outer layer, thereby improving the stove's heat transfer efficiency. The figure does not represent a real stove.

gases can linger and cool, giving up their heat through the walls into the room.

STOVE DESIGN FEATURES

SECONDARY COMBUSTION

Some stoves have two distinct air inlets. The primary air inlet feeds air to the wood, usually near the bottom of the fire. The secondary air can be admitted above the wood, or into a separate secondary-combustion chamber located either above or beside the primary-combustion chamber (Figure 6-7).

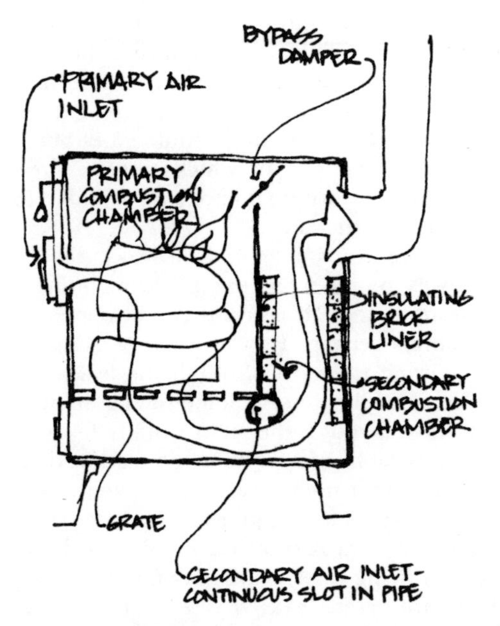

Figure 6-7. A schematic of a stove with a secondary air inlet and a secondary combustion chamber. In this stove, the secondary air is delivered in a pipe at the bottom of the wall at the back of the primary combustion chamber. It is added to the combustible gases through a continuous slot in the pipe. The secondary combustion chamber is lined with insulating brick to help assure the high temperatures necessary for complete combustion. This is not an illustration of any existing stove.

A substantial portion of the chemical energy in "raw" wood is converted to combustible gases and airborne tar droplets — one-half to two-thirds, depending on wood type, its moisture content and how quickly it heats up in the fire. Substantial energy is lost if none of the combustible vapors are burned. But this is rarely the case. The large, yellow flames in a wood fire are the burning gases. Most fires include flames, indicating that at least some of the gases are being burned. Only in a totally flameless (smoldering) fire is all of the energy in the combustible gases lost.

Complete combustion of wood is a desirable objective, and this *does* require the burning of both the evolved gases and the solid wood. This requires a certain minimum amount of air, sufficiently high temperatures, and adequate distribution and mixing of the air with the combustible materials. But feeding air into the stove through two different inlets is not the only way to achieve this. In many stoves with only one inlet, plenty of air is available to burn the gases as well as the wood.

However, many stoves which have primary and secondary inlets need the second inlet. The need generally arises from an attempt to achieve a steady burning rate for a long time with a large fuel charge. To prevent the fuel from burning too quickly, the air supplied to the wood (the primary air) is limited. Yet the large load of wood usually gets hot enough to generate considerable amounts of combustible gases. The primary air supply is inadequate to completely burn the gases, and so secondary air and a secondary combustion chamber are necessary.

In practice, the secondary-air inlets on many stoves do not usually increase combustion efficiency. The principal reason is that the air is admitted at a point where the temperature of the flue gases is already below the ignition temperature (roughly 1100° F.). Thus, no additional combustion takes place and the net effect is the same as an air leak at that location, increasing the amount of excess air and hence decreasing the efficiency of heat transfer. A rough rule of thumb is that secondary air will help only if admitted into a region where flames already exist. This is likely to be the case for any kind of secondary air inlet in a stove operated at very high firing rates, for then flames fill the combustion region and may even extend up the stove pipe a little way; in this case even most air leaks become effective secondary air supplies. But at medium and low firing rates, secondary air must be admitted very close to the burning wood to have a chance of being effective.

Preheating of the secondary air can help, but the heating must be substantial (at least a few hundred degrees Fahrenheit) to be very significant. Three basic situations can arise: (1) If the secondary air is being added to already burning, but fuel-rich gases, then the preheating can make the secondary combustion a little more complete by increasing the average temperature of the burning mixture. But the effect of preheating is unlikely to be large in this case since the burning gases are already so hot. Flame temperatures are roughly 2000 degrees Fahrenheit; thus, preheating by even a few hundred degrees of the relatively small secondary air flow is unlikely to have much effect. (2) If the secondary air is being added to flue gases which are not burning because the temperature is below the ignition point, the secondary air would have to be heated to well over 1100 degrees Fahrenheit to help initiate ignition. (3) Finally, if the gases are not burning only because they contain insufficient oxygen (but they are hot enough), ignition will occur without preheating of the secondary air, and then the situation is similar to case (1). The conclusion is that only if preheating is substantial (at least a few hundred degrees Fahrenheit) will it have significant benefit. In many stoves with this feature, the preheating is not enough to be of much use.

In designs where secondary combustion is concentrated in a small volume, lining the region with insulating brick might be a practical and effective way to help maintain high temperatures in the secondary combustion chamber.[10]

Another reason many secondary air inlets do not perform well is that the air they admit does not mix adequately with gases to be burned. A localized single inlet which admits air smoothly into the flue-gas stream is usually not optimal. Mixing, of course, eventually takes place, but often not until the gases have cooled below their ignition temperature. The inlet should distribute the air as evenly as possible in the flue gases, and/or turbulence should be high. Experiments with coal heaters indicate that one good system is to admit air through an essentially continuous slot across the region where secondary combustion is desired.[11]

The amount of secondary air admitted is also critical. If it is too little, not much of the combustible gas can be burned. If it is too much, its cooling and diluting effects may inhibit combustion, and excess air is detrimental to heat transfer. The amount needed is not predictable and not constant. It depends on how much of the oxygen in the primary air is left and on how much of the combustible gases is unburned. This depends on many details, such as the exact location of the primary air inlet, the amount, direction and speed of the air as it enters the interior of the stove, how close the wood is to the primary air inlet, how much wood is in the stove, the average size of the pieces, and how densely piled it is. On the other hand, the limits of flammability of the gases are fairly wide (Table 4-1), and mixing of air with the gases is never perfect. Thus secondary combustion is possible over a moderately wide range of secondary-air flows. However, the flow will not be optimal much of the time due to the variability of combustion in stoves.

It is not possible to generalize about either the need for, or the effectiveness of, secondary air. Some stoves achieve essentially complete combustion without it. Others do not. Those stoves that have secondary air inlets do not necessarily benefit from them — whenever the extra air is not used in combustion, it is detrimental to overall energy efficiency, and this is the case at least part of the time in any such stove. The amount of secondary air needed varies in time. It is highest when a new large charge of wood is placed in an already hot stove and the primary air supply is small, and lower when the amount of fuel in the stove is small. The net effect cannot be known through analysis alone. Careful testing is necessary. To the best of my knowledge, this has not been done on wood stoves.[11]

AIRTIGHTNESS

How important is it that a stove be "airtight," that no air be able to get into a stove except at the explicit air inlet(s)? The immediate effect of small cracks where various parts of a stove join together is that air leaks into the stove. The effect of this extra air inside the stove depends on where it enters and the condition of the fire.

No stove is literally airtight — no stove is able to hold a vacuum or contain pressurized air. But many stoves are tight enough that their performance as stoves would be no different if they were absolutely airtight. Leakage can occur through cracks around the loading door and around other movable exterior parts, and through cracks where the various fixed parts of a stove join together. Steel-plate stoves with welded seams rarely have significant leakage except possibly around the door, because good welds seal well and permanently. Cast iron stoves are assembled from separate cast parts and rely upon compounds such as furnace cement to make the joints airtight. Most furnace cements tend to shrink and crack, and can fall out, leaving small cracks where air can leak in.[12] In good quality cast iron stoves, this does not happen because the parts fit so closely together that the cement is held in place.

Air leaks are almost always detrimental. Efficiencies are usually lower, and it can be difficult to obtain low (overnight) firing rates. Much of the air which gets into a stove through leaks is not likely to be used by the burning fire because the air enters where there is no wood or any combustible gases to be burned, or it enters where there are combustible gases but they are too cool to burn. Any air which enters a stove but is not used for combustion is excess air and decreases the stove's energy efficiency.

Controllability is decreased in leaky stoves. In airtight stoves, no air enters except through the air inlet, and the amount of air entering can be regulated by the inlet-damper setting. In a leaky stove, some air is always entering the stove through the leaks no matter how the controls are set. Some of the air which gets into a stove through the leaks usually feeds the fire. This can make it difficult to achieve a low firing rate; even with the air inlet(s) completely shut, enough air may leak in to make a hotter fire than desired (as well as less efficient heat transfer, as explained above).

At high firing rates, leaks may not matter so much, and in rare cases, leaks may even increase a stove's efficiency. With a large fire, more of the leaked air is likely to be used in combustion, for the same reason that secondary-air inlets are more likely to work with a hot fire.

The use of a stovepipe damper or built-in flue-gas damper can improve the efficiency of non-airtight stoves by decreasing the amount of air leaking through the cracks without decreasing the flow through the air inlet. A stovepipe damper is a rotatable metal plate inside a section of stovepipe which can be oriented to partially close the pipe if desired (see Figure 6-3 in Chapter 6). Suppose a leaky stove is burning steadily at the desired heat output rate with the stovepipe damper wide open. If the damper is adjusted towards the closed position there is a decrease in the suction (draft) inside the stove — the buoyancy of the hot gases in the chimney is less effective when acting through a partially closed damper. The result is that less air is drawn into the stove through all openings, both its air inlet and its cracks. With less air coming in the air inlet, the combustion rate is decreased. If now the air inlet damper is opened to the point where the same amount of air is coming through as before the damper was closed, the fire will be burning at the same rate. The net effect will be a decrease in the amount of air coming through the leaks, and hence an improvement in energy efficiency. The further closed the stovepipe damper is, the better; the practical limits are determined by when the stove starts to smoke, or by when the stove is not putting out enough heat, even with its air inlet wide open.

FLOW PATTERNS

Updraft? Downdraft? Crossdraft? S-draft? What do they mean and what is the significance? These terms refer to the basic flow pattern of the gases in a stove. The term "draft" here does not refer to pressure difference (or suction), which is its technical meaning, but to air flow. There is virtually no limit to the number of possible stove designs; thus, any classification scheme is somewhat arbitrary and incomplete. Nonetheless, it is useful to discuss broad categories of stoves. The flow pattern of air through most (but not all) stoves falls in one of 5 categories: up, diagonal, across, down, and "S" (Figure 6-8).

In updraft stoves the combustion air enters through grates under the wood and leaves the main combustion area near its top. Examples of updraft stoves are many of the antique (and contemporary) pot-bellied stoves and the Shenandoah. Additional (or secondary) air may be necessary above the wood to make complete combustion possible, particularly if the wood charge is thick and completely covers the grate, for then very little oxygen will get through the wood without being consumed, and yet the heat from the burning wood at the bottom will cause the wood at the top to evolve combustible gases. If the fire covers only a portion of the grate, ample air may flow around the outside of the pile of wood. Thus, whether or not secondary air is needed depends on both details of the stove's design and how it is fired. Natural locations for a secondary air inlet, if needed, are anywhere in the upper portions of the combustion chamber, or at the stovepipe collar. If this were to result in large flames inside the stove pipe, it would be safer and more efficient to change the design and add a secondary combustion chamber above the primary chamber.

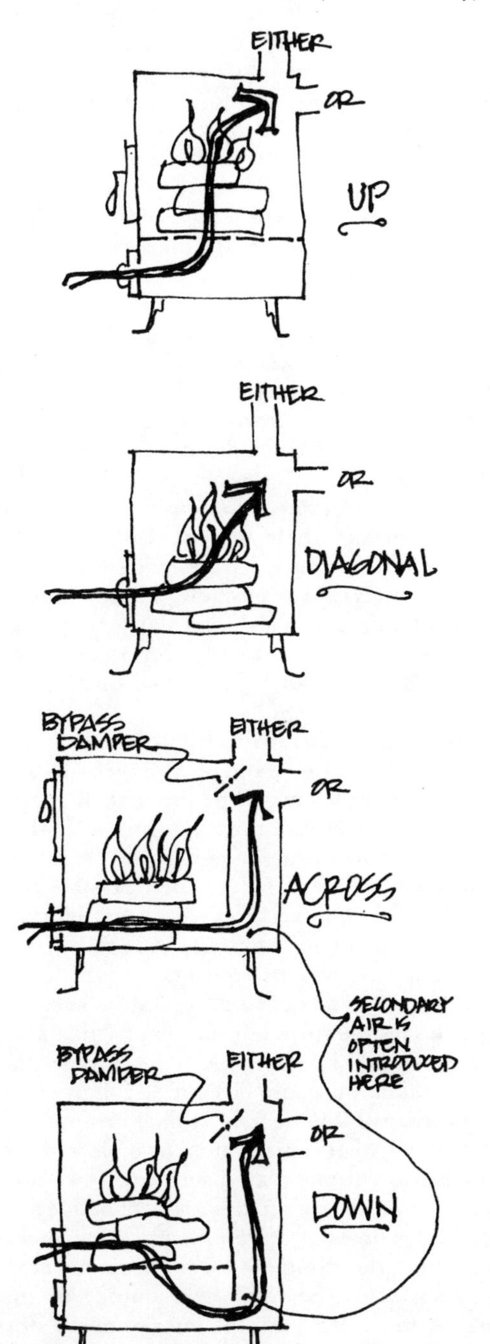

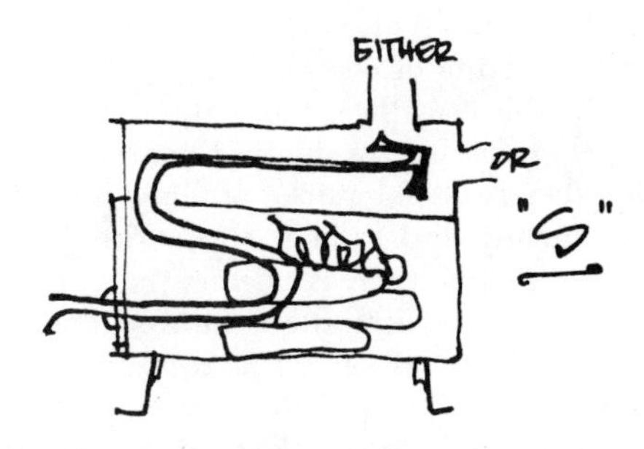

Figure 6-8. Schematic sketches of some flow patterns of air through closed stoves. Only the up-flow and down-flow stoves are shown with the grates because passage of air through the grates is essential to these flow patterns. The other types may have grates and an ash tray without affecting their flow classification as long as there is no significant flow through the grates. Similarly, the difference between a flue collar on the top of the stove, or at the top of the back or a side, does not affect the classification.

The diagonal flow pattern is typical of most plain box stoves. Air enters near the bottom of the front and flue gases exit near the top of the back. Examples are the Ashley 25HF, Fisher and drum stoves. A second air inlet above the wood, again, may or may not assist combustion depending upon, among other things, how much air travels around rather than through the logs.

In stoves with a cross flow pattern, air enters the main combustion chamber near the bottom and exits near the bottom of the opposite side. However, the air and wood gases do not flow neatly across the stove; the heat of the fire causes strong vertical currents so that there is substantial swirling convective motion in addition to the average horizontal motion. In any case, all gases, in leaving the primary combustion chamber, must pass through or near the hot coals and/or flames at the back of the chamber. Any unburned gases may burn as they travel through this region if enough oxygen from the main air inlet remains. Whether or not it does depends on the details as previously discussed. A natural place to add secondary air is at the bottom of the rear baffle plate under which the wood gases must pass, for the wood gases should always be relatively hot here. The abrupt change in direction the gases must undergo here should also help by inducing turbulence and hence, mixing of combustion gases and the air.

Stoves of this cross flow (and the down flow) type usually require a bypass damper near the top of the fuel chamber to avoid smoke spilling into the room during refueling and to help a new fire start. When the damper is open, there is a direct route from the top of the fuel chamber into the chimney. The flow pattern is then diagonal, and the resistance to flow is usually decreased. Thus, more air is supplied to help a new fire get started, and smoke tends to be cleared out of the chamber.

In downdraft stoves, the wood rests on grates. The main combustion air supply enters at some point above the grates, and the gases leave by passing down through the grates. An example is the Downdrafter stove. If the entire grate area is covered with hot coals or burning wood, then this design guarantees that all generated combustible gases must come into intimate contact with burning material on the way out, and in this respect the downdraft flow pattern is better than the crossdraft design. The high temperatures will assist the burning of any combustible gases *if* enough oxygen is present. Secondary air can be introduced in a number of places — at the bottom of the baffle plate as suggested for the crossdraft design, or under the grate, or even within the grate.

Many contemporary stoves with crossdraft or downdraft flow patterns have large fuel capacities and are airtight. They are designed to operate for long periods of time between refueling by filling the stove and restricting the primary air supply. Since the fuel generally gets quite hot and hence evolves combustible gases, secondary combustion is generally important in these designs.

The most widely known examples of the "S" flow pattern are the Jøtul 118 and 602 stoves from Norway, although many other imported and American stoves incorporate the same design feature (e.g., the Independence, the Ram Woodstove and the Sunshine). A horizontal baffle plate extends from the back of the stove towards the front. It butts up against the back and sides so that gases can only exit from the fire by passing up and around the front edge of the baffle plate. Wood is placed under the plate and air enters through the loading door.

Particularly if the stove is filled with wood up to the baffle plate, burning tends to proceed from front to back, somewhat like a cigarette. The reason is that not much oxygen can get to the wood in the back of the stove until the wood in the front has burned. (Recommended operating procedure is to rake all coals as far forward as possible before adding wood, which also encourages the front-to-back burning pattern.) This feature contributes to the steadiness of the stove's heat output.

Another consequence of this design is that combustible gases evolved from the wood must pass through the region of most active combustion, which should favor their burning. Most stoves of this type do not have a separate secondary air inlet to help burn the wood gases because the one air inlet supplies air to the front of the stove where this need could arise. Much of the wood waiting to be burned may remain cool enough that it does not evolve gases until the combustion zone approaches it. The result could be that the gases are generated relatively steadily throughout the burning cycle rather than mostly near the beginning. Thus, there may never be the intense need for secondary air which usually is present in crossdraft and downdraft stoves after a new load of wood is added. Only careful measurement can determine how important each of these effects is.

Flow patterns can affect heat transfer efficiency as well as combustion efficiency. Little can be predicted with certainty because of all the complexities and interactions. The larger the exterior surface area and the higher its temperature, the more heat will be transferred out of the stove. Stoves with up or diagonal flow patterns may have higher average surface temperatures because combustion can be occuring in the entire volume of the stove, whereas in the cross and down draft stoves, little combustion probably take place in the upper half of the fuel magazine. On the other hand, the total exterior surface area of the crossdraft and downdraft stoves may be larger than others (for the same heating capability) because of the internal baffle, and/or because they tend to be built with larger fuel capacities. Of course, the less heat is transferred out of the stove, the more will be transferred out of the exposed stovepipe since then *it* will be hotter. The longer stovepipes are, the more they tend to equalize the heat transfer efficiency of wood heating systems.

There are many stoves which are hybrids of these flow patterns, and others which are quite different from any of them. The total variety is immense. Flow patterns can affect many important properties of stoves, such as efficiency, duration of burn, steadiness of heat output and creosote generation. However, there are few *necessary* relations between the type of flow pattern and these various stove characteristics. Valid arguments can often be made both for and against the effects of various flow patterns. But the total process of burning wood and extracting heat is so complicated that the relative importance of each factor by itself cannot be predicted theoretically.

STOVE AND STOVE PIPE FINISHES

The color of a stove (and of stovepipe) can effect its heat-transfer efficiency. The amount of energy radiated by a given surface area depends on both its temperature and its infrared "emissivity." A surface with an emissivity of 1.0 is the very best possible emitter of infrared radiation. Surfaces with a flat black appearance are usually good emitters. Very shiny bare-metal surfaces are the poorest emitters.

Roughly 70 percent of a normal stove's useful energy output leaves the stove as radiation (Table 2-3). If an entire stove and its stovepipe were chrome plated almost no radiation would be emitted; shiny chrome has an emissivity of 0.02 to 0.04, which means that only 2 to 4 percent of the maximum possible radiation actually leaves chrome-plated surfaces. It would not be possible to warm one's hands and feet beside a chrome-plated stove unless one were within a few inches so as to be within the rising hot air immediately next to the stove's surface. The total heat output of the stove would be much less, and the interior of the stove, and the flue gases, would be much hotter. Restricting the outer surfaces from radiating much energy holds the energy inside the stove. (This effect might be a useful way to increase the temperature in combustion chambers and get more complete combustion.)

Most surfaces on stoves and stovepipe are moderately good emitters. The emissivities of black paint and carbon (the black ingredient in many stove-blackening fluids) range from 0.90 to 0.98.[13] Blue-oxide steel stove pipe is probably nearly as good.[14] Paints and enamels of any color except metallic colors have emissivities between 0.70 and 0.95. Many colors, including white and other light colors, are equivalent to black paint as radiators. Only metallic colors such as bright aluminum paint have lower emissivities, from 0.40 to 0.60. Bright galvanized steel (e.g., galvanized stovepipe) has an emissivity of 0.2 to 0.5.[15]

It is impossible to tell from the color exactly what the infrared emissivity is. It depends on exactly what pigments and binders are used, and on how smooth the surface is on a microscopic scale. Light colored finishes which can withstand the high temperatures of stoves exist at least in the form of baked-on enamels. Thus stoves and stovepipe need not be dark in color to be efficient radiators.

EXPERIMENTAL EVIDENCE AND DISCUSSION

Between 1823 and 1826, Marcus Bull performed some very careful experiments on the heating efficiencies of a fireplace and various stoves.[16] The burners were inside a room within a room, and a constant temperature difference was maintained between the two rooms. Bull compared the amounts of time that each device could maintain the higher temperature of the inner room, while burning an equal amount of fuel. The results indicated that an open masonry fireplace is the least efficient heating device. Bull's most efficient system was a closed "sheet iron cylinder stove" with 42 feet of 2 inch stovepipe. The details of the stove designs are not described in his paper, but his results clearly indicate the higher energy efficiencies obtainable 1) with closed versus open stoves, and 2) with long lengths of exposed stovepipe.

Seeley and Keater[17] investigated the energy efficiency of wood stoves using the stack-loss method (Appendix 4). Since only 2 stoves were tested, and neither appears to be available today, the results have little direct and practical applicability today.

More research has been done on coal stoves than on wood stoves.[18-20] Landry and Sherman used a calorimeter room (Appendix 4) to measure the heat output directly; some of the design ideas they investigated are probably applicable to wood stoves, particularly in the areas of secondary combustion and back puffing.

The British measured the performance of a variety of stoves burning coal and coke around 1948 to 1952.[21,22] The heat output was determined by separate measurements of radiation and convection. Many stoves included water heaters or boilers, and all stoves were of the openable type, designed to be operated with the doors either open or closed. Although wood was not used as a fuel, some of the results are probably applicable to wood stoves. For instance, efficiencies were not much affected by closing the doors. Closing the doors always decreased the amount of radiation given off in the forward direction, but the amount of heat in the air rising around the stove was usually increased, as well as the heat in the circulating water. The net effect of closing the doors was very small. This result also appears to be valid for Franklin stoves burning wood.[23]

More recently, experiments have been conducted by Shelton on wood stoves, using a calorimeter room to measure the heat output directly.[23] Many of the major stove types available today were tested. From this research and that done earlier on both wood and coal stoves, I draw the following conclusions concerning energy efficiencies and stove design.

The total range of efficiencies covered by all kinds of stoves is probably 40 to 65 percent[24] (excluding heat transferred from extra long stovepipes and from interior chimneys). Many stoves are grouped in the middle of this range. Fine distinctions are difficult to make between them because the performance of most stoves is significantly affected by how they are operated, as well as by their designs. However, three design features do stand out as having a significant effect, and they are openability, significant leakiness and smoke chambers.

Open stoves tend to have the lowest energy efficiencies. If the doors do not seal well, as is true in most Franklin stoves, then closing the doors does not have a strong effect on efficiency when the flue damper(s) is (are) open. However, if the built-in flue gas damper and/or a stovepipe damper are kept as closed as possible without making the stove smoke, the efficiency is improved. When the doors are closed, considerable damping is possible.

Most non-airtight stoves with large air leaks have efficiencies similar to that of a Franklin with the doors shut. Use of a flue gas damper can increase the efficiency, but if the leaks are large, the efficiency is usually less than most airtight stoves.

Some openable stoves have tightly fitting doors (e.g., the Morsø 1125, Jøtul No. 4, and the Defiant). In such stoves, flue gas damping is not necessary when the doors are shut; as in any airtight stove, the air inlet damper is the only control needed. These openable stoves have the potential of being as efficient as an airtight closed stove, although airtightness is no guarantee of high efficiency.

For maximum energy efficiency, efficient heat transfer is essential. Among the most efficient stoves available today are those with some kind of smoke chamber (e.g., the double-drum stove, the Sevca and the Lange 6303). This type of stove is roughly 5 to 10 percentage points more efficient than most other stoves. This means that 10 to 20 percent less wood would be burned to produce the same amount of useful heat. Adding extra exposed stovepipe, or use of an interior, exposed, uninsulated chimney can have a similar effect on the system's net efficiency.

The evidence from experiments conducted thus far indicates that other stove design features have less impact on energy efficiency. In particular, completeness of combustion is often not a dominant factor in practice. The reason is most clearly illustrated with an example. If a small airtight stove like the Jotul 602 is filled with wood and very little air is let in, a flameless smoldering burn results. The charcoal burns, but none of the gases do; combustion is very incomplete.

Yet the energy efficiency of such a burn is about the same as any other type of burn, including one where a maximum amount of air is admitted, ensuring relatively complete combustion. The overall efficiencies are about the same because the heat transfer efficiency is higher during the smouldering burn; the flue gas velocity is much slower, allowing more time for heat to be transferred out of the gases into the room.

This trade off between completeness of combustion and completeness of heat transfer is one reason stoves of radically different design often do not have very different energy efficiencies.[25] Combustion is almost always made more complete by admitting more air, and admitting more air is always detrimental to heat transfer. Efficient use of air is important.

Conflict between combustion and heat transfer efficiencies is not a necessary one. Even when there is excess air both diluting the temperature and increasing the velocity of the flue gases, heat transfer can always be nearly complete if there is sufficient effective heat transfer surface area. With enough exposed stovepipe attached to a stove, the flue gases will cool down to room temperature before leaving the building, no matter what their velocity. The result is very high heat transfer efficiency under all burning conditions.

Another common side effect of admitting more air (in order to achieve more complete combustion) is an increase in the stove's combustion rate. In most stoves, the heat output is controlled by limiting and adjusting the air inlets. Much effort in stove design is devoted to achieving complete combustion despite the limited air supply to the wood. Use of secondary combustion chambers and air is the most common technique. The secondary air is intended to have no effect on the primary combustion rate, and to burn the gases evolved in primary combustion. In well designed stoves, secondary combustion can make it possible to burn a large charge of wood both slowly and completely.

A completely different approach to achieving steady heating without frequent and small additions of fuel to the fire is to use heat storage between the stove and the house to control the heating rate to the house. This permits the stove to burn very hot and completely with plenty of air without overheating the house.

A wood furnace designed by Prof. Richard C. Hill of the University of Maine, Orono, Maine, for the Maine Audubon building in Falmouth incorporates this storage principle, as well as complete physical separation of combustion and heat transfer. The furnace is heavily insulated which helps make combustion more complete by keeping temperatures high. The combustion gases are also recirculated through the combustion region an average of 10 times before leaving the furnace. Most of the heat is carried out of the furnace in the flue gases, which are sent through a 140 foot long, 8 inch diameter pipe in a gravel heat storage bin. The gases enter as hot as 1000° F. and cool down to as low as 150° before leaving the storage area, indicating very efficient heat transfer. Condensation occurs but combustion is so complete that the condensate is almost pure water. Heat for the building is extracted as needed from the hot gravel by circulating air through it. The furnace is fired up once every day or two, as needed, with a few hundred pounds of wood which burns in a few hours. The fast burning hot fire with plenty of air assures relatively complete combustion, the long flue pipe assures relatively complete heat transfer, and the gravel bin heat storage controls the heating rate to the house. The flue gases are so cool that the natural draft of the chimney is inadequate, so a draft-inducing fan is used.

Heat storage and heat transfer can also be built into or onto a stove, but the result would not resemble typical stoves in the least. A stove built of tons of masonry materials with many internal passageways to transfer heat from the flue gases into the masonry could be fired with occasional hot fires with high energy efficiency and little creosote generation. When the fire is nearly out, the air inlet would be shut tight to prevent heat losses from the masonry up the chimney. The stored heat would gradually come out of the masonry into the room, or air could be blown through another set of passageways to extract more heat as needed. European tile stoves[26] and a conventional steel or iron stove surrounded with rocks or gravel (Chapter 8, pp. 64) are two examples of stoves which utilize these ideas.

Systems incorporating large scale heat storage are awkward for many applications because of the very large mass and volume of heat storing material needed. In most applications, the conventional approach of limiting primary air and striving for secondary combustion of the gases is more practical; such systems have the potential for high efficiency and steady heating although most current designs are probably not optimal.

Chapter Seven Operating Characteristics

Though energy efficiency of a stove is important, there are other characteristics which are equally significant. The two most important stove properties other than energy efficiency are heating capability or "power," and the ability to generate heat at a steady average rate without frequent refueling or adjustment for an extended period of time (8-12 hours).

POWER AND STOVE RATINGS

If one were trying to decide what size stove to buy, the promotional literature of manufacturers and importers would not be very helpful. To state that a stove can heat 15,000 cubic feet is an unusual way to rate a heating appliance. Virtually every other heating appliance (furnaces, wall heaters, portable electric heaters, etc.) are rated not in terms of volume of air they can heat, but in terms of power, the rate at which they can produce heat. Fossil fueled heaters in the United States are rated in "British thermal units" (Btu) per hour. Electrical heaters in America and all heating systems in the rest of the world are rated in watts. (One watt is the same as 3.41 Btu per hour.)

The reason "cubic feet" is inappropriate is that the need for heat in a house (or room) is determined not only by its size, but also by how well insulated and sealed it is, how many windows it has, and how cold it is outdoors. An old, leaky house without insulation or storm windows needs much more heat than a new, snug house of the same volume and in the same climate. Likewise, identical houses (including the same cubic feet) in Maine and Southern California will not need the same amount of heat because of the difference in climate. Volume by itself just does not matter. Stoves, like any other heaters, should be rated in units of power, such as Btu per hour, or watts.

Most conventional heaters, in a sense, have only one power; the device is either on or off (e.g. most furnaces, water heaters and portable electric heaters). When a furnace's burner is on, heat is being delivered at the maximum rate, and ratings are based on this maximum output. Usually less than the maximum power is desired, and the device then continuously cycles on and off, controlled by a thermostat. The *average* power can thereby be varied from a maximum value all the way down to zero. Rating conventional heaters by their maximum powers tells essentially the whole story.

It is not easy to give a stove a power rating, because unlike most conventional heaters, stoves do not have clearly and easily defined upper or lower limits to their possible heat-output rates. A hot stove can always be coaxed to be a little hotter either by putting in more or smaller size pieces of wood, or by letting more air in. The minimum practical heat output rate is determined by the minimum possible combustion rate without risk of the fire going out.[1] Both the maximum and minimum powers may also depend on particular installation details through the effects of chimneys on draft and the effects of exposed stovepipe length on energy efficiency. (A more efficient installation burning wood at the same rate has a higher heat-output rate.) Thus, both maximum and minimum powers may depend on other things than the stove itself.

The human effort required to operate a stove is also critical information. For some users, a stove which requires refueling every hour when operating at the required power may not be practical. For other people, the need for frequent refueling is not a liability. But in order for prospective purchasers to be able to make wise decisions, this kind of information is required.

Ideally, I think that manufacturers (or some consumer testing organization) should supply at least the following information: a minimum and a maximum average power output, and at each of these powers as well as at an intermediate power, 1) the time between fuel loadings (and damper adjustments, if any are necessary), 2) the stove's energy efficiency, 3) the tendency towards chimney creosote deposits, and 4) some indication of the steadiness of the stove's heat output. Although the species, moisture content and sizes of wood burned, and installation details such as length of exposed stove pipe and chimney properties all may affect these characteristics, at least reliable relative data could be obtained through standardized testing. This would be of immense benefit to consumers. Such testing would be a major undertaking and has not been done. At this time, manufacturer's estimates (not measurements, in most cases) of power (or cubic feet) are about the only available information.

Determining the needed heating rate of a room or house is the other half of the problem, and it also is difficult to do precisely. Standard practices for estimating heat losses from houses (Appendix 5) are just as applicable to wood heated houses as to houses with conventional heating systems, particularly if the wood heating is done with a central, wood-fueled furnace. Such estimates are useful despite their having a large uncertainty [due mostly to the uncertainty in estimating the contribution of air exchange (or infiltration) to the total heat loss.]

If the stove is intended to heat one room only, and if the rest of the house will be maintained at a comfortable temperature by some other heating system, then the heating needs of the one room can be computed using the method in Appendix 5 by calculating the heat losses through only the "cold" surfaces, such as exterior walls (no heat is conducted through a wall, floor, or ceiling, both sides of which are equally warm). But if the stove is intended to heat as much of the house as possible and if portions of the rest of the house will be less warm than the room with the stove in it, then computing the usable heat output is more difficult. Relevant factors include how easily heat can move into other rooms (see Chapter 8), and how large a temperature difference between rooms is acceptable. There are no standardized and readily available guides for computing heating needs under such circumstances because it is not customary to have temperature differences and natural-convection heat distribution in houses.

Because of the lack of reliable stove-performance data and the difficulty of accurately predicting heating needs, the best practical procedure at this time for selecting the appropriate size stove is to seek advice locally from people with wood-heating experience. Stove retailers should be a good source of information. Friends and neighbors with similarly constructed houses (in the same climate) who already have stoves will have some experience with stove sizing.

Most stoves can operate over a wide range of heat outputs; this is a necessity in order to meet the varying need as the seasons and the weather change. If a stove is undersized, it may be inadequate only on the coldest days. However, if a stove operates near its peak power most of the time, it will require more frequent refueling than if operated in other portions of its range. If a stove is oversized, it can usually be operated in the lower portion of its power range, although its practical range may not extend low enough to handle times when just a little heat is needed. (The most common complaint concerning stove size comes from owners of oversize stoves which put out too much heat even at their lowest powers.) Proper sizing optimizes a stove's usefulness, but sizing is not overly critical — considerable leeway exists for error.

CONTROLLABILITY OF HEAT OUTPUT

For those people interested in serious heating with wood, a stove should have a large capacity for fuel to lengthen time between refuelings and should be capable of putting out heat at a fairly steady rate without frequent adjustment of the controls. Since the needed heating rate is variable, ideally a stove should be easily controllable over a range of powers. In the spring and fall, in a particular house, one might need only about 10,000 Btu per hour, whereas in the winter 40,000 Btu per hour might be needed. Some stoves need more attention than others to achieve steady power. For example, a Franklin-type stove, operated with grates and the doors open, quickly burns whatever fuel is in it because of the abundant supply of oxygen to all of the wood. One consequence is that the only way to get low heating rates from such stoves is to add small amounts of wood frequently, an inconvenience under most circumstances. A large chunk of wood will sometimes slowly smolder, but there is always the chance that it may go out or suddenly flare up. Ideally, for the most carefree heating, a stove should be able to take a large amount of wood at a time, and be able to burn it at a steady rate, selected by the operator, without further attention.

All closed stoves (which excludes, for example, Franklins operated with open doors) are controllable to a reasonable degree. The almost universally used principle is to limit the supply of combustion air. With no air, there is no oxygen to react chemically with the wood, converting its stored energy into heat. Limiting a stove's air supply can be done manually with air inlet and / or flue dampers, or automatically with a thermostat.

A number of stoves are equipped with thermostatically controlled air inlets. The thermostat works on the same principle as do many wall thermostats for conventional heating systems. It consists of a bimetallic coil or strip which tends to wind up tighter (or in some cases, unwind) as its temperature decreases. It is made of two strips, of two different kinds of metal, bonded together. Both kinds of metal expand (or contract) as they are heated (or cooled), but one expands more than the other. Since they are bonded, the only way one can change its length more than the other is for the shape of the bonded pair to change as illustrated in Figure 7-1. If one end is held fixed, as the temperature of the bimetallic strip changes, the movement of the other end can be used to open or close an air-inlet damper. In a wall thermostat, the movement opens or closes an electrical switch.

Figure 7-1. Bimetallic strips and coils as thermostatic controls for air inlets.

In most stoves, the coil's temperature is about the same as the air in the room being heated.[2] When the room warms to the desired temperature, the movement of the coil closes the air inlet, and combustion stops (or slows, if some air can still get in). Heat stored in the stove walls and in the fuel continues to enter the room, but little or no new heat is generated inside the stove. The room then slowly cools, which makes the bimetallic strip open the air inlet, allowing combustion to increase again. A knob on the control can be turned to select the desired room temperature. The automatic control on Riteway stoves has a small magnet which tends to pull and hold the air damper shut. When the force of the bimetallic strip exceeds that of the magnet, the damper pops open. As a result, the stove tends to get either a substantial amount of air, or none. The control on Ashley stoves and the Downdrafter is more continuous - the damper gradually opens and closes and can remain in any position. It is not clear which system, if either, is better all-or-none control versus gradual control.

Figures 7-2 to 7-4 illustrate the effects of each kind of control on heat output, for two particular stoves, including one with a thermostatically controlled built-in heat extractor. Both types of control, if well designed, are capable of regulating room temperature without uncomfortable oscillations. (The oscillations in room temperature would be much smaller than the oscillations in heat output shown in the figures, due to heat storage effects of the room.)

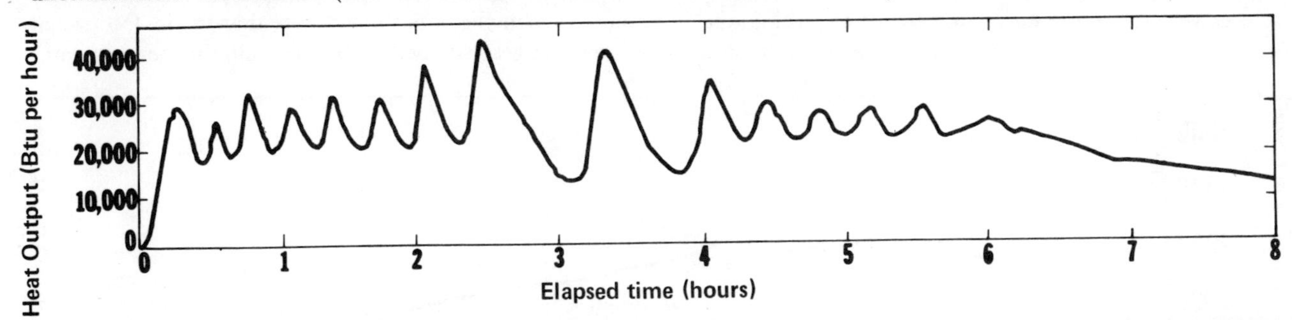

Figure 7-2. The heat output of a Riteway Model 2000 during a particular run, illustrating its on/off-type thermostatic control. Near each peak, the thermostat suddenly shuts off the primary air completely; in each valley, the primary air damper pops open. The result is a fairly sharply modulated power output around an average of about 25,000 Btu per hour in this case. The rate of the decrease in power after the primary air is stopped is determined partly by the duration of flaming in the secondary combustion chamber, and partly by the rate at which stored heat leaks out of this relatively massive stove. After 6 hours most of the initial charge of 50 pounds of wood had been consumed and the damper stayed open from then on. The end of the burn is not shown in the graph.[3]

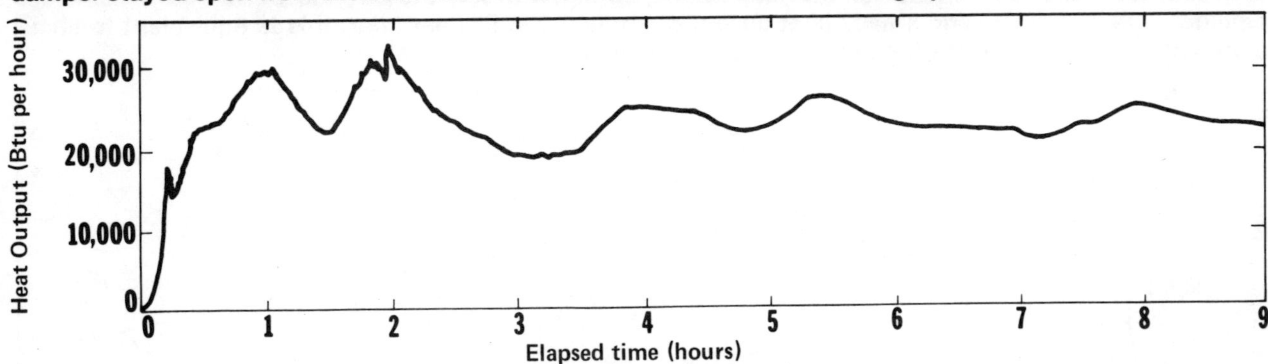

Figure 7-3. The heat output of a Downdrafter stove during a particular run, illustrating the effect of its continuous or proportional thermostatic control. The combustion air supply is never fully cut off; it is adjusted slightly and gradually in response to the thermostat. The result in this stove is a relatively steady power output. The graph depicts most of a burn which started with a 50 pound load of wood. (The Downdrafter has a blower which did not come on during this test.)[3]

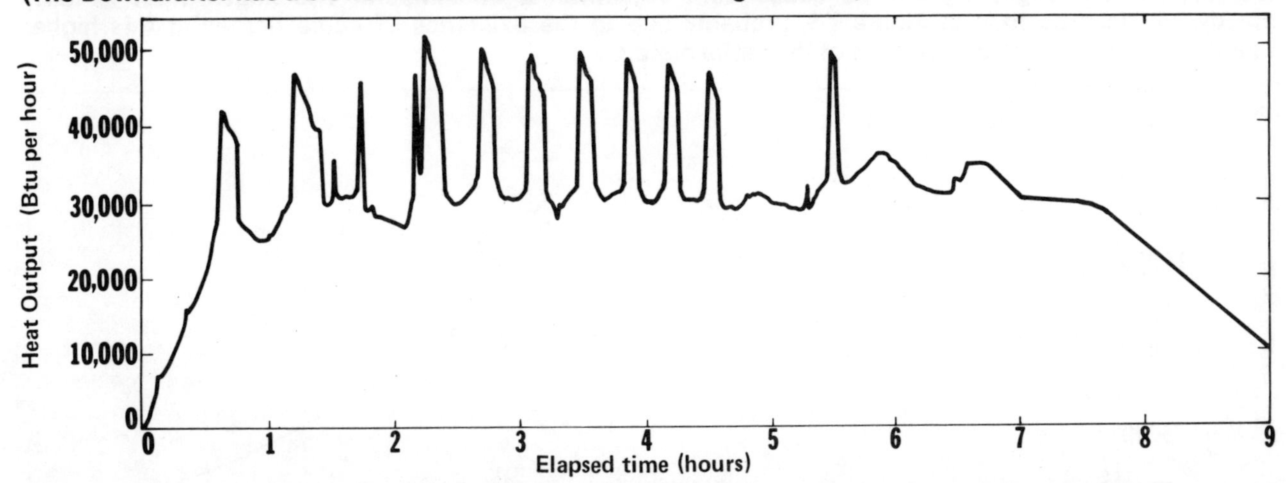

Figure 7-4. Heat output of a Downdrafter stove during a particular run, illustrating the effect of a built-in electrically powered and thermostatically controlled heat extractor. When the flue-gas temperature exceeds a preset value (about 400 degrees Fahrenheit in this case) a thermostat turns on a small blower, forcing room air to circulate through passages built into the stove. Substantial additional heat is extracted while the blower is on, and as a result the flue gases are cooled; the thermostat then turns the blower off. Combustion air is also thermostatically controlled, resulting in a constant average heat output. The Downdrafter has grates, so that wood combustion ends quite abruptly when the fuel is exhausted, after about 7½ hours in this case. The subsequent heat output is due mostly to stored heat in the stove coming out into the room.[3]

Most stoves do not have thermostats to automatically adjust the air-inlet damper. The damper is set at fixed positions, depending on circumstances and as learned from experience. Some stoves are capable of putting out heat at a remarkably constant rate under these conditions. Figures 7-5 to 7-7 illustrate some heating rates of airtight stoves without thermostat control. In each case a little kindling was burned before the stove was filled with wood; no further adjustments were then made. Occasional small surges can occur, but overall the heating rates

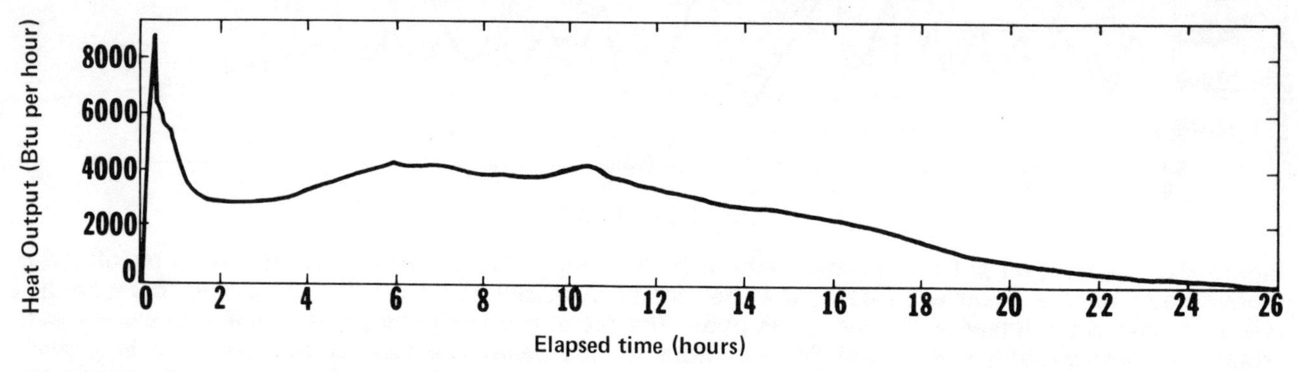

Figure 7-5. The heat output of a Jøtul 602 in a very low-power burn during a particular run. The initial sharp peak is due to the burning of kindling to generate some coals. After 20 minutes the stove was filled with about 18 pounds of red oak. The air inlet damper was then nearly shut, and no adjustments or wood additions were made for the rest of the burn. With so little air, the burn was a flameless smolder with a very low and steady heat output of about 3,000 Btu per hour; this is equivalent to about a third of the heat output of a top burner in a cooking range.

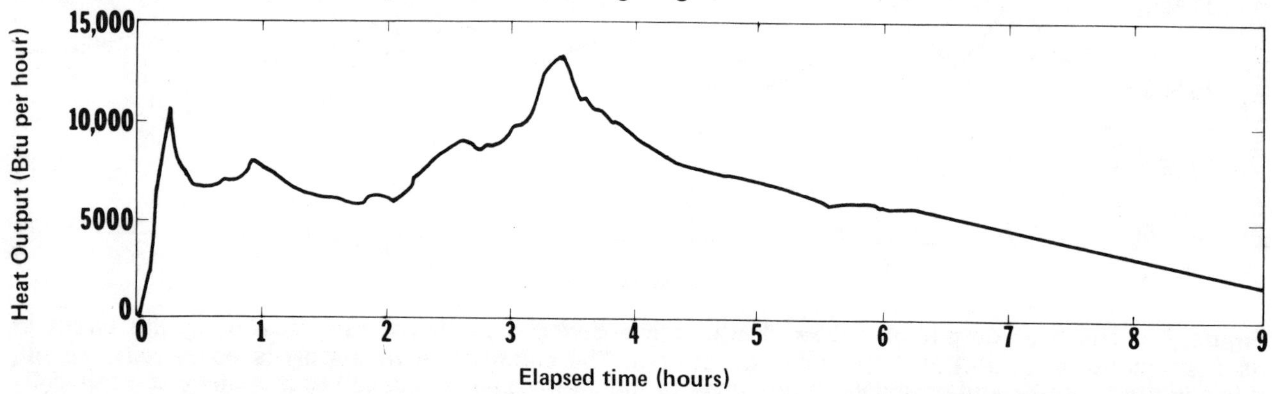

Figure 7-6. The heat output of a Jøtul 602 without its baffle plate during a particular run with one full fuel load at the beginning and no subsequent adjustments of dampers. The heat output is fairly steady, but less so than in Figure 7-5, probably due to the existence of some flames at this higher heating rate, and due to the absence of the baffle plate.[3]

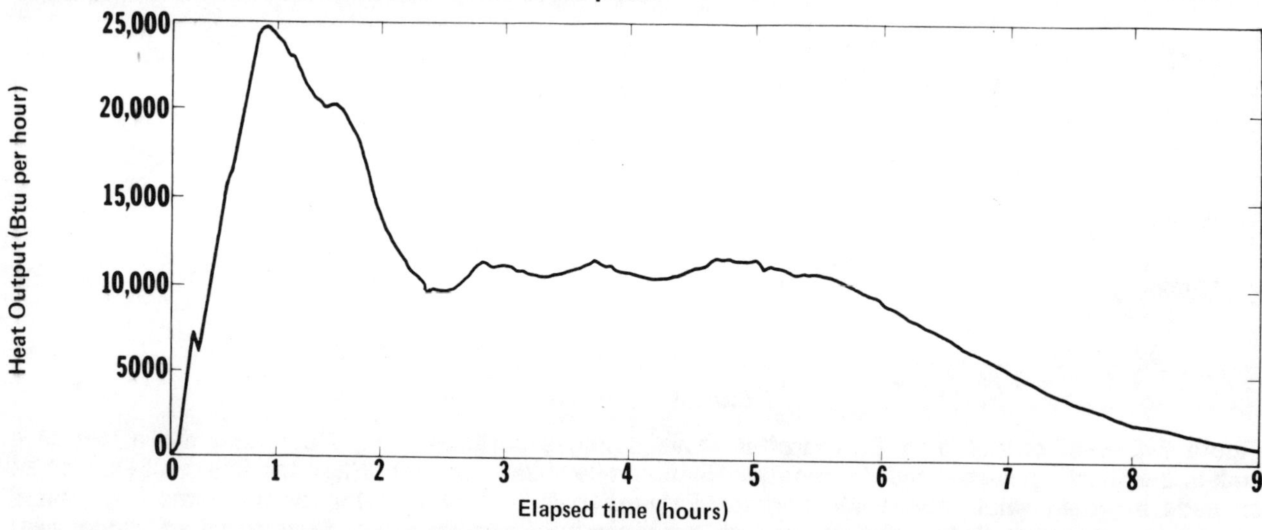

Figure 7-7. The heat output of a Morsø 2B stove during a particular run involving one full loading with fuel at the beginning and no subsequent adjustments. The surge at the beginning is typical when non-automatic stoves are operated at medium power, and probably corresponds to the burning of combustible gases, which are most abundant in the early portions of a fire.[3]

are satisfactorily steady.

In addition to adjustment of air inlets, there are two other ways to limit the amount of combustion air entering stoves. Both involve regulating the draft (suction) at the air inlets, rather than the size of the openings. The most common is a flue-gas damper, usually placed in the stovepipe above the stove, but in some cases, built into the stove. Its use is discussed in Chapter 6. The other draft-regulating mechanism for controlling combustion air is called a draft regulator. Whenever the amount of draft in the chimney exceeds a preset value the draft regulator opens, admitting air to the chimney and thus relieving some of the pressure difference. Although commonly used on fossil-fuel furnaces, draft regulators may not be safe to use with wood stoves because of the chance, however remote, that a spark or other burning matter might come out of the regulator. A draft regulator would also make chimney fires more intense by feeding air to the fire.

Neither a stovepipe damper nor a draft regulator is of much use in an airtight stove. Complete control of the amount of air entering the stove is achieved by adjusting the air-inlet dampers.[4] But non air-tight stoves benefit from draft control. If the air inlet itself is too leaky, draft regulation is the only way to limit combustion air enough to achieve low power operation. If leaks are so located that their air is not likely to contribute to combustion, then decreasing the draft will cut down on this excess air flow, which will improve the stove's energy efficiency.

In an open stove (or fireplace) or a very leaky closed stove, there is no practical way to control the heat output by controlling the amount of combustion air. If air is always abundant, limiting the fuel supply is the only alternative. This can be done by controlling the frequency and size of refuelings. Figure 7-8 shows the heat output of a Franklin stove with its doors closed. Each surge corresponds to one refueling. A fairly steady average heating rate can be maintained with frequent refuelings.

Although a thermostat is not necessary to achieve steady heating, thermostatic control can have some advantages. In any stove the combustion rate can suddenly increase, such as when smoke ignites, or some of the wood settles. A thermostat can compensate, keeping the heat output steady, by decreasing the air supply. In stoves without thermostats a surge of heat results. Such surges can be substantial, resulting in sudden large increases in room temperature.

Thermostatic control can have an advantage when the need for wood heat changes. The switching on and off of heat-producing appliances, such as a cooking range, can change the amount of heat needed from a wood stove. Sudden changes in weather can change the house's total heating need. Since a thermostat responds to room temperature, it can automatically compensate for such changes. However in practice this is only a small advantage.

HEAT STORAGE EFFECTS

Another completely unrelated way to help obtain steady heating from wood stoves is to have a large heat storage capability either in the stove, or in the room in which the stove is installed. Such "thermal mass" or heat capacity moderates temperature fluctuations by temporarily storing heat. For instance, if there is a combustion surge in a light thin-walled stove, there will be an almost immediate surge of heat into the room, which can make the room uncomfortably warm. The same combustion surge in a massive stove will be much less noticeable because much of the heat generated in the surge will temporarily be in storage in the stove walls and structure. Most of the heat will still come out, but it will be delayed and spread out over a longer time. Similarly, as combustion rates decline, the heat output of a heavy stove will fall more slowly than that of a light stove, as it gradually gives up its larger quantity of stored heat.

The only possible detriment of a massive stove is that it is impossible to get an intentional surge of heat out of it as quickly as with light stoves. When a house is cold and one wants heat in a hurry, less massive stoves are better. They start putting out significant amounts of heat within a few minutes after lighting. Even though the fire starts just as quickly in a massive stove, its heat output into the room does not reach full power for 10 to 30 minutes.[5] Thus in applications where wood heating is not wanted continuously or for long periods of time, (e.g., in occasionally used cellar or garage workshops) thin-walled stoves may be more satisfactory.

A comparison of Figure 7-9 with Figures 7-2, 7-3, 7-4 and 7-8 illustrates this effect. A thin-walled drum stove reaches 30,000 Btu per hour of heat output in six minutes after ignition (Figure 7-9). The Riteway

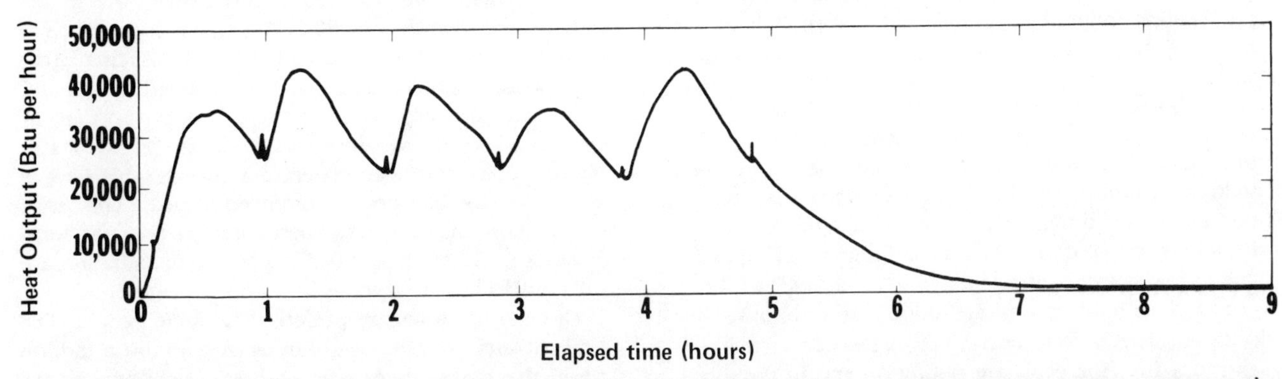

Figure 7-8. The heat output of a Franklin stove with its doors shut. Each surge corresponds to the burning of a load of fuel. Fuel was added 4 times after the initial loading, in each of the major valleys in the graph. The small sharp spikes represent the bursts of heat entering the room while the doors are open for refueling.[3]

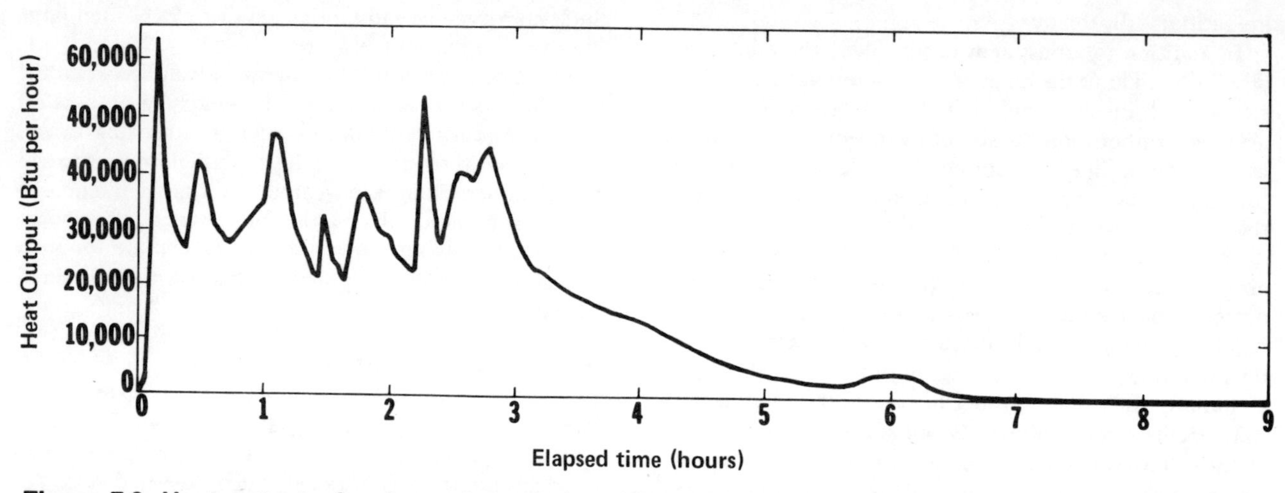

Figure 7-9. Heat output of a drum stove during a particular run, illustrating the quickness with which the heat output can change in a relatively thin-walled light-weight stove. Only 6 minutes after ignition the stove was heating the room at a rate of 30,000 Btu per hour, and 2 minutes later, just before the inlet damper was closed somewhat, the power was 60,000 Btu per hour. This is a much faster buildup of heat output than in the much heavier stoves shown in Figures 7-2 to 7-4 and 7-8. Wood was added twice during this run, and the inlet damper adjusted four times. Much of the variation in heat output was spontaneous, resulting from shifting and settling of the wood.[3]

Downdrafter and Franklin stoves are all very heavy, and they take 15 to 30 minutes to reach the same power, although in two of these stoves part of the delay is due to slower initial combustion (related to air flow patterns). The Fisher Papa Bear is a comparably heavy diagonal-draft stove. It takes 10 to 15 minutes to reach 30,000 Btu per hour under similar conditions.

Thermal mass in the room has the same effect. If a stove is placed in a corner surrounded with masonry materials, the heat output of the stove is similarly moderated. Because masonry materials are poorer conductors of heat than steel or iron, they absorb and let out heat more slowly. Stone will not absorb much of the heat from a 10 minute surge. However if a stove operates at higher-than-needed power for an hour, masonry surroundings can take up some of the excess, preventing the room from getting too warm, and then let it back out as the stove's output diminishes, thus evening out the temperature in the room.

OVERNIGHT BURNING AND RETENTION OF HOT COALS

Many stove users appreciate a stove which can put out heat all night long and is still hot enough in the morning that added wood will ignite spontaneously without the need to use kindling and matches. Most stoves can operate only at the low end of their power range during such a long time without restoking. The total amount of heat that a stove can give off between stokings is limited by the amount of wood that can be put in at one time. So the ability to hold a low fire for a long time is determined by the stove's fuel capacity, and its low-power controllability.

Stoves without grates are more likely to have hot coals left after a long burn. This is because at the very end of a fire, the coals are resting on and in the ashes. The ashes perform two useful functions. They impede oxygen from reaching the coal's surfaces and hence slow the rate of combustion. They also thermally insulate the coals, helping to keep them from cooling off and going out. The decline in heat output of the Franklin stove in Figure 7-8 is considerably faster than that of the barrel stove in Figure 7-9. The Franklin has grates and the barrel does not. (This result is despite the Franklin's larger mass, which gives it the advantage of more heat storage capacity.) The slow declines in heat outputs in Figures 7-5 and 7-7 are due both to the absence of grates in these stoves and to the fact that these stoves are relatively airtight.

The nature of the wood used affects the duration of a low-power fire. Larger (especially thicker) pieces burn more slowly because of their relatively smaller surface areas for a given weight of wood. Wood with more moisture in it also burns more slowly. Thus many people put some larger and/or greener pieces of wood in the stove before going to bed. It takes a little practice to do this successfully, since, if over done the fire will die in the night without burning up all the wood.

CIRCULATING STOVES

Stoves can be classified as either radiant or circulating depending upon the dominant mechanism by which they transfer energy into the room. Most stoves are of the radiant type, transferring 60 to 70 percent of their energy output as infrared radiation (Table 2-3). A circulating stove is essentially a radiant stove surrounded by an outer jacket with openings at the bottom and top so that air circulates between the stove and the jacket (Figure 7-10). Natural convection can generate a considerable flow of air, but small blowers are often employed. The result is that more of the stove's energy output is in the form of hot air, although significant amounts of radiation are still emitted by the jacket.

Circulating stoves are preferred by some people. The exposed surfaces are not as hot as they are on a radiant stove; this makes them safer with small children in the house, and allows people, furniture, and walls to be closer to the stove without discomfort or damage. (See Table 9-1 in Chapter 9 for safe clearances from combustible materials.) The jackets are often designed to

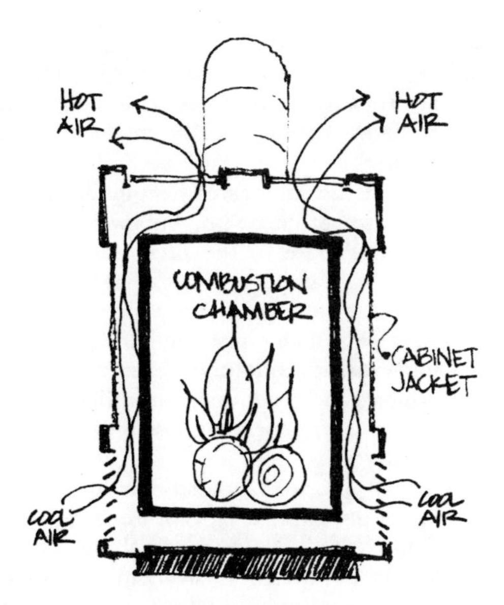

Figure 7-10. The Ashley Imperial (Model C-60), a circulating stove.

disguise the fact that they house a wood stove.

There are some differences in thermal comfort between rooms heated mostly by radiation and rooms heated mostly by hot air. Since the radiation from a stove comes from one direction only, a person's side facing the stove will be slightly warmer than the opposite side. A piece of furniture or a person can block the radiation from reaching parts of the room. But in a room with high levels of infrared radiation, the air temperature can be cooler without causing any thermal discomfort, and some people feel better breathing and being in cooler air.

There is no fundamental reason why either type of stove would necessarily have a higher energy efficiency.

WOOD LOADING EASE

Stoves vary considerably in wood-loading ease. Fuel is added to most stoves through a side opening door which is placed near the bottom of the primary combustion area. One advantage of the low location for the loading door is that smoke seldom comes out of the stove into the room during refueling. However, it is sometimes difficult to get large pieces of wood in, partly because of weight and leverage problems, and partly because of the danger of burning one's hand. The difficulty is most accentuated with longer and narrower stoves. One does not have to use pieces of wood that are so big that there is any loading problem, but in that case the wood must either be split into smaller pieces and/or cut shorter. In addition, then, one cannot take advantage of the slower burning properties of larger pieces of wood.

There are stoves that are loaded through a door in the top or near the top. Such a loading scheme usually eliminates the possible awkwardness of handling large pieces of wood, for they can just be dropped in (gently). It is also often easier to completely fill the fuel magazine from above. Larger doors, wherever they are located also make fuel loading easier. Some front-loading stoves have doors that are considerably smaller than the stove's fuel chamber, which makes it a little awkward to fill the stove with wood, especially with medium and large pieces of wood.

Top loading doors are somewhat more likely to let some smoke out during loading, but proper procedure can largely eliminate this possibility. The air inlet should be opened fully a few minutes before opening the loading door. If there is not much smoking fuel left, this should clear the air in the stove. If there is some smoking fuel, the additional air will usually ignite it, eliminating the smoke. In addition, the resulting combustion surge will warm the fuel gases, increasing the chimney's draft, thus providing larger suction at the loading door. This will tend to keep the combustion gases from escaping when the door is opened.

BACK PUFFING

It is wise to follow this same procedure before opening a door on any relatively air tight stove in order to avoid the danger of back puffing. The combustion rate in air tight stoves is usually air-limited (as opposed to fuel limited). Under this circumstance, if air is suddenly admitted there can be a combustion surge of the gases which is so quick that the resulting pressure can force hot gases out any available openings, such as the loading door, the air inlet, and cracks or leaky joints in both the stove and stovepipe. The phenomenon could be termed a slow, very small explosion. Back puffing is somewhat dangerous because the emerging gases can be burning, and thus can cause burns to people nearby. Therefore, all openings in operating stoves should be opened slowly. The air inlet should be fully open for a few minutes before opening the door. The operator's face should also be kept well back from the stove until the door has been open for a few moments. The possibility of back puffing is one reason children should be cautioned about touching a wood stove.

Back puffing can occur in most stoves. It can sometimes be induced by adding a large amount of new wood on top of a very hot bed of coals; if the air inlet is then closed for a few seconds, the stove will become filled with combustible gases. Then if the air inlet is opened, the gases may suddenly ignite, creating a substantial back puff. Back puffing can also occur spontaneously at any time. On rare occasions the pressure within the stove is sufficient to lift off heavy metal cooking and warming plates. A more common occurence is pulsating back puffing - small puffs occurring consecutively, sometimes separated by only half a second, sometimes separated by as much as five seconds.[6] Each puff consumes the available oxygen and pressurizes the stove, momentarily preventing more oxygen from getting in. The next puff will occur when enough oxygen has re-entered the stove.

The phenomenon of back puffing ought to be carefully investigated since it can be hazardous. There may be particular kinds of stove designs which are especially susceptible, and others which are immune.[7] I have never heard of a stove exploding, but the blowing off of a cooking plate clearly leaves a stove in a more dangerous condition, and serious burns are possible if a back puff occurs while the door is being opened. Fortunately, injuries and serious accidents resulting from back puffing appear to be very rare.

HOT HANDLES

A minor but annoying aspect of many stoves is that

the handles and knobs needed in their operation are often too hot to touch with bare fingers. Some stove manufactures supply small metal rods with hooks to operate the handles. Many stove users keep a glove nearby. However, two stoves I am aware of have handles which never seem to be too hot to touch; they are the Jøtul 602 and 118. Other things being equal, a black handle will be cooler than a chrome handle because it radiates much more heat. A chrome surface tends to hold the heat in, making a chrome handle significantly hotter.

MATERIALS AND DURABILITY

Most stoves are sufficiently durable that durability need not be a prime consideration in purchasing a stove. But there are exceptions, and there are conditions which are especially wearing on particular types of stoves.

Most stoves are made of steel (plate or sheet) and/or cast iron. There are good and bad aspects of both materials. Mechanically, cast iron is much harder and stiffer, and less susceptible to distortion, which makes it preferable for doors and door frames where small distortions could result in significant air leakage. But also because of cast iron's stiffness, it is susceptible to cracking. Cast iron cannot "give" very much. If the center of a cast iron stove is much hotter than the rest because the fire or a hot coal is against it, the thermal stress can crack it.

Steel is relatively soft and malleable. If stressed, it bends, often permanently. Some steel stoves develop slight distortions in their walls due to thermal stress; they can be aesthetically displeasing, but the stove's functioning is rarely impaired. Cylindrical and oval shapes are more resistant to distortion than the flat sides of rectangular stoves.

Both steel and cast iron are susceptible to corrosion. Some oxidation (rusting) of the stove walls from the inside due to the fire is unavoidable. The rate of oxidation at very high temperatures ("red hot") is much higher than at normal stove temperatures. Thin-walled stoves operated at very high temperatures have been known to burn out in one season. (A quarter-inch of steel can be oxidized completely away if it is held at red-hot temperatures for about a day.)

Firebrick or metal liners lessen all three of these stove ailments - cracking of cast iron, distorting of steel and corrosion of both materials. The liners keep the main stove body from getting too hot. The liners are easily replaceable.

Thick stove walls are also less susceptible to these problems. Thicker walls have less need for the protection of a liner. Being stronger, they are less likely to crack or distort in the first place, and being thicker, they will last longer with respect to corrosion even if the corrosion rate were the same. But in fact, both the thermal stresses and the corrosion rate are less. A thick stove wall cannot have such intense hot spots in it as thin walls can; the thickness makes it much easier for heat to spread out sideways or laterally within the wall. Thus, even though a fire or hot coal is against the inside of a thick wall, it cannot get as hot as a thin wall. The increased conductance within the wall also makes the more distant portions (corners) hotter than they would be in a thin-walled stove. The wall's temperature is more uniform, and therefore the thermal stresses are less.

Stainless steel is much less susceptible to oxidation. If there are thin metal parts inside a stove which are red hot at times, they should be made of a material like stainless steel.

With normal use and care, heat-induced oxidation is the only source of corrosion, but two kinds of carelessness can contribute to the premature demise of a stove. Storage in damp conditions has always been a common fate for stoves, but the modern way to ruin a stove, stovepipe and chimney is to burn trash, including plastics. Some plastics contain the elements chlorine and fluorine, which can be converted in a fire to extremely corrosive acids. Stovepipe may last only a year under such circumstances.

The thermal properties of steel and cast iron are virtually identical (Table 2-1). Cast iron's reputation for "holding heat better" is the result of the thickness of most cast iron stovewalls. A steel stove with walls as thick as a cast-iron stove will have just as much heat storage capability. Cast iron and steel stoves of identical design, including wall thickness, would be indistinguishable in their thermal performance. Overall, both steel and cast iron are suitable materials for stoves. They have some different properties, but neither can justifiably be claimed far superior.

Soapstone, clay tiles and other masonry materials are sometimes used in stove construction. Their thermal properties are roughly the same as firebrick's (Table 2-1). Compared to cast iron and steel, the most pronounced difference is their low thermal conductivity; they take significantly longer to warm and to let their stored heat out. Their specific heats are a little higher than those of iron and steel, but their lower densities result in about half as much heat storage capability for the same thickness. Most masonry materials are susceptible to cracking under thermal stress.

ASHES

Other things being equal, the frequency with which ashes need to be removed varies considerably. With heavy use, some stoves require ash removal every two or three days, others every few weeks or even less frequently. Four factors are probably most important.

The volume of ashes the stove can accumulate without impairing operation is the most obvious factor. In the case of stoves with ash drawers or trays under grates, the drawer's volume is the limit. For stoves without grates, the ashes gradually accumulate, and, if desired, the stove can still be operated with a large amount of ash.

In stoves with grates, the ashes are quite loose - their weight in a given volume (density) is very small. In stoves without grates, where the wood rests directly on the ashes, compaction is possible; in addition, the heat of the fire on top of the ashes can fuse them into a semi-solid (in the form of a crust). More ash (in terms of weight) can be stored in the same volume. For the same volume of storage space, ash clean-outs can therefore be more infrequent in grateless stoves. However, such stoves rarely, if ever, have ash drawers; ashes must be shoveled out, which is somewhat less convenient than using a built in ash drawer.

In some stoves a considerable amount of ash can be blown out of the stove and up the chimney during normal operation. This seems to be especially true of stoves where the wood rests directly on the ashes, for then there can be fairly high air velocities over the ashes which can cause them to be carried along. Natural dispersal and fallout of ashes which come out a chimney is fairly even and helps to fertilize the surrounding land.

Finally, different wood types leave different amounts of ash residue when burned. Ash contents of most woods types vary from 0.1 to 3 percent. The ash content also depends on the age and the part of the tree from which the wood comes (branches versus trunk, and how much bark is included). Although there are significant differences in ash contents, rarely does this property of wood affect choices of what type to burn.

The total amount of actual residue left in a stove after a fire may be quite different from the wood's ash content. The amount may be less because the ash has gone up the chimney, or apparently more because small bits of incompletely burned wood are mixed with the ash.

PRACTICAL RECOMMENDATIONS

The following points should be kept in mind when purchasing a stove:

1. A stove's heat-output range should match the heating needs of the room or house as closely as possible.

2. If occasional quick heating is important, such as in a cabin, garage, or basement, a lightweight thin-walled stove is appropriate. Examples are drum or barrel stoves, the Ashley models 23 and 25, and the Yankee.

3. For continuous heating, ease of operation is probably the most important consideration. Thus a stove that holds a relatively large charge of wood and burns it steadily is most desirable. A good thermostatically controlled air inlet system guarantees steady heating between refuelings. A large automatic stove like the Riteway is capable of very long burns with substantial heat output. S-draft stoves like the Jøtul 118 and the Independence need more frequent refueling and are probably somewhat less steady in their average heat output, but have simpler and more rugged designs.

4. If the availability or cost of wood fuel is the principle concern, a system with a high energy efficiency should be selected, such as the Sevca, Lange 6303, or double drum. The most expensive stoves are not necessarily more efficient.

5. The airtightness of a stove is important for steady heat output at very low rates, as is needed in the spring or fall, or overnight. The only practical test for airtightness is a visual inspection for gaps and cracks.

6. Heating a mobile home with a wood stove may work best with a small circulating stove because of the space limitations, and with direct feeding of outdoor air to the stove (for combustion) because of the small natural air infiltration rates. Stoves made especially for mobile homes are available (e.g., the Suburban Mobil-matic).

7. Cast iron and steel plate or sheet are both suitable materials for stoves. Steel can warp, but this does not usually impair the functioning of the stove. Cast iron can crack, but rarely does. Both materials can rust and / or "burn out." The thicker the material, the longer the life-expectancy. I see no strong overall superiority of either cast iron or steel as a stove-construction material.

Chapter Eight Installations— Usefulness for Heating

Where and how a stove is installed in a house can significantly affect its usefulness. A stove's location in a house can influence its energy efficiency, the steadiness of its heating, and the distribution of its heat.

STEADINESS

In houses or rooms heated with stoves, temperatures usually fluctuate more than in conventionally heated spaces due to the irregularity of the heat output from most stoves. The temperature fluctuations are usually not a serious drawback of wood heating, but most people are more comfortable the smaller these fluctuations are.

The location of a stove affects the size of the temperature fluctuations it causes. The larger the heat capacity (also called "thermal mass") of its surroundings, the smaller will be the fluctuations. The mass in furniture, walls, floors and ceiling near a stove contributes to thermal mass but in most types of houses it helps significantly to add heavy heat-storing material near the stove, such as brick or stone under and behind the stove. These heavy materials can soak up and store some of the stove's heat output when it is operating at high power, and release it when the fire subsides, thereby making the air temperature in the house more constant.

A considerable mass is required for the effect to be significant. For example, suppose that a stove is placed in the corner of a room, and that both walls behind the stove are lined with four-inch thick bricks from floor to ceiling which extend 5 feet from the corner. Suppose that the stove also rests on a four-inch thick, 5 x 5 feet brick hearth. This constitutes a total of 35 cubic foot or a little over 2 tons of masonry. If use of the stove can raise the average temperature of all this mass by 20 degrees Fahrenheit, the stored heat would be about 16,000 Btu[1]. The typical heat output rate of stoves ranges from 10,000 to 50,000 Btu per hour. Thus the stored heat is equivalent to roughly 20 minutes to an hour and a half of stove output. The actual time it would take for a substantial amount of this heat to enter or escape from the masonry into the room is roughly a few hours.

This same amount of thermal mass is even more significant in keeping house temperatures from going too low at night and at times when the stove is putting out little or no heat. If the masonry starts at 85 degrees Fahrenheit (20 degrees above a room temperature of 65 degrees), and if it were to cool all the way down to 45 degrees the heat that would be released would be 32,000 Btu. This amount of heat can be a substantial contribution, which might, for instance, prevent water pipes in the house from freezing on a cold night.

Installing a stove in front of or inside a fireplace has some steadying effect on temperatures, but may result in a decrease in energy efficiency, especially if the fireplace and chimney are in an exterior wall of the house.

Some people pile rocks all around and on top of a stove (except the front). The rocks can be constrained by a wire cage to make the pile neater and more compact. The rocks can get much hotter than a masonry wall would, and the amount of stored heat per pound of rock is thus greater. The temperature-leveling effect will be strong, but along with that comes sluggishness - the system would not be able to heat up a room quickly. Rock-covered stoves might be particularly useful for heating a greenhouse; the rocks would also decrease the intensity of the heater's radiation, allowing plants to be closer.

The greatest temperature-leveling effect is achieved if the whole house is built of masonry materials.[3] Both the tonnage and surface area of heat storing material is then very large. Large amounts of heat can then be absorbed, stored or released. Wood or any kind of heating in such a structure is especially even. (There are interesting disadvantages to such masonry houses. For instance, nighttime thermostat setbacks do not save as much energy as in wood-frame houses because the house does not cool down very quickly. See Appendix 6.)

HEAT DISTRIBUTION

Wood stoves are "local" sources of heat. A stove can maintain comfortable conditions throughout most of a typical-size room, but temperatures are usually significantly cooler in the other rooms. This is an inherent property of any kind of "room" heater. Some people accommodate to temperature differences from room to room. Others have room heaters in more than one room. Of course, the norm in America today is a central furnace distributing heat to all rooms. Wood furnaces are available, as well as furnaces which can burn both wood and conventional fuels. Whenever wood stoves are used, it is often desirable that a considerable portion of the heat be distributed to other rooms - in an all-wood-heated house, temperatures will be more uniform, and in wood-supplemented houses, more use can be made of a stove and thus more oil, electricity or gas can be saved.

The useful energy output of a stove is delivered to the room both as infrared radiation and as heated air. Radiation travels in straight lines in all directions from a stove until absorbed or reflected by a solid surface.[4] Of common surfaces, only shiny metals can reflect very much of the radiation; all other surfaces absorb most of it, reflecting or scattering only a small part. The radiant energy which is absorbed on surfaces becomes sensible heat there, some of which conducts into the solid, some of which is conducted to adjacent air, and some of which is reradiated out into the room. Air is heated directly by contact with the stove, and indirectly by contact with other surfaces in the room which have been heated by the stove's radiation.

Air which is warmer than the immediately surrounding air rises. If rising warm air hits a flat ceiling, it spreads out. Hot air at a ceiling is not useless for heating. It warms up the ceiling, which then radiates energy downward, in significant amounts. If there is a room above, the heat which conducts through the ceiling is also used. In some cases air currents (or forced convection) bring some of the warmer air down, tending to even out ceiling and floor temperatures.

The extent of a wood stove's heat distribution is affected by its location, and can be improved by some simple schemes. Placed in a large opening between two rooms, a stove can heat both. Placed in one room considerably more heat will get to an adjacent room if there is unobstructed air flow at both the ceiling and the floor; ceilings in particular are often not continuous from one room to the next - warm air at the ceiling often must flow down to get through a doorway and this impedes heat distribution. (Figure 8-1). The warm air at the ceiling behaves like water in an upsidedown lake - the ceiling is the lake bottom and the walls are sides or dams; a doorway which is not open all the way up to the ceiling is like a spillway - the air will not flow out until its level is "down" to the lip of the spillway. If the doorway does provide clear passage at ceiling level, warm air at the ceiling will always flow out to the next room. If doorways are not open at ceiling level, large registers at the tops of walls will serve the same purpose. Lateral movement of heat can also be assisted with a fan. For instance, placed at the ceiling in the lake of warm air, it can force the air down through a doorway.

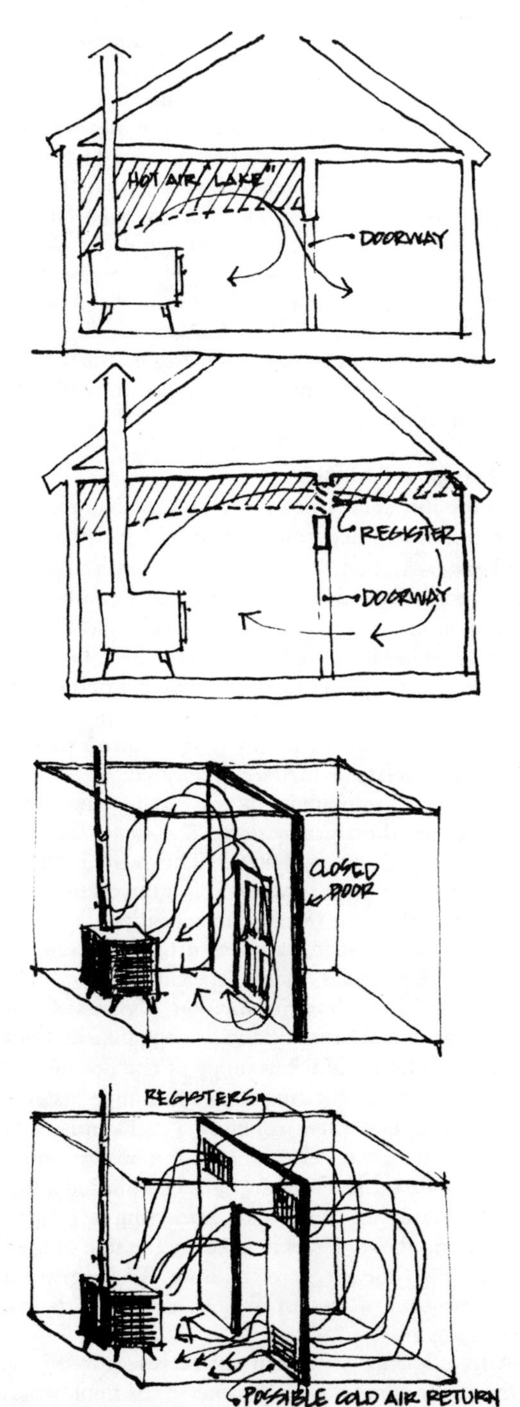

Figure 8-1. The effect of ceiling-level registers on the distribution of heat from the stove room to an adjacent room connected by a doorway which is not open to the ceiling. If the door is closed, floor level registers are also needed to provide a return path.

Vertical heat distribution comes very naturally. Stairwells provide efficient passage for warm air to go up and cooler air to return. The two currents do not interfere with each other - the warm flows up along the ceiling, and the cooler air flows down over the stairs. A register installed in the ceiling of the room with a stove in it is very effective for heating the room above as long as the air in the upstairs room can get out to make way for the incoming air. In very old houses, enough air may leak out through the walls and ceiling; in tighter houses, there must be an air return route to the stove room, such as open doorways, or a register at the bottom of a wall of the upstairs room, providing passage to a stairwell.

A wood stove can also affect the temperature distribution by its interaction with a conventional, thermostatically-controlled heating system. If the stove is located in the same room as a thermostat, the conventional heating system will not come on as long as the stove keeps the room above the thermostat's setting. This of course saves conventional energy - it is essentially equivalent to a very large thermostat setback, with one room kept warm with a stove. It is quite possible for this to result in serious damage to pipes in other rooms if they get so cold that water in the pipes freezes.

If a stove is located far from thermostats there is little such danger of freezing, since the thermostats will control temperatures in their normal fashion. Essentially the same conventional-energy savings as in the case above can be achieved with explicit thermostat setbacks which may be necessary anyway to keep the stove room from getting too warm. The advantage of not having a stove affect a thermostat is the independent controlability of the two heating systems. Aside from the danger of freezing damage to pipes, either arrangement is satisfactory.

It is possible to take some advantage of an existing forced hot air distribution system to help distribute the heat output of a wood stove to the rest of the house. If the stove can be installed in the same room as a cold-air return register, then, whenever the furnace is on, the return air can be somewhat preheated by the stove. Of course, if the thermostat for the furnace is in the same room as the stove, the furnace may not come on when the stove is in use. Independent controls for the blower alone can be installed. A manually operated switch can turn on the blower even when the furnace burner is off; or the switch could be controlled by a thermostat in the room with the stove, so that the blower comes on whenever the temperature in the room exceeds a certain value. However, if the basement or crawl space is cold, a significant amount of the wood stove's heat may be dissipated out through the walls of the ducts and plenum. Thus, unless the space around the ductwork is warm from frequent use of the furnace, some of the stove's heat will be wasted when circulated with the furnace's blower.

ENERGY EFFICIENCY

How and where a stove is installed can affect its net energy efficiency. Since a significant amount of heat is given off by hot stove pipe, increasing the pipe's length increases the energy efficiency. The 4 or 5 feet of stove pipe in typical installations contributes 15 to 25 percent of the total useful heat and hence contributes about 8 to 13 percentage points of a typical 50 percent efficient stove. Adding an additional 5 feet of stovepipe increases the system's energy efficiency by 3 to 8 percentage points.[5] Decades ago in buildings heated by wood only, the stove was often placed very far away from its chimney, so that very long lengths (often more than 50 feet) of stovepipe could be used. This not only extracted more of the heat from the flue gases, but also helped distribute the heat throughout a large room. If properly installed and maintained, such installations do not constitute an undue safety hazard, in my opinion, but the National Fire Protection Association safety standards prohibit it. Similarly, use of an exposed interior brick chimney whose heat leakage contributes useful heat to the house can improve net efficiency 5 to 15 percentage points.[6]

Chrome stovepipe decreases the efficiency of a wood stove heating system. Shiny metal surfaces emit much less radiation than do most other surfaces at the same temperature. The net effect for a typical chrome stovepipe installation is probably a 10 to 15 percent decrease in heat output (a decrease in energy efficiency of 5 to 8 percentage points). Galvanized pipe also restricts radiation, although not as much as chrome. Flat black and blue-oxide stovepipe are both good radiators.

Locating a stove near an exterior wall results in larger heat loss through that wall. For a normally insulated wall, the effect on net efficiency is probably less than a percent under most circumstances. For a single-pane glass wall, or other uninsulated wall with no confined air space, the decrease in net efficiency can be a few percentage points. (The glass may also crack). Placing a stove near an interior wall is not detrimental unless the extra heat conducted into the wall is not fully used, because for instance, it convects up into an unused attic, or because it is conducted to an unheated pantry or laundry room. A stove placed close to an asbestos board or sheet metal wall protector probably is slightly less efficient, particularly if the wall protector has a shiny metal surface.

Installing a stove in front of a fireplace, so as to use the fireplace flue, need not, but often does cause a decrease in efficiency. It is very important to prevent room air from escaping. Normally the smoke pipe from the stove is inserted up into the chimney a few feet through the open fireplace damper. A carefully fitted metal sheet and/or packed fiberglass insulation can be used to seal the rest of the damper opening.

Placing a stove inside a fireplace will normally decrease its net efficiency. Much of the stove's direct radiation will be intercepted by the masonry, and not all of this energy will eventually be used, particularly if the fireplace is built into an exterior wall. The warm masonry environment will also result in less net heat transfer out of the stove. In addition, with no exposed stove pipe, much of the heat output from a normal installation will not be available, although some of this heat may be conducted through the chimney walls into the room and thus be used.

BASEMENT INSTALLATIONS

A stove installed in a basement is a satisfactory way to heat the basement, but it is usually very inefficient for heating any other part of the house, even if there is free air circulation between the basement and the rest of the house. The reason is, it takes a huge amount of heat to keep an uninsulated basement warm. A wood stove can deliver heat to the basement; but because basement walls and floors soak up so much heat, a significant amount of the heat will never rise to other floors of the house. This is especially true of radiant woodstove heat, most of which leaves the stove as radiant energy. The radiant energy passes through the air and only becomes sensible heat when absorbed by surfaces such as the walls and floor; but by then, most of it is well on its way out into the ground. Circulating stoves would be better since most of their energy output is hot air, which can rise directly to the first floor of the house. Basement wall and floor coverings also help when they have some insulating value. But in general, having a stove in a basement is not an efficient way to help heat the rest of the house.

CENTRAL HEATING

Wood furnaces are appropriate in basements. The basic difference between a stove and a furnace is that the heat output of furnaces is collected and ducted to other rooms. Wood furnaces and stoves can be identical on the inside - the same combustion techniques and control mechanisms are used. A stove becomes a furnace if a sheet-metal jacket or box is placed around the stove which allows air to enter near the bottom and circulate up around the stove. If it is not required to heat the space in which the furnace is located, it is helpful if the jacket is insulated—increasing the heat transfer efficiency to the duct system. The heated air is then guided through ducts system. The heated air is then guided through ducts to where it is wanted and eventually returned back to the furnace for reheating.

Factory-built wood furnaces usually include blowers to force the air, as is done in typical gas or oil furnaces; in this case the whole installation is virtually identical to ordinary fossil-fuel furnaces except that fuel is added by hand. The natural buoyancy of the hot air can also be used as the circulating force. A simple system using buoyancy is illustrated in Figure 8-2. Decades ago such "gravity" hot-air systems were common. A wood stove can be made into a crude furnace by building a fireproof jacket around the stove and connecting it to a register in the floor above. Holes or doors in the side of the jacket are necessary for access to the stove. The register should be a few square feet in area so as not to restrict the flow of hot air.

A free flowing air return system is required in gravity furnaces. In some houses registers in the floor around the perimeter of the house can let the return air into the basement. Although it is not essential, it is usually best to then duct the return air to the furnace rather than allow it to meander across the basement.

Furnaces are available which burn both wood and conventional fossil fuels. There are also multi-fuel boilers for steam heating systems, hot water systems, and simply

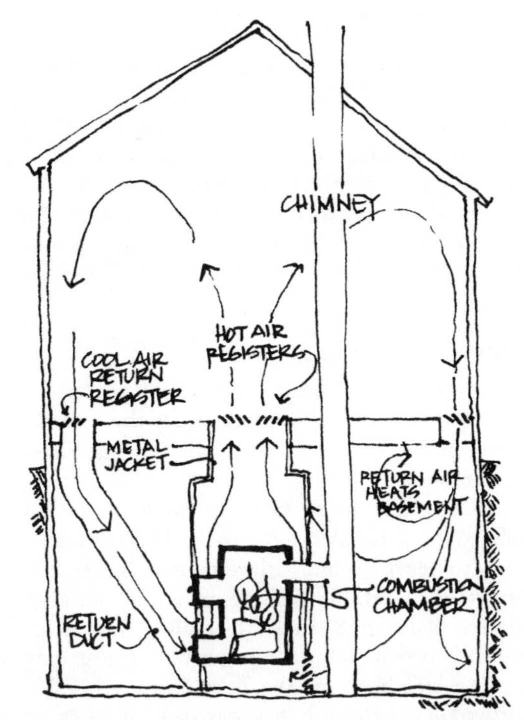

Figure 8-2. A gravity or natural-convection wood furnace. Return air can be ducted directly to the furnace as shown on the left, or simply let into the basement as shown on the right.

for making hot tap water.

A good wood or multi-fuel furnace or boiler with conventional contemporary heat circulating systems provides the ultimate in even heat distribution throughout a house and offers the convenience of remote temperature control. More fuel will usually be consumed to warm a house with central wood heating than with stoves because 1) central systems can heat the entire house (and that takes more energy), and 2) the basement will absorb some of the furnace heat. But wood furnaces and boilers can provide even, uniform heat to a large house, and with proper duct and register design, heat to unoccupied rooms may be turned off. In addition to space heating, furnaces and boilers can heat all of or a substantial part of a home's domestic hot water and all this can be done with only one to three fuel reloadings per day.

FLOOR-LEVEL DRAFTS

Rooms heated by stoves often are uncomfortably drafty at floor level. An operating stove creates a natural-convection pattern of air rising around the stove, spreading out at ceiling level, descending beside cooler exterior walls, and returning to the stove along the floor (Figure 8-3). Air flow along the floor towards the stove is also needed to satisfy the stove's combustion-air requirements. This combustion air is not recirculated but must be a net influx of air from the outside. It normally leaks into a house through many small cracks around windows, doors, sills, etc.

Most people do not make special provisions to limit cool floor-level air currents. But if desired, both contributions to the currents can be diminished - combustion air can be ducted directly from outdoors or from a basement to the heater, and some of the floor-level portion of the natural-convection circulation can be guided under the floor (Figures 8-3 and 8-4).

Where feasible, the simplest way to get combustion air directly to the wood heater is through an opening in the floor under or in front of the stove (or fireplace), which provides an air passage from a basement or crawl

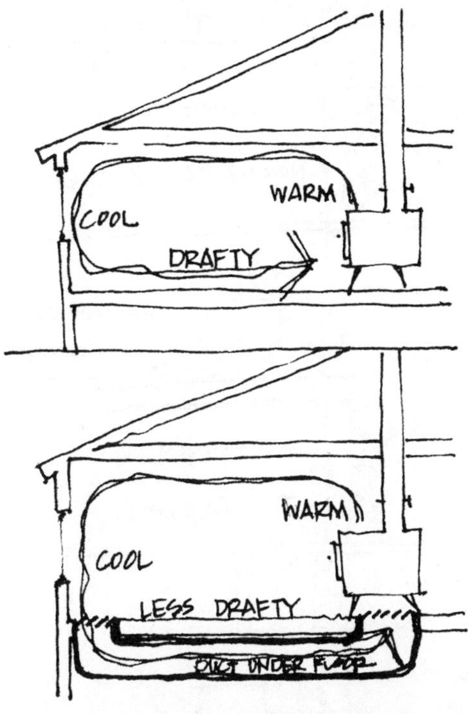

Figure 8-3. A method of decreasing the discomfort of floor-level drafts. Floor registers along the inside of an external wall permit some of the descending cooler air to return to the stove via the basement and a register under the stove.

space. The size (area) of the passage should be roughly the same as the flue. Alternatively, air can be ducted directly from an outside wall of the house to the heater, either via the basement or through a wall behind the stove. It is probably best to actually attach an outside-air duct directly to the stove's air inlet, but this is frequently inconvenient.

Feeding outdoor combustion air to a wood heater will usually decrease the net energy efficiency of the system, due both to decreased heat transfer from the stove[7] and to the larger heating needs of the house due to the new hole in it. A damper should be installed to help limit the outside air flow when it is not needed, but some leakage is virtually inevitable. The increase in comfort from eliminating part of the floor draft may be worth using a little extra energy.

Guiding the natural-convection circulation under the floor is a little harder to do in an existing house; but when building a new house it is not hard to provide for. If there is a continuous gap half a foot wide between the floor and the exterior walls, or an occasional larger floor opening, much of the cold air which flows down the inside of exterior walls will not flow across the floor but will drop down into the basement or into ducts provided for it under the floor. The cold-air ducts (or the basement itself) allow the cold air to flow into an opening under the stove. The openings are covered with registers. This system is similar to many "gravity" hot-air distribution systems, except that the heater is located

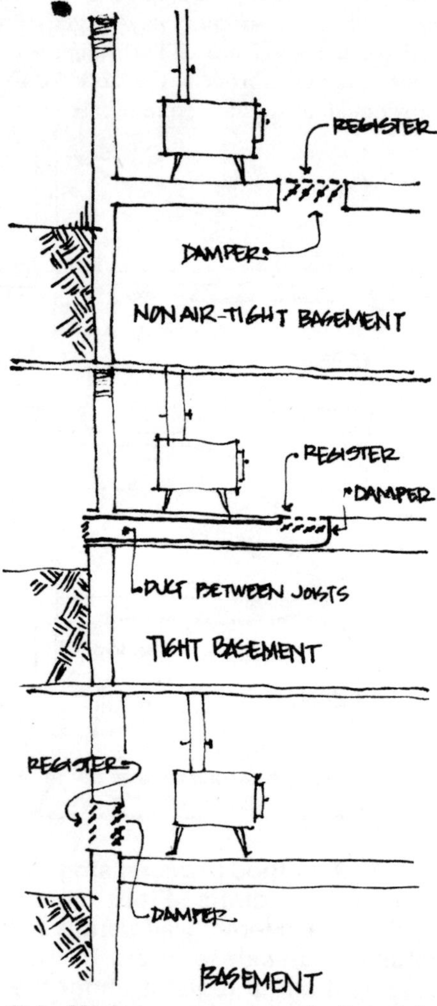

Figure 8-4. Various ways to admit outdoor air directly to the vicinity of a stove.

in the living space rather than in the basement. Both have an air return system around the perimeter of the house, alleviating cool, floor-level drafts. However, in this system neither the rising hot air nor falling cool air is contained in a duct during any significant part of its vertical motion. The driving forces for the air circulation under the floor are weak. Thus it is important that the floor registers have large areas. If furniture such as a book case is placed in front of (but not covering) perimeter registers, it can help guide the cold air down into the basement.

PRACTICAL SUMMARY

A stove's location should be most influenced by where one wants heat and, if applicable, by the location of an existing usable chimney. The following are some other considerations:

1. If it is desired to heat as much of the house as possible, a central location is best. Heat should be able to flow to nearby rooms. Placement near a staircase will result in considerable heating of the floors above. Installation of large (2 square feet) floor and ceiling level registers between rooms can improve heat distribution.

2. The steadiness of the heat from a wood stove is improved a) if there is a great deal of mass near the stove, such as masonry walls and hearth, or a fireplace and b) if an interior and exposed masonry chimney is used.

3. The energy efficiency of the system is improved a) by using a long length of exposed stovepipe to connect the stove to its chimney, and b) by using an interior, exposed chimney.

4. Installing a stove in a basement is usually not an efficient way to heat the rest of the house.

5. Wood furnaces and boilers can provide the most uniform and steady wood heat, and in so doing, they usually burn more wood.

Chapter Nine
Safety

SAFETY

Above all else, stove installations should be safe. There is little doubt that in actual practice, fires are more likely in buildings heated with wood than with conventional heating systems. It is also clear that most building fires associated with wood heating are due to unsafe installations, not inherent dangers of heating with wood. By complying with safety standards and building codes, virtually all danger is eliminated except operator carelessness.

The potential fire hazards of heating with wood are fires started by 1) radiation from the stove, stovepipe or chimney, 2) sparks and glowing coals getting out of the stove, 3) flames leaking out of faulty chimneys, 4) conduction of heat from chimneys into combustible material, and 5) burning or glowing material coming out of the top of the chimney. There also is a small danger of chimney flow reversal, leading to either flames or smoke coming out of the stove's air inlets.

The National Fire Protection Association has formulated installation standards[1] intended to prevent these occurrences. The standards include recommended clearances between various hot surfaces (stoves, stovepipe, chimneys) and combustible materials, safe ways for stovepipe to pass through a wall, and safety aspects of chimney design. The standards are intended to ensure that under the worst conditions (such as continuous high power operation, and perhaps an occasional, intense chimney fire), the building will be safe, so the only accidents which are likely to occur would be due to extreme operator carelessness.

Safety standards are always somewhat arbitrary. They usually represent a compromise between rationality and simplicity — between taking into account the many different types of stoves, varying combustibilities of house construction materials, etc., and trying to keep the standards sufficiently short and simple so they are reasonably easy to follow. As a result, standards can be overly conservative. On the other hand, observing standards is no absolute guarantee of safety, even without human carelessness in the operation and maintenance of the system.

What follows is a presentation and discussion of the major National Fire Protection Association standards for safe wood stove installations. I have reworded the standards and in a few cases I have made minor additions. For the original standards, see the references in footnote 1. The comments about the standards are my personal opinions about their importance and adequacy. I do not recommend deviating from the NFPA standards or from local building codes.

STOVE CLEARANCES

STANDARD

Radiation from large hot surfaces can be intense enough to cause spontaneous ignition of nearby combustible materials. The NFPA minimum recommended clearances for stoves and stovepipe are given in Table 9-1. A typical (radiant) stove should be no closer than 36 inches from walls made of or containing wood, unless the wall is protected in an approved manner, as described in the Table.

The protective wall coverings for radiant-type stoves should be large enough and so placed that no unprotected

TYPE OF PROTECTION[3]	STOVES[4]		STOVEPIPE[5]
	Radiant[6]	Circulating[6]	
Unprotected	36″	12″	18″
¼″ asbestos millboard	36″	12″	18″
¼″ asbestos millboard spaced out 1″	18″	6″	12″
28 ga. sheet metal on ¼″ asbestos millboard	18″	6″	12″
28 ga. sheet metal spaced out 1″	12″	4″	9″
28 ga. sheet metal on 1/8″ asbestos millboard spaced out 1″	12″	4″	9″

TABLE 9-1. NFPA Recommended Minimum Clearances From Combustible[1] Walls and Ceilings[2]

[1] Combustible materials include wood, cloth, vinyl, paper, etc. Wood covered by plaster is also considered combustible. Thus, of the most common wall and ceiling constructions, only masonry walls are excluded.
[2] Adapted from National Fire Protection Association Standard No. 89M, "Heat Producing Appliance Clearances (1971.)
[3] Protection should extend over enough of the wall or ceiling that no uncovered part of the surface is closer than the "unprotected" clearance.
[4] These are clearances from the sides and rear of a stove. Slightly different standards exist for the tops and fronts of stoves, but they are almost always met in practice (stoves are rarely placed close to ceilings and the space needed in front for loading is normally more than ample for safety). See NFPA 89M for details.
[5] These are clearances from a parallel wall or ceiling. See text for case where stove pipe passes through a wall or ceiling.
[6] Most stoves are classified as "radiant." "Circulating" stoves have outer jackets all around the stove with openings, usually grilled, to allow air to circulate. An example is the Ashley C-60 console model.

part of the wall is within the standard 36 inches of any part of the stove. Similarly, for convective stoves and stovepipes, the protection must cover all combustible surfaces within the standard unprotected distances of 12 and 18 inches respectively.

A "combustible wall" in this context includes walls containing wood which is not exposed. For example, sheetrock over wood framing is considered a combustible wall. Thus these clearances are applicable to virtually all kinds of construction except masonry.

COMMENT

A very important fact about wall and ceiling protection is the vital importance of ventilated air space. Figure 9-1 illustrates two common but unsafe installations. A sheet of ¼ inch asbestos millboard placed in contact with a wall is rated to have no protective value. The

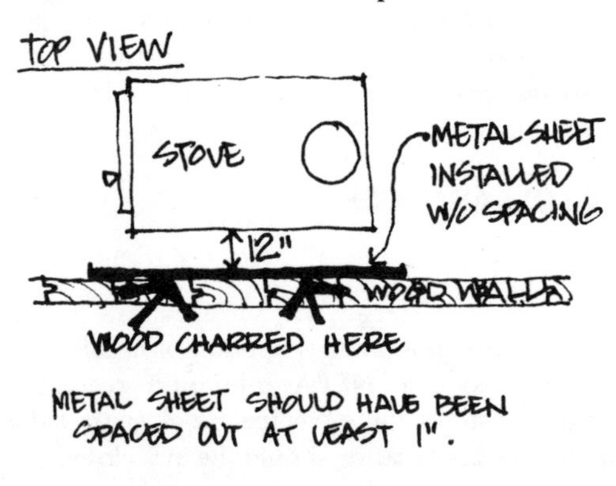

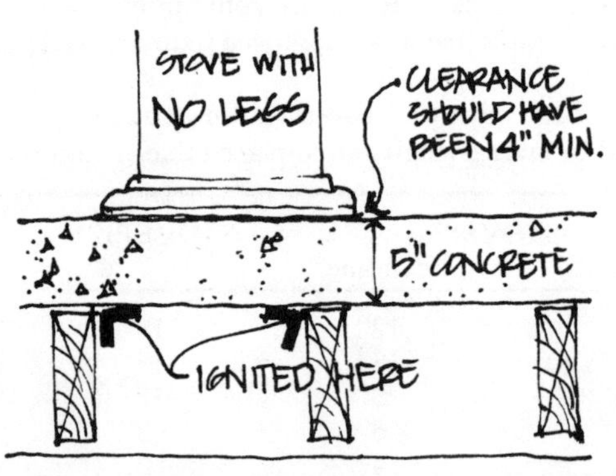

Figure 9-1. Typical improper stove installations of kinds which have resulted in fires. [Adapted from "Fire Protection Handbook, 13th Ed., pp. 9-47, National Fire Protection Association, Boston (1969).]

material is such an excellent conductor of heat that its back side, which is in contact with the wall, is at nearly the same temperature as the side exposed to the stove. But the same material, spaced 1 inch from the wall with (non-combustible) spacers so air can freely circulate behind it (Figure 9-2) permits placing the stove half as far from the wall (18 inches).

A shiny metal surface on a wall actually does give it some protection, even without an air gap behind it. The protection is not related to the non-combustible nature of metal but to the fact that highly polished metal surfaces reflect almost all infrared radiation. As a consequence, the surface, and whatever is behind it, do not get as hot. The NFPA standards do not make allowances for this kind of wall protection because if the surface loses its shine, its protection is lost. This

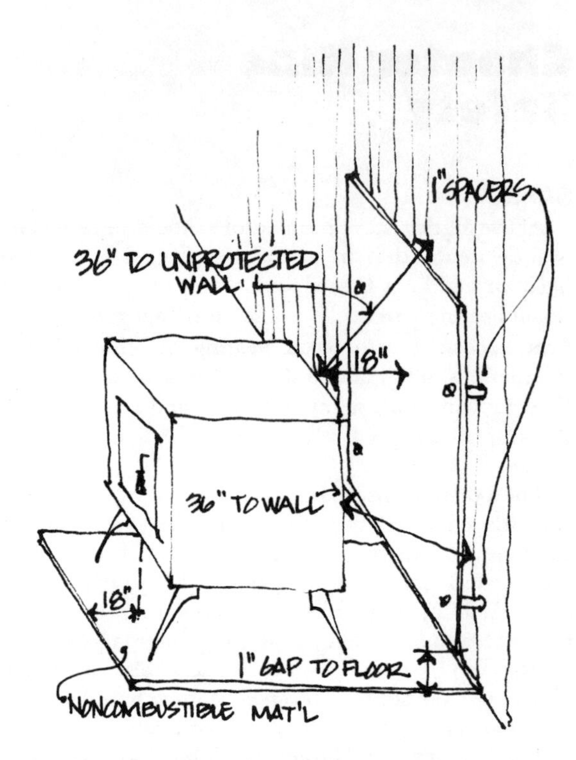

Figure 9-2. A safe way to place a stove closer than 36 inches to a combustible wall.

can happen all too easily; corrosion and paint are two likely agents.

It may seem surprising that the NFPA standards for stove clearances require a stove to be as far as 3 feet from an unprotected wall (note that plaster over wood framing is considered an "unprotected wall" in this context — see Note 2 in Table 9-1). It was stated previously in Chapter 4 that the spontaneous ignition temperature of wood is about 700°F. Could a stove 3 feet from a wood wall possibly make the wall's surface that hot? The answer in "No." Yet the standard is reasonable. The ignition temperatures usually measured in laboratories (and quoted in Chapter 4) involve heating wood for a few minutes or perhaps an hour, and under these circumstances, wood does not ignite until the temperature of its surface is about 700°F. But if wood is heated for days and weeks and months, it is found that ignition, or at least dangerous conditions, can occur at much lower temperatures — around 200-250°F.[2] Left at these temperatures for months, wood will char (combustible gases and tar droplets slowly evolve and a deepening layer of charcoal is created).

If a stove is fired the same way every day for months, exposed wood surfaces nearby will reach higher and higher peak temperatures each day. Two properties of the wood are gradually changing during prolonged exposure to infrared radiation from a stove. The wood gets drier and darker. Drying makes it harder for heat at its surface to be conducted into the wood and beyond into the rest of the wall. As the surface darkens, it becomes a more efficient absorber of infrared radiation. Thus, since the surface absorbs more energy from the stove, and since heat is less easily conducted away, the surface gets hotter. If the wood actually starts to char, both effects are even stronger.

The intensity of radiation which, over a period of months, will result in the surface of exposed wood being at 250°F is about 600-700 Btu per hour per square foot — only about 3 times the intensity of direct (unfocused) solar radiation at the surface of the earth (200 to 300 Btu per hour per square foot).

How intense is the radiation from a hot stove on a nearby wall? As discussed in Chapter 2, the three most important factors are the area of the stove's side which is facing the wall, its average[3] temperature, and its distance from the wall. Theoretical calculations show that a large stove, such as a Riteway would have to have an average temperature of 800°F or more on its side facing a combustible wall, 36 inches away, to constitute a danger. (The surface area of the Riteway's side is about 5.5 square feet.) Since this is an average temperature over the surface, parts of the surface would no doubt have to be over 1,000°F, which is glowing red hot. This may be possible, but only if the stove is pushed well beyond normal operating conditions. Thus the standard 36 inch clearance appears to be about right for a large stove, although perhaps not conservative enough for some of the very biggest stoves operated at their very highest powers.

The back of a small stove such as the Jotul 602 would have to have an average outside surface temperature of about 1,400°F to constitute a hazard to a wall 36 inches away (the Jøtul's back has an area of only about 1 square foot). At this temperature the stove would appear a bright cherry red. At least with wood as the fuel, this is impossible — energy would be radiating away from the stove at many times the rate it could be released from the burning wood in the stove. The maximum possible average exterior surface temperature of wood stoves is probably about 900°F, at which temperature it would be faintly glowing. At this temperature the back of the Jøtul 602 can be 18 inches from a wall without the radiation intensity on the wall ever exceeding the safe limit of about 600 Btu per hour per square foot. Thus for small stoves, the standards for clearances are considerably more conservative than necessary.

In addition to its size, the design of a stove can also affect the clearance necessary for safety. Sides with metal or firebrick liners cannot get as hot as unlined sides, and hence, can safely be placed somewhat closer to a combustible wall.

The 18 inch minimum clearance for stovepipe is half that for stoves, and yet stovepipes can get much hotter than stoves. The reason is stovepipe presents a smaller area to a wall than do most stoves (more precisely, its angular area (p. 13) is less when viewed from the same distance). During a chimney fire, stovepipe can glow with a cherry-red color, suggesting a temperature of about 1,400°F. Another indication of the high temperatures stovepipe may achieve is that chimneys which are in compliance with Underwriters' Laboratories Standard 103 are suitable for use where the flue-gas temperature is as high as 1,000°F on a continuous basis, 1,400°F for periods of an hour, and 1700°F for 10 minutes.)

In summary, the NFPA clearance standards are generally conservative. Only very large and very hot stoves may require more clearance. Smaller stoves and stoves with protected sides do not require as much clearance as the standards specify. Thus following the standards almost always guarantees a safe installation.

FLOOR PROTECTION

STANDARD

The standards for floor protection are different for two reasons: 1) stove bottoms are almost always cooler than stove sides, and 2) in addition to the danger of radiant ignition, glowing embers falling from the stove can start a fire. 18 inches is considered adequate clearance to prevent ignition by radiation. But no stove should be mounted on an unprotected combustible floor no matter what the clearance, because of the danger of hot and/or burning material falling from the stove. This danger is greatest during reloading and ash removal, but even with the door shut, it is possible for sparks to come out through the air inlets of some stoves. The NFPA standard for protection against this kind of danger is a sturdy noncombustible floor covering, which extends at least 18 inches beyond the stove on the sides with doors or other potential openings. Suitable materials are 24 gauge or thicker sheet metal by itself, ¼ inch or thicker asbestos millboard covered with 24 gauge sheet metal, mortared bricks or stone, concrete, etc. For stoves with less than 18 inches, but at least 4 inches, of clearance to the floor, these same floor coverings meet the standards for preventing radiant ignition as well.

COMMENT

These floor protection standards presume stove bottoms are cooler than sides, which is virtually always the case. Some stoves have double bottoms or ash trays. Most manufacturers recommend that a layer of sand or ash always be in the stove. All of these features and procedures help insulate the fire from the outside of the stove's bottom. Even if a stove is fired without any ash or sand in its bottom, ash very quickly appears, so the bottom will not be hot for long.

The proper amount of floor protection on the sides of the stove without opening doors is not specified explicitly by the NFPA standards. I would suggest the floor protection extend at least 12 inches beyond all sides of the stove and at least 18 inches beyond the front.

Apparently, floors can also be ignited by the high temperatures of the legs which support the wood burner. A particular free-standing metal fireplace was supported above the floor by stiff metal straps which were screwed or bolted to the floor joists under the finished floor. Enough heat was conducted through the straps to ignite the underside of the floor.

CHIMNEY CONNECTOR: SAFETY

Usually a few feet of stovepipe are used to connect the stove to the chimney, and in such cases, this section of the venting system is called the chimney connector. A great number of fires associated with wood stoves are due to unsafe connector installations.

The NFPA connector standards are designed to prevent two hazards — ignition of the surroundings, principally by radiant heating, and inadequate draft, which could result in serious smoke concentrations in the house.

STANDARD

The NFPA connector standards for fire safety with minor modifications, are as follows. If steel stovepipe is used,[4] the steel should be at least as thick as 24 gauge.[5] Each stovepipe joint should be secured with at least 3 sheet metal screws or the equivalent. Outside mechanical support should be used if the connector is more than about 6 feet long. The stovepipe should be accessible for inspection, cleaning and possible replacement.

If the stovepipe connects to an existing masonry chimney, the pipe should penetrate through to the inner surface of the masonry wall, but not beyond (not into the flue space itself). The connection should be made physically secure such as through the use of high temperature cement.

The clearance between any part of a chimney connector and any combustible material must be at least 18 inches, except as noted below and in Table 9-1.

A chimney connector may not pass through a ceiling, but may pass through a wall if the wall is protected in one of the following ways:

1) A metal ventilated thimble is used whose diameter is at least 12 inches larger than the stovepipe's diameter (Figure 9-3) giving at least 6 inches of ventilated, metal-lined clearance around the pipe;

2) A metal thimble, or a burned fire-clay thimble is used and is surrounded by fireproofing materials (such as brickwork) extending at least 8 inches beyond the thimble;

3) No thimble is used, and no combustible material is within 18 inches of the pipe. For a 6 inch diameter pipe, this requires a 6 + 2 x 18 = 42 in. diameter hole through a combustible wall. The hole may be closed in or covered with non-combustible materials such as masonry, asbestos millboard, or sheet metal.

COMMENT

These standards are designed to make installations safe under the worst possible conditions — a chimney fire — during which the stovepipe is glowing red hot and may be subjected to violent forces and vibrations. The physical strength of the pipe is important — sheet metal screws at every joint, even if it feels tight without them, and additional external support (e.g., hangers) for long segments of stovepipe, are essential. Many house fires have resulted from stovepipe joints vibrating apart during a chimney fire.

The 18-inch clearance between the stovepipe and any combustible material assumes the pipe is an efficient radiator — that its surface has an emissivity of close to 1.0. Thus, if pipe is used which is a poor emitter of radiation, it could safely be closer than 18 inches to a wall or ceiling. New shiny galvanized pipe emits about half as much radiation as black pipe, and new chrome-plated pipe emits about 1 / 30 as much. However no type of (single-wall) stovepipe is generally recognized as being safe with a clearance smaller than 18 inches. The emissivity of most shiny surfaces changes with time; emissivities increase as surfaces become dirty, scratched or corroded. The zinc coating on galvanized pipe becomes permanently much duller (and it is a much better emitter) if the pipe's surface ever exceeds about 800 degrees Fahrenheit, the melting point of zinc. Because of these problems with the durability of the low emissivity surface, it is not recommended that any kind of stovepipe be placed closer than 18 inches to combustible materials.

The prescriptions for safe passage of stovepipe through a wall are important. Many common ventilated metal thimbles are not adequate protection by themselves because their diameters are too small. For 6 inch pipe, a hole at least 18 inches in diameter should be cut through the wall, and the thimble should then be supported in the middle by sheet steel or asbestos millboard covers on either side of the hole.

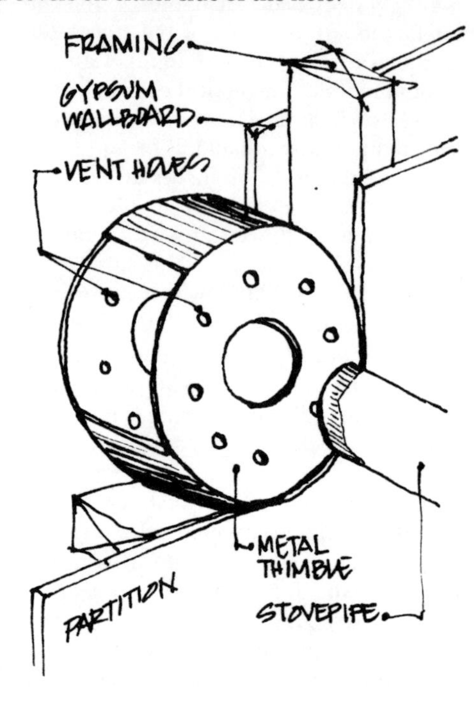

Figure 9-3. A ventilated thimble — a device permitting a stovepipe to pass safely through a combustible wall.

In using these smaller-than-recommended thimbles, the question of whether or not to insulate or ventilate the space beyond the thimble is unanswered by the NFPA standards. I can only guess what would make these smaller thimbles equivalent to the recommended size in terms of safety. In my own home I would either line the exposed framing with heavy aluminum flashing, or I would fill the space with loosely packed fiberglass insulation (without any paper or foil backing!).

Ventilation through the thimble is an essential aspect of its design; the ventilating holes on either side should not be blocked.

NFPA standards do not permit stovepipe to pass through a floor or ceiling under any conditions. The reason is not that ignition is more likely — with equal protection, both ceilings and walls are equally unlikely to ignite. Rather, in the improbable event that the protection fails and there is a fire, the fire can spread much more quickly through the building, if it pierces through a hole between floors than if it is in a wall.

The most common 6-inch stovepipe used in houses is gauge 28, considerably thinner than gauge 24. If pipe is available which is thicker than gauge 28, it should be used. If not, considerable experience indicates gauge 28 will serve well. Its principle disadvantage is that it will need more frequent replacement. Corrosion will eventually eat through almost any stovepipe; if the pipe is wet with creosote most of the time, annual replacement may be necessary. Properly installed and in good condition, gauge 28 stovepipe has adequate strength.

CHIMNEY CONNECTOR: DRAFT

STANDARD

The NFPA chimney-connector standards which are designed to assure adequate draft are as follows. The diameter of the connector pipe should be no smaller than the stove's fuel collar diameter, and the length of the connector should be as short as possible (or, the stove should be placed as close as possible to its chimney). The connectors' horizontal portion should be no more than 75% as long as the vertical portion of the chimney (above the point of connection). The horizontal portion of the connector should slope upwards towards the chimney with a slope of at least ¼ inch per foot. There should be no more than two 90° turns and they should be gentle, sweeping corners.

COMMENT

As indicated in Chapter 5 on chimneys, some of these standards are rather arbitrary. Both the total length and the number of elbows (which induce flue-gas turbulence) contribute to an installation's energy efficiency. The maximum safe length and number of corners depends on other aspects of the venting system — its diameter, height, and insulating properties. Similarly, sharp (mitered) corners offer more resistance to flue-gas flow than gentle corners, but they can be used in many systems with no ill effects. The slope requirement is particularly arbitrary. Hot flue gases will flow downhill. A downhill portion of smokepipe will decrease the net draft of the whole system, but as long as adequate draft is developed in other portions of the chimney system, the flue gas will flow downhill with no difficulty. Some people have even used a vertical loop of stovepipe as one way to increase the pipe's length; the pipe leaves the stove, rises to about 18 inches from the ceiling, turns through 180 degrees and extends down to near the floor, travels a few feet horizontally near the floor and then rises up to the chimney connection. No flow problems (smoking) were encountered in the case best known to the author, due, no doubt, to the chimney's being somewhat overdesigned (20 feet of 7" insulated double-walled metal chimney with no corners). Chimney residues (ash and creosote) will tend to accumulate at the low point(s) in such a system and could block flow. Thus frequent inspection is advisable.

In any case, inadequate chimney flow is rarely a safety hazard. An inadequate venting system will result in some smoke getting into the living-space of the house and/or the stove will not be able to operate at full power. These are annoyances more than safety hazards. The only kind of serious accident which inadequate chimney capacity could cause is asphyxiation of sleeping occupants, if very heavy smoking should occur. This appears to be very rare, and is more likely due to flow reversal than inadequate capacity. A smoke detection alarm, which is a wise investment in general for wood heated houses, would warn the occupants before the situation was serious.

CHIMNEY

STANDARD

Factory-built metal chimneys should be of a type designed for wood fuel. Because of the possibility of chimney fires, and the corrosiveness of creosote, such chimneys are built to withstand more severe conditions than other standard types. They are typically called "all-fuel" or "Class A" chimneys. The appropriate clearances of these chimneys from combustible materials should be given by the manufacturers — they may differ because of different insulating characteristics.

Masonry chimneys should have fire-clay flue liners, and the space between the liner and the chimney wall should not be filled. In general, combustible material (wood, in practice) should be no closer than 2 inches to the outside surface of the chimney. However, ends of wood girders may rest on a shelf of the chimney if the chimney wall is 8 inches thick at that point. Also, exterior chimneys may be placed against the building's sheathing.[6]

A chimney connector should not be connected to a flue which also serves a fireplace unless the fireplace or its flue is permanently sealed shut. In general, connecting two or more appliances to a single flue is not advisable, but if done, the flue should have adequate capacity to serve both appliances under all conditions, and the connectors should not enter the chimney at the same level but should be separated by a few feet.

Chimneys should terminate at least 3 feet above the roof where they penetrate through it, and at least 2 feet higher than any part of the roof within 10 horizontal feet (Figure 9-4).

COMMENT

Most prefabricated (factory-built) metal chimneys require minimum clearances from combustible materials. The fact that they may be insulated does not mean their outer surfaces remain cool, only that they are less hot. A section of prefabricated chimney is not a safe way to pass flue gases through a wall or ceiling unless it is installed in accordance with the manufacturer's instructions, which will often specify 2 inches of clearance to any combustible material.

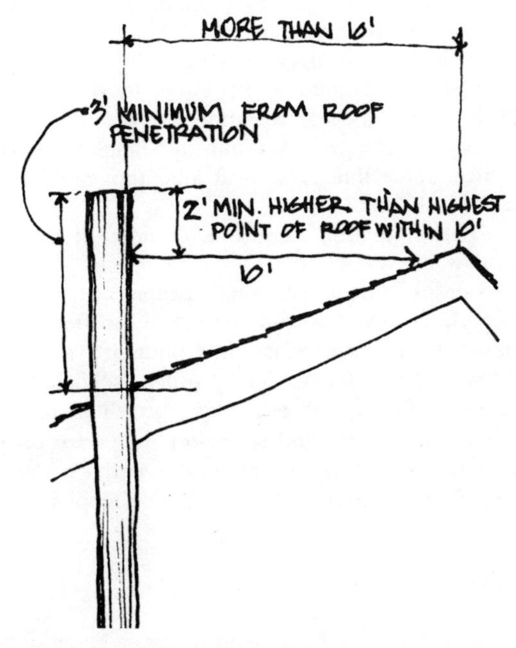

Figure 9-4. Minimum chimney heights above a roof, according to National Fire Protection Association Standard No. 211.

Existing masonry chimneys without fireclay liners are often used without apparent problems, but they are more dangerous. Liners provide additional corrosion and temperature protection, and may be used to create additional flues if more than one appliance is vented into the same overall chimney structure. In some cases fireclay liners have been retrofitted by lowering liner sections (with mortar applied on the upper edge) down from the top of the chimney. Another remedy for an unlined chimney is to insert a stovepipe inside the masonry chimney. This will serve most of the same functions as a fireclay liner, but is much less durable and hence will require periodic replacement. (Uninsulated stainless steel pipe may be available and probably would be a good investment in this situation.)

Close visual inspection and a smoke test will indicate the soundness of an existing chimney. A smoke test is performed by lighting a *small* smoky fire or setting off a smoke bomb in a connected stove or fireplace, and then closing or partially closing the top of the chimney. If there are leaks or cracks in the chimney, smoke will be seen coming out of them. If there is a local chimney expert, let him or her help perform this test.

In regions where fire hazards outside the house are very high, a spark arresting screen over the top of the chimney may be advisable. The screen should be approximately a ½-inch mesh and should be formed into a rounded hill, cone, or can shape over the chimney to ensure free enough flow of the flue gases through it. However spark screens over wood-fire chimneys will require occasional cleaning, as they will almost inevitably become clogged with an accumulation of creosote, tar or soot.

There are three, potential hazards from venting two appliances into one flue. First, if either of them is a fireplace or an open-type stove, it is possible that sparks from the operation of the other appliance may find their way through the open appliance into the house. Second, either or both appliances may vent improperly due to draft and flow interactions, and dangerous fumes may back up and accumulate in the house. Third, if a chimney fire were to occur, it is less likely that it could be suffocated by cutting off the air supply if more than one appliance is vented through the same flue. The draft adequacy problem is discussed in Chapter 5.

The spark-migration hazard is minimal if normal precautions are taken, but it is the intent of safety codes to ensure safety even when people are careless. Most fireplaces have both flue dampers and firescreens or doors. When there is no fire, both are normally closed, and sparks from a stove which is hooked up to the same flue could not come out. The danger arises if both the flue and screen or doors are open. Then a falling chunk of burning matter, perhaps ignited chimney deposits, could fall through the fireplace and out beyond the hearth.

SMOKE DETECTORS

A number of different warning devices are available today which can detect smoke. The device can then sound an alarm, or notify the fire department, or do anything else which can be triggered by an electrical signal from the detector. The least expensive devices require external electric power (they plug into an electrical outlet) and sound an alarm in the house. They cost as little as $50. I recommend having such a device. Smoke will almost always precede any wood-stove induced fire, and with the warning provided by these smoke detectors, preventive action can almost always be taken in time; at least the lives of the occupants can be saved. Smoke detectors are increasingly available at hardware and building-supply stores. Local fire departments may offer advice and information.[7]

INSURANCE

Most insurance companies require policyholders to disclose major changes in a home which might affect safety. There is no evidence that insurance companies are charging higher premiums for houses heated principally with wood, but the company may insist that the installation meet certain safety standards. If the company is not notified by the policyholder that wood heating has been installed and/or if the system does not meet safety standards, insurance policies may be void. Thus checking with the insurance company or agent is important. It is alleged that some companies have refused to pay damages resulting from a fire caused by unsafe wood stove installations.

VIGILANCE

Installation standards do not guarantee safety. Chimneys and chimney connectors require regular inspection and cleaning to remain reasonably safe. Thin gauge stovepipe can corrode through in one year under some conditions. Clearances must be vigilantly watched after the initial installation. Furniture, wood, newspapers, matches, etc., if placed and left too close to a stove, can ignite. Combustible materials must also be kept the required safe distance from the chimney. Placing green or wet wood on top of a stove is an effective way to dry it quickly, but also an effective way to ignite it; such a practice is very dangerous. Stove surfaces are hot enough to ignite any combustible material and melt all plastics. Apparently cool ashes often contain hot coals and should not be dumped carelessly, but should be put into a metal container, or dumped where there is no danger of fire. Using flammable liquids such as kerosene or even charcoal lighter fluid to light fires is dangerous, as is storing such fluids near stoves. Complacency is risky even if an installation meets safety standards or codes; carelessness can still result in the house burning down.

Chapter Ten
Stove Accessories

A number of devices are available which are intended to improve or extend the capabilities of a wood stove. These include heat extractors to heat both air and water, stovepipe dampers, and draft regulators. Chemical chimney cleaners and so-called creosote inhibitors are discussed in Chapter 12.

HEAT EXTRACTORS

Air heat extractors (also known as heat exchangers, heat savers, heat economizers or heat robbers) are intended to improve the heating efficiency of the system by extracting additional heat from the flue gases, over and above what would normally come out through the stove and stovepipe. Some are passive, operating on radiation and natural convection; others are active, employing a blower or fan to help transfer the heat by forced convection.

Typical active heat extractors are shown in Figure 10-1. In these devices, the flue gases enter a closed box with pipes running through it. The gases flow around the outsides of the set of pipes while room air is blown through the pipes. Heat is conducted from the flue gases through the pipes and into the moving air, warming it. The heated air usually enters the room although it can be ducted to other parts of the house. However, the blower or fan that comes with such units is usually of insufficient capacity to move much air through ducts. Variations on this design including reversing the functions of the box and the tubes; the flue gases can be directed through a set of vertically oriented tubes while air is blown around the outside of the tubes. Heat extractors are also available employing heat pipes[1] to convey the heat to the outside of the stovepipe, where a blower or fan may then help to extract the heat from the heat pipe. The blower or fan on most heat extractors is thermostatically controlled to be on only when the flue gas temperature is above a given value.

Passive heat extractors include a heat-pipe type without a blower, somewhat larger scale vertical pipe arrays, and various types of metal fins attached to stovepipe. (See Figure 10-2.)

There is no doubt that most of these devices work — they do improve the heating efficiency of the system. Typical energy efficiency increases are probably in the range of from 3 to 20 percentage points which can result in from 6 to 40 percent less wood burned for the same useful heat output.

The range is very large due to differences in design of the extractors, and because the effectiveness of an added heat extractor depends very much upon the heat-transfer efficiency of the stove.

Any simple up-draft or diagonal-draft stove operated at high power will benefit greatly from a heat extractor — in such cases flue gas temperatures at the stove's flue-collar are roughly 800 to 1200°F, indicating large amounts of available heat. At the opposite extreme are stoves such as the Sevca or a double drum stove (see Figure 6-4). These stoves have so much surface area relative to their combustion rates that flue gases are rarely above 500°F even with a large hot fire. At medium and low firing rates, typical temperatures are under 300°F. Since the stove extracts so much more of the available heat in these cases, an additional heat-extractor device is relatively superfluous.

Measuring the net effect of a heat extractor accessory can involve some subtleties. In addition to the heat output of the device itself, three other factors should be included in an assessment of the device's *net* effect. First, the more heat the device extracts, the cooler will be the stovepipe that comes after the device; hence use of the device decreases the amount of heat given off from the down-stream stovepipe. There is still a net gain, but it is smaller than the heat output of the

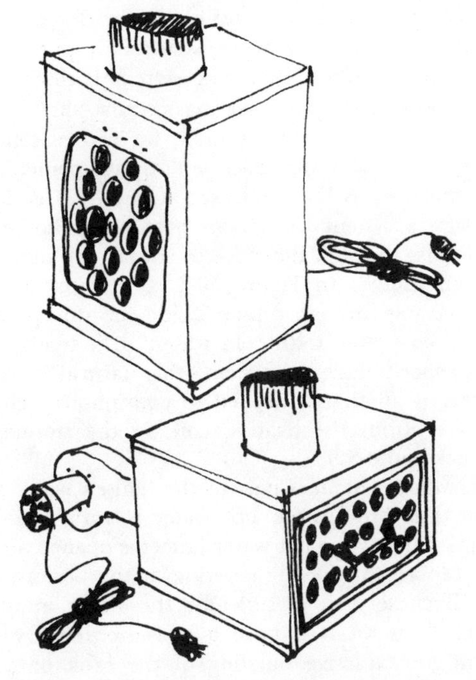

Figure 10-1. Electrically powered flue-gas heat extractors.

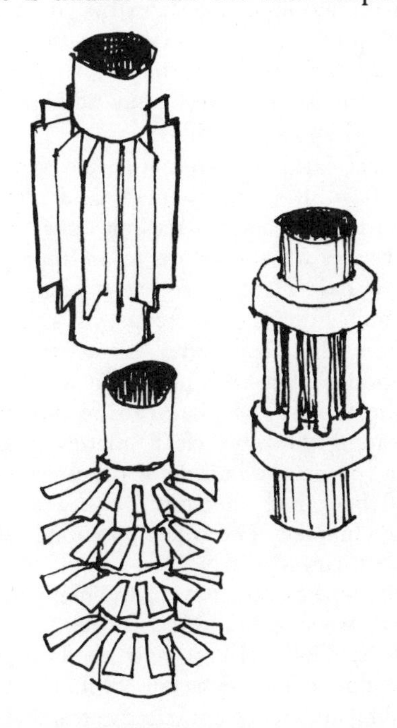

Figure 10-2. Some passive (or natural-convection) flue-gas heat extractors.

extractor itself. Second, if the extractor is not present, the ordinary stovepipe which would be in the same place transfers considerable heat to the room by itself. Thus some of the decrease in temperature of the flue gases passing through the extractor would have occurred without the device. Third, all the electric energy used by the active heat extractors is ultimately converted to heat. Consequently a small amount (50 to 100 watts, or 170 to 340 Btu per hour) of electric heating is a side effect of active devices but should not be included as part of the extracted wood heat. The net extraction effect of a heat extractor is its total heat output, minus the output of the length of stovepipe the extractor replaced, minus the decrease in heat output from the stovepipe (and chimney) beyond the extractor, minus the device's electrical power consumption.

Active (electrically powered) heat extractors are generally more effectivie than the passive types and can also significantly improve the heat distribution in a wood heated room or house because of the air mixing and circulation the blower induces. In some case the unit can be oriented to induce a general air circulation pattern around a room or even a small house. Two possible disadvantages of active heat extractors are the noise the blower or fan makes, which some people find annoying, and the electric energy they consume. The motors used are typically from 50 to 100 watts. The net increase in stove heat output is roughly 1,000 to 8,000 Btu per hour, which is equal to about 300 to 2,500 watts of heating. Economically, the value of the extra heat virtually always exceeds the cost of the electricity used in extracting it.

A major consideration for all types of heat extractors is the ease with which they can be cleaned on the inside. The heat extractors get more heat from flue gases by creating cooler surface areas inside the flue. All kinds of chimney deposits from wood smoke preferentially build up on cooler surfaces. These deposits are generally good thermal insulators; thus the buildup decreases the amount of heat extracted by the device. The buildup of deposits is inevitable; to maintain their effectiveness, heat extractors require frequent cleaning, typically about once a week, but perhaps as often as daily, or as seldom as monthly.

Many heat extractors are not easy to clean, which is a serious flaw. It is inconvenient and often messy to have to partially disassemble a unit or sections of stovepipe, and carry the dirty parts to an appropriate place and brush them clean. I feel it is vital to have easy cleaning built into heat extractors, as is the case in the Magic Heat and Heat Saver extractors. These units have a movable plate with holes in it which fits snuggly around the tubes. The plate can be moved back and forth, scraping the tubes clean, merely by pulling and pushing a rod on the outside of the unit. The deposits either fall back into the stove or are carried up and out of the chimney. The whole operation takes seconds and creates no mess.

The heat-pipe extractors are particularly inappropriate to use with wood stoves. Precisely because of the very high heat capabilities of heat pipes, the heat pipe surfaces inside the flue will be especially cool, resulting in very quick fouling and hence lost effectiveness.

In some installations use of a heat extractor may cut the power output of the stove while at the same time increasing its energy efficiency. An electrically powered and thermostatically controlled extractor will turn on whenever the flue gas temperature exceeds a certain value. When it is on, the device may extract enough heat to significantly cool the gases, thus reducing the draft developed by the chimney. With less suction inside the stove, less combustion air will be drawn in and hence the combustion rate will fall. With cooler gases coming from the stove, the heat extractor's thermostat will turn off the device. With the heat extractor off, the gases in the chimney become warmer again, increasing the draft which increases the combustion rate, which increases the temperature of the heat extractor, which turns the extractor on again. The cycle can keep repeating until the fuel is consumed. The net effect is a suppression of *extended* heat surges; the heat output oscillates up and down fairly quickly, but the average heat output is steadier than without the heat extractor. (This is not a universal effect of all heat extractors in all installations.)

There are some alternative ways to accomplish the same basic objective of heat extractors. Increasing the length of stovepipe from the typical 4 to 5 feet to about 10 feet will improve the heating efficiency of most systems by 3 to 8 percentage points. A fan blowing air over a stove will increase its efficiency by roughly a similar amount. The fan can also serve to help improve the distribution of the stove's heat throughout the room and to other rooms.

WATER HEATING

Heating water with wood can serve two purposes — to supply the inhabitants with hot tap (or domestic) water, and to help heat the house via a hot-water heat distribution system. There are boilers designed to do nothing but heat water. In this chapter, heating water with wood is considered a possible additional feature of a stove whose principle function is direct space heating. Whether a system is designed to heat tap water or house-heating water, many of the components and problems are similar.

Considerable variety is possible in the design of components and in the way whole systems are put together. A particular tap-water heating system is illustrated in Figure 10-3. Tap-water heating systems almost always employ a hot water storage tank, for two reasons. The demand for hot water does not always coincide with times the stove is hot, and even if it did, it is difficult to design a system which can heat water hot enough in one pass through the stove to supply the demand of an open faucet. In Figure 10-3, the water is heated while passing through piping inside the stovepipe. Hot water is less dense than cold water, so it tends to rise. In a properly designed system, this natural convection of water is all that is needed to continuously circulate the water from the heating coil up the storage tank and back to the coil.

The whole system is part of the house's water system and is thus pressurized; hot water is withdrawn from the tank whenever a hot water faucet is opened and cold water replaces it, ideally entering at the bottom of the tank. Because it is possible for the water in the coil to boil, it is vital to have a pressure relief valve to prevent an explosive bursting of the tank or a pipe. These valves automatically vent the system when the pressure exceeds a certain amount. Such valves are part

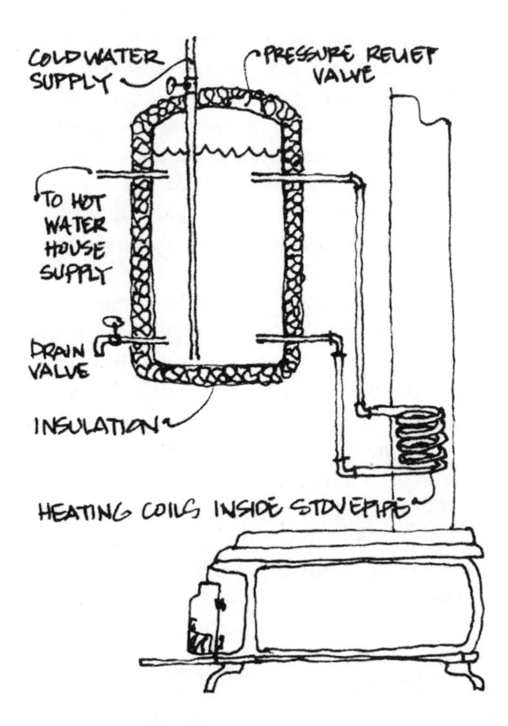

Figure 10-3. A wood-stove tap-water heating system. In this case, the water circulates naturally from the heating coils to the tank and back.

of all conventional water heaters, and are typically set for about 150 pounds per square inch.

A particular house heating system is shown in Figure 10-4. This system is designed to distribute some of the stove's heat to an adjacent room via hot water. The water heated by the pipe in the stove is pumped through ordinary hot-water baseboard units. The pump is controlled by a thermostat so that it only comes on when the stove is hot. An expansion tank absorbs small changes in volume of the water in the pipes as its temperature changes, and its loose cover serves as a pressure relief valve to prevent the system from exploding. The tank also serves as a water reserve for the system. Replenishing evaporation losses from the tank is occasionally necessary.

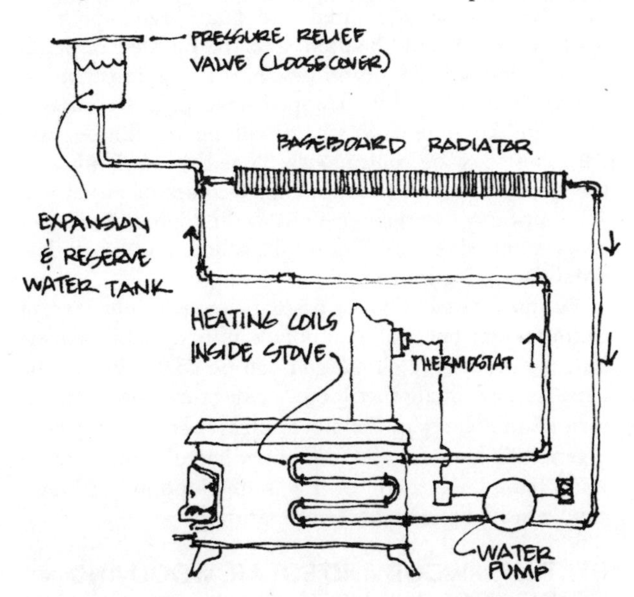

Figure 10-4. A wood-stove water heating system for distributing some of the stove's heat to other rooms.

HEATING COIL OR TANK

Three effective ways to extract heat from a wood stove and transfer it into water are illustrated in Figure 10-5. Piping inside a stove can be located wherever it is most convenient, commonly against one side. A few feet of 1-inch pipe provides a substantial amount of hot water. For systems with pumped circulation, the details of pipe-array shape do not matter much. If natural convection is to be relied upon, air traps must be avoided or flow-stoppage will occur (see p. 79). The three most common geometries for natural convection systems are a

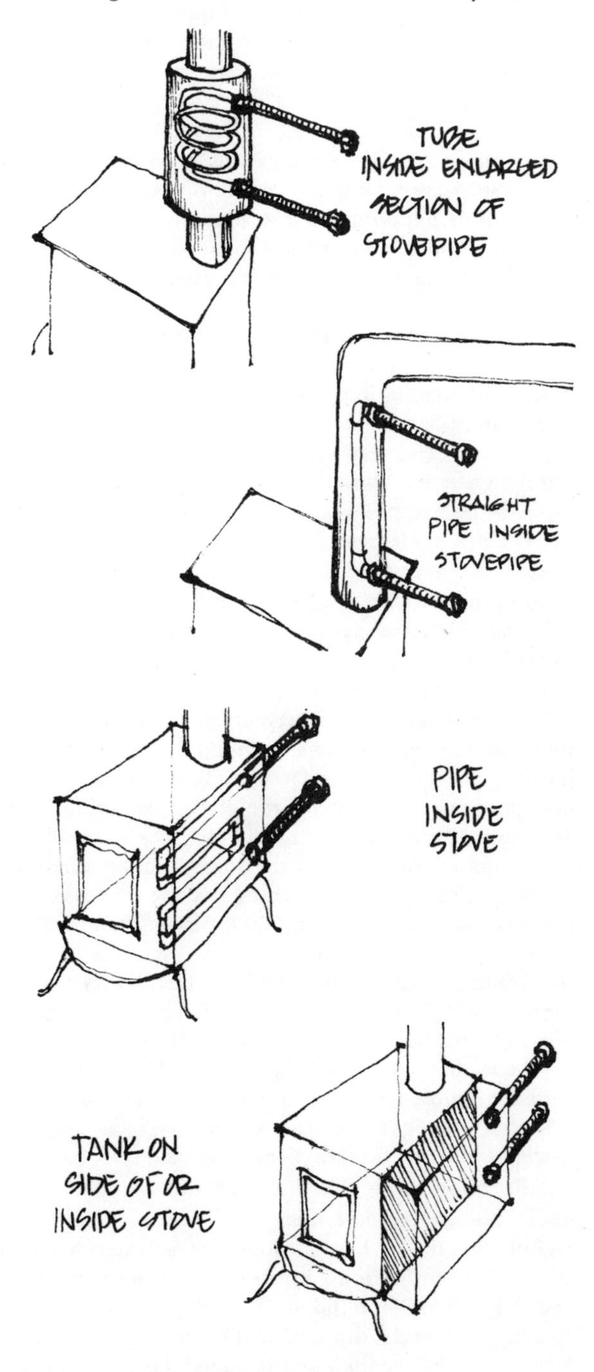

Figure 10-5. Some systems for heating water with a wood stove.

serpentine curve, parallel pipes with headers, and a helical coil. (The latter is usually more suited for use inside a stovepipe than in a stove.)

Materials used include brass pipe, copper tubing, and various more exotic alloys and composite materials. The heat transfer characteristics of these materials are essentially equivalent in this application. Copper is often used by people who make their own water heaters because it is easy to work with. Since ordinary 50/50 solder melts at about 420°F, well below typical interior stove temperatures, either there should be no soldered joints inside the stove or stovepipe, or the system should be designed and operated so that the pipes always have water in them. Liquid water inside the tubes ensures that the metal cannot be heated far above the boiling point of water. (It is virtually impossible to unsolder a copper-tube fitting with water in it.) One way to avoid soldered connections is by bending the copper tubing. Copper tube comes in both rigid and flexible forms. The flexible tubing can be bent easily, but it will often crimp unless special bending tools are used, or unless it is first packed with dry sand. A lot of tapping is necessary to pack the sand adequately, but once thus filled, the tubing can be bent with no danger of crimping. ½-inch tube can be coiled to fit inside 6 inch pipe as in Figure 10-5.

Ordinary threaded brass pipe and fittings have been used in wood water heating applications, although brazing or welding of joints which are exposed to high temperatures is probably wise.

Unfortunately, very few stoves on the market (in 1976) are designed for convenient addition of water heating pipes. (One exception is the Portland Foundry Atlantic 121 B Kitchen Heater.) But many stoves are fairly easily adapted. The most difficult task is usually cutting the holes in the stove body for the piping to enter and leave, and this job is perhaps best done by a skilled metal worker.

Inserting pipes or coils in the stove pipe avoids marring the stove, but may create a need for more frequent and very difficult cleaning. The pipes or tubes have relatively cool surfaces and hence accumulate deposits from the flue gases at higher than normal rates. Even fairly thin deposits can have a significant insulating effect, destroying much of the water heating capability of the system. Frequent cleaning may be necessary, and the task can be much more difficult than cleaning stovepipe — the colder water pipe surfaces seem to accumulate a much harder deposit which is very resistant even to a stiff wire brush. A coil inside a stovepipe also has some relatively inaccessible surfaces.

Not all stovepipe water heaters will have this fouling problem. Stoves with relatively complete combustion and hot flue gases at the stovepipe collar are least susceptible. The further away from the stove the unit is, the more severe the problem will be, due to the cooler flue-gas temperatures.

But by far the best location for a water heater is inside the combustion chamber of the stove, for there the deposits never build up to the point where water heating is seriously impaired. The reason is the fire is there to keep the deposits trimmed back by burning them. Cleaning is never, or at least very seldom, necessary.

Stovepipe coils can restrict flue-gas flow, and increasingly so as deposits build up on them. Thus unless the venting system has considerable excess capacity, the coils should be in a somewhat oversize section of stovepipe.

A tank built onto one side of a stove can be an excellent water heater because of the large surface area in contact with the fire. The inner wall of the tank should also be the combustion chamber wall — only a single piece of metal should separate the water from the fire. Separate water-heating tanks inside a stove are another effective means which is very common in British coal stoves.

Wrapping or coiling copper tube around the outside of the stovepipe works, but requires much more tube than a coil inside the pipe or in the combustion chamber to achieve the same amount of heating, because the coil is not exposed to such high temperatures. It is not at all easy to achieve a good and lasting thermal bond between the tube and stovepipe. Wrapping the assembly in insulation helps. Cleaning flue deposits is easier because the stovepipe does not have the hidden inaccessible surfaces of a coil inside a pipe. But in general, the best location for water heaters is in the combustion chamber of the stove.

CIRCULATION

Either natural convection or a pump can be used to circulate water through the stove and to storage, in the case of a tap-water heater, or to radiators or other heat distributors in the case of a house-heating system. Natural convection has the advantage of being silent, simple and inexpensive both to buy (no pump or controls) and to operate. Water flows whenever it is hotter in the heating pipes or coils than in storage and the amount of flow increases with the difference in temperature. The major limitation is geometrical — hot water cannot be moved downward or over long horizontal distances by natural convection. In practice it is most successful if the destination (storage tank or heat distributor) is located somewhere above the stove. Distributing heat to a room one floor above the stove can easily be done with natural convection of water (or air). Proper detailed design is also critical. Since the motive forces are small, resistance to flow must be minimized. Adequate size piping (at least ¾ inch) should be used, and the number of bends and corners should be minimized. There must be no places where air can get trapped, (except at the highest point in the system) or there will be no flow at all. All pipes must be continuously sloped upwards towards the system's highest point so that bubbles of air (which will inevitably form) can get out of the piping. Examples of air traps where bubbles would collect and block flow are shown in Figure 10-6.

Pumped circulation imposes no constraints (except during power failures). Not only is any geometry possible but, if desired, the circulation can be controlled by the need for heat in another room, rather than the temperature of the water in the stove. Heat transfer efficiency is generally higher with the pumped circulation — more water can be heated, or the same amount of water can be heated to a higher temperature.

SYSTEMS - WOOD AND SOLAR, WOOD AND CONVENTIONAL

A wood water-heating system is capable of supplying most of the hot water needs in a typical household only during the cold winter months when the stove is in

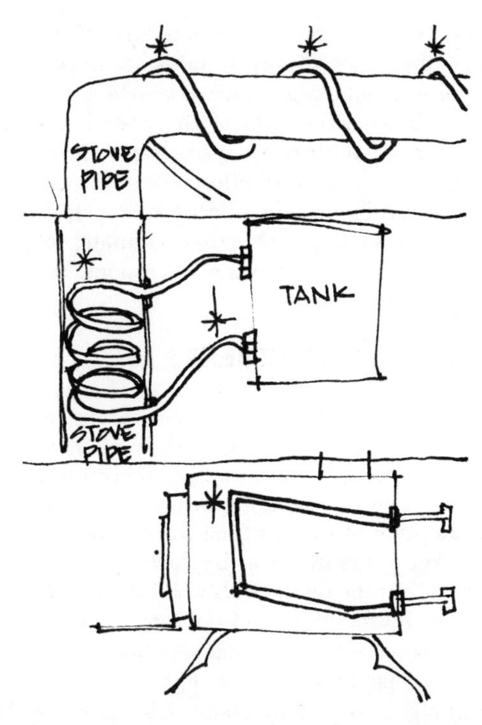

Figure 10-6. Some water heating systems which will not work under the influence of gravity or buoyancy alone, but would require pumping. The asterisks mark locations where air pockets will form, blocking the flow of water.

fairly constant use. In the fall and spring, its hot water production is more irregular, and in the summer, non-existent. A natural way to achieve a more dependable supply is to add a solar water heater. In many climates, the solar and wood system will complement each other. If the sun is shining and the weather is not very cold, the solar system will be heating water; if it is cold, the stove will be in use. The total hot water output can be fairly constant. The same hot water storage tank can be used by both systems, so there is very little additional complexity in the plumbing.

A further or alternative step to increase the dependability of the hot water supply is piping the output from the storage tank to the input of an ordinary water heater, i.e., to treat the wood or wood-solar system as a pre-heater. The conventional water heater's thermostat will sense the temperature of the incoming water and add heat if necessary, or just let the water pass through if it is already hot enough. With this system, a hot water supply is assured. The savings in energy for the conventional heater will depend on how much heating the wood and/or solar system can do. Since most people want a dependable supply of hot water and already have a conventional water heater, this is generally the best way to couple a wood or solar or combination heater to the house's plumbing.

Chapter Eleven Fireplaces

The open fireplace once was the only source of heat in all American homes, but now is treated mostly as a contribution to interior decoration, designed to cheer the heart and soul, but not to warm the body. At least in its older versions it was clearly capable of heating a house, or at least parts of a house. Historically the British have always championed the open fireplace as a heating method, not because of its energy efficiency, which is low, but because of the value placed on "fresh" air. Fireplaces in use induce high ventilation rates. During much of the 19th century it was believed to be most important to avoid having any air breathed twice; "used" air contained "carbonic acid gas" (carbon dioxide), which was thought to be bad for health.

Today in this country, the emphasis is on preventing excess ventilation because of its energy cost. If interior air is pushed or drawn out of a house to make room for fresh air, a great deal of heat is lost. As mentioned in Chapter 6, even in a typical new snug house, about half of the heating need is due to ventilation, most of which is unintentional air leakage.

Because of the large air consumption of ordinary fireplaces, the distinction between efficiency and net efficiency is important (see Chapter 6). The efficiency of any heating device is the fraction of its energy input (e.g., wood) which emerges as useful heat. Net efficiences take into account possible changes in total heating needs of a house due to the presence and operation of the appliance, and possible heat contributions from the appliance's chimney.

The amount of the extra heating burden when a fireplace is in use depends on two factors: the increase in outdoor air flow into the house, and the temperature difference between indoors and outdoors. The colder it is outside the more energy it takes to warm up the exterior air to room temperature. Thus the extra heating load depends on details which vary considerably from house to house and from day to day. (Closed stoves use so little air that this effect is usually negligible for them). Even though the effect for fireplaces is quite variable, it is so important that estimates of it are included in all of the following discussions. See Table 11-1 for one estimate.

EFFICIENCY MEASUREMENTS

Very few good measurements have been made of fireplace efficiencies. One of the most careful and accurate studies was done by the University of Illinois Engineering Experiment Station, using a highly instrumented test house.[1] The investigators measured the change in gas consumption of the gas-fired forced hot air furnace as a function of the use of a fireplace. Coal, not wood, was burned, but the results with wood should be roughly similar. Careful monitoring of outdoor conditions made it possible to correct for varying weather. The fireplace damper was opened and a fire lighted at 4 p.m. and kept burning briskly until about 7:30 p.m. By 10:30 it was reduced to glowing coke. Only a few glowing embers were left at 11 p.m. The fireplace damper was closed at 7 a.m. This sequence was then repeated.

The results were very interesting. With the furnace thermostat located in a different room than the fireplace the net effect on temperature was that the room with the fireplace was a few degrees warmer than the rest of the house. Gas consumption was slightly less (than without the fire) during the evening, but slightly more during the rest of the night until the damper was closed, due to loss through the chimney of heated air from the house. The net effect was a slight (3 to 4 percent) increase in gas consumption i.e., a negative net efficiency for the fireplace. Use of a tight fitting

	PERCENT OF WOOD ENERGY CONSUMPTION	POWER (Btu per hour)
Wood consumption 20 pounds per hour, @ 8600 Btu per hour	100	172,000
Direct or primary radiation	10	17,200
Indirect or secondary radiation	5	8,600
Chimney conduction contribution — 10% of sensible heat entering chimney, for a 2 story house and an interior chimney with exposed sides[1]	7	12,000
Air heat loss 1500 pounds per hour, 40°F colder outdoors than indoors	-8.4	-14,500

Net efficiency 10 + 5 + 7 - 8.4 = 13.6%
while operating (not including overnight losses):

Net heat gain 17,200
+8,600
+12,000
-14,500
23,300 Btu per hour

TABLE 11-1. Rough estimate of fireplace energy flows.

[1] This estimate is extrapolated from data in W.S. Harris and J.R. Martin, "Heat Transmitted to the I = B = R Research Home from the Inside Chimney," Trans. of ASHVE, (1953), pp. 97-112.

metal cover over the fireplace opening from 10:30 p.m. until 4 p.m. the next day decreased the nighttime air loss enough so that the overall gas consumption was about the same as without the use of the fireplace (zero net efficiency).

With the furnace thermostat located in the same room as the fireplace, all the rest of the house dropped below the temperature setting when the fireplace was in use. (The heat from the fire kept the thermostat from turning on the furnace). Less gas was consumed, but probably about the same amount less as if the thermostat had been set back to the cooler temperature of the other rooms.

It is apparent from these test results that the net efficiency of a fireplace is positive when it is in use, and is negative when it is not and the damper is open. Thus the net efficiency is highest if a fireplace is used continuously, as it was in colonial America and least (probably negative) for very short fires, due to the large fraction of the time the damper must be open with little heat coming from the fire. The net efficiency of a fireplace is also always less the colder it is outside (this was verified in the above experiments). The most efficient way to use a fireplace is to use it a great deal of the time during cool but not cold weather, such as in the fall and spring, and to make sure the damper is closed whenever possible. If the damper has more than one open position, it should be used at the narrowest setting possible without making the fireplace smoke.

More direct measurements on fireplaces have been made, where either the heat output itself is measured or the energy it loses up the chimney is measured.[2] Coal or natural gas were the fuels in most cases, but these results should still bear some resemblance to the case where wood is used. These results suggest that the steady-state net efficiency of an operating fireplace is probably substantial (from 10 to 30 percent). This is the net efficiency while a large fire is burning. The average net efficiency in most homes is lower (perhaps from -5 to +10 percent) because with a small (or no) fire and the damper open, there is a net loss of heat.

A direct experimental comparison between a closed stove and a fireplace was conducted by Tatnall Kollock in 1911.[3] Equal amounts of wood were burned, either in a fireplace, or a stove. Three different rooms and two fireplaces were involved. The stove was built of sheet metal and was called "Hot Stuff." Room temperatures were measured constantly, and the appliances were compared on the basis of how long each could maintain the temperature above 65°F. Although not all precautions were taken or corrections made to ensure a fair comparison, the results clearly indicated the superiority of the stove, which kept the room warm from 4 to 10 times longer than either fireplace. Since stoves are known to have efficiencies in the range from 40 to 65 percent, the above comparison suggests fireplace efficiencies are about 5 to 15 percent, which is consistent with other estimates.

FIREPLACE THEORY AND DESIGN

Fireplace designs which are intended to optimize heating effectiveness must take into account the three main ways a fireplace influences the energy balance in a house; (1) radiant energy from the fireplace heats the room it is in, (2) heat in the hot flue gases may conduct through the chimney walls into the house, and (3) warm house air is pulled into the fireplace and up the chimney. Very rough estimates of the size of each of these effects while a steady fire is burning are that for every 100 units of wood energy put into the fire, 15 units enter the room as radiation from the fireplace, 7 units conduct through the chimney walls and become useful heat if the chimney has all its walls exposed to the living spaces, and 8 units of energy leave the house in the air which goes up the chimney (Table 11-1). The ideal energy-efficient fireplace should maximize the amount of radiation it emits, and minimize the amount of air escaping up the chimney. The excess air is not only a direct loss of heat from the house but its diluting effect decreases the amount of heat which will conduct through the chimney walls.

It is useful to think of there being two kinds of radiant energy coming out of a fireplace; primary radiation coming directly from the coals and flames of the fire into the room, and secondary radiation which is reflected or emitted from the fireplace walls. Direct radiation is maximum for fireplaces which are especially shallow, tall and wide (Figure 11-1), since with this shape less radiation is intercepted by the fireplace walls. Taken to the extreme, this geometry becomes a fire built against a flat wall, with a smoke-gathering hood far above it. The idea is to bring the fire into the room as much as possible.

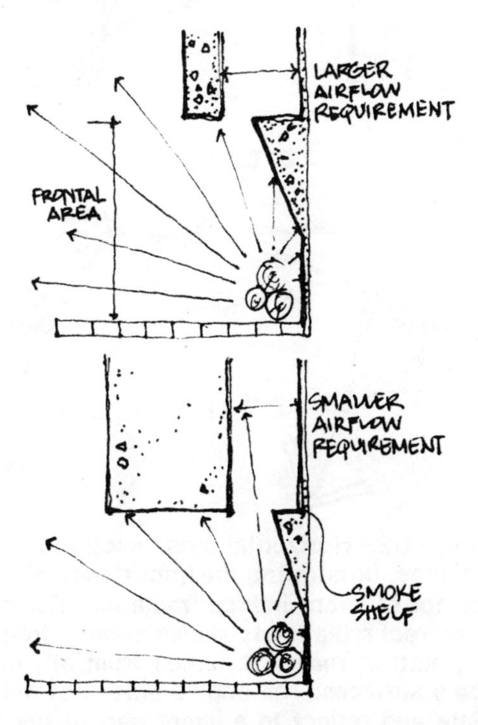

Figure 11-1. A relatively deep and a relatively tall fireplace. More direct radiation gets out of a tall, shallow fireplace than out of a deep, cave-like fireplace.

The secondary radiation is mostly emitted radiation, not reflected primary radiation. Bricks are highly absorbing for infrared radiation, just as black objects are for visible radiation. Thus most of the direct radiation which is intercepted by the fireplace walls is absorbed by them. Parts of the walls are also heated by direct contact with flames and hot gases from the fire. The hot bricks then radiate some of their energy; the rest of their energy is either conducted away from the fireplace through the masonry or is conducted into the air and / or gases that are moving through the fireplace.

The radiation from a surface travels in all directions and its amount increases both with the surface's temperature and its area. To maximize the secondary-radiation contribution to heating the room, without diminishing the primary radiation, two things can be done. The upper part of the back of a fireplace should slope forwards and be tall (Figure 11-1), and the sides should not be perpendicular to the back, but angle outwards. (Figure 11-2). In addition, the temperature of all the surfaces will be still higher if highly insulating materials are used in the fireplace's construction. This makes the surfaces hotter by decreasing the heat conduction into the structure of the fireplace. Since the bricks which are better insulators are also physically weaker (they are more porous), they are not generally used in exposed locations but are covered with a layer of more durable firebrick. Loose, high temperature insulating material can also be used behind firebrick.

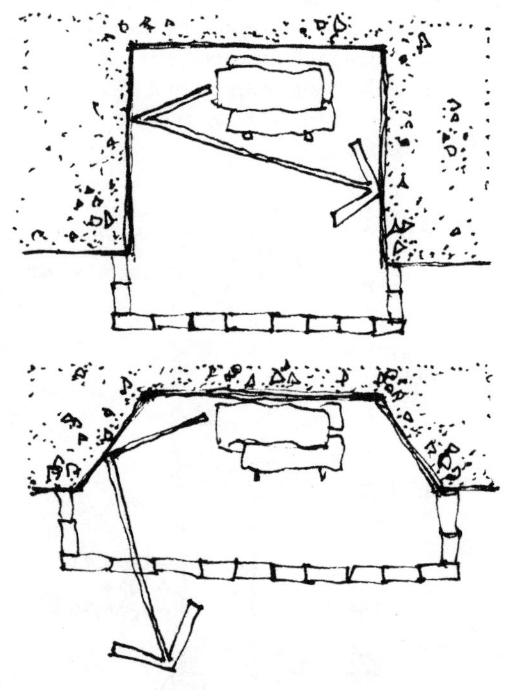

Figure 11-2. Horizontal cross sections through fireplaces, illustrating the importance of angled sides for efficient indirect radiation. For clarity, the indirect radiation is shown coming from only one point; in reality it comes from all the fireplace's surfaces. The angled sides not only can radiate and reflect to a larger part of the room, but, being closer to the fire, they are hotter and thus emit more intense radiation.

To minimize the loss of heated room air up the chimney and to maximize the heat conducted out through the chimney walls require cutting back the air flow into the fireplace as much as possible. This is usually done, in part, by having a narrowing or a constriction in the throat of the fireplace before it enters the chimney. A damper with adjustable settings is ideal to meet the varying conditions of fireplace use. The more challenging problem in fireplace design is to shape the fireplace so the constriction can be as substantial as possible without making smoke come into the room. The turbulence and eddies of the flames in a fireplace tend to eject smoke unless countered by an inward flow of air through the fireplace opening. The required average velocity of this air is about 0.8 ft / sec.[4] This implies an air flow, for each square foot of fireplace opening, of 48 cubic feet per minute or 216 pounds per hour. Thus the most obvious way to decrease the amount of air needed to prevent smoking is to reduce the frontal area of the fireplace. This requires making the fireplace narrower or less tall, which in turn inhibits the radiant energy transfer to the room.

A central compromise in fireplace design is between very open and shallow fireplaces, which are the best radiators but may require relatively large total airflow to prevent smoking, and a more closed box-like design with less frontal area for the same size fire, which decreases total air requirements but intercepts much radiation. It is also probable that the necessary inward frontal air velocity to prevent smoking is not the same for all fireplace shapes. As is the case in stove design, the number of design variations and interactions is so large that thought and prediction alone are inadequate — experimental testing is essential. To my knowledge, no careful testing has been done on these issues of fireplace design. Various people, most notably Count Rumford[5], have given supposedly ideal shapes and dimensions of fireplaces; I see no reason they are not optimum designs, but there seems to be little quantitative evidence.

Most fireplaces are built with a "smoke shelf" as indicated in Figure 11-1. It is my belief that under most circumstances the shelf itself serves no function at all, but is an incidental side effect of fireplace designs with throat constrictions and forward sloping backs (Figure 11-1). Smoke shelves are often thought to be important for preventing downdrafts in chimneys.[6] However, smoke shelves are not used in the venting systems of any other kind of fuel-burning appliances (free-standing fireplaces, stoves, furnaces of all kinds, water heaters, etc.) Many all-metal prefabricated fireplaces have no smoke shelf. Thus a smoke shelf cannot be generally necessary.

Smoke shelves *may* be useful in oversize chimneys, or when a fireplace has only a very small fire in it, particularly if the chimney has a side exposed to the outdoors. The upward velocity of the flue gases may then be quite small. This can permit cool outdoor air to enter the top of the chimney. The air may or may not survive as a distinct entity all the way down to the bottom of the chimney, for as it descends it will mix with the rising warm flue gases, and it will be warmed by the chimney walls. If it does reach the chimney shelf; the shelf may help keep it, along with some smoke, from coming out into the house. Exterior masonry chimneys are especially susceptible; the relative coolness of their exposed side(s) can assist or even induce downdrafts against the cold flue surfaces.

Oversize chimneys are likely to cause smoking whether or not there is a smoke shelf. The low velocity of the flue gases results in more cooling than is usual, which decreases the draft. The low velocity also makes a chimney more vulnerable to wind-induced draft problems. If there are down currents in the chimney (whether or not they reach the smoke shelf) they decrease the draft by their own cooling effect, and they physically impede the upward flow of the flue gases. Thus many of the problems of an oversized chimney are not solved by a smoke shelf; yet, the only problem a smoke shelf might help prevent only occurs if a chimney is oversized. Thus the only general solution is a good chimney, not a smoke shelf. The best chimneys are appropriately sized, are not on an exterior wall of the house, and have wind caps.

FIREPLACE ACCESSORIES

Many varieties of fireplace heat extractors and heat savers are available. All are designed either to extract and use some of the heat, which would otherwise rise up the chimney, or to decrease the warm air loss from the house up the chimney. Since a fireplace's normal efficiency is so low, there is considerable room for improvement.

The most common and least expensive type of fireplace heat extractor is the tube grate (Figure 11-3). The tube grate replaces ordinary grates or andirons and fulfills their functions, as well as extracting additional heat. The units are all of some sort of hollow construction — most commonly, open pipes bent roughly into a "U" shape. Room air enters the lower ends, is warmed as it passes under, around the back and over the fire, and then reenters the room through the upper ends of the pipes. Some units work by natural convection, others are equipped with blowers and manifolds to increase the speed of the air. Some do not use round pipes but rectangular passages and chambers.

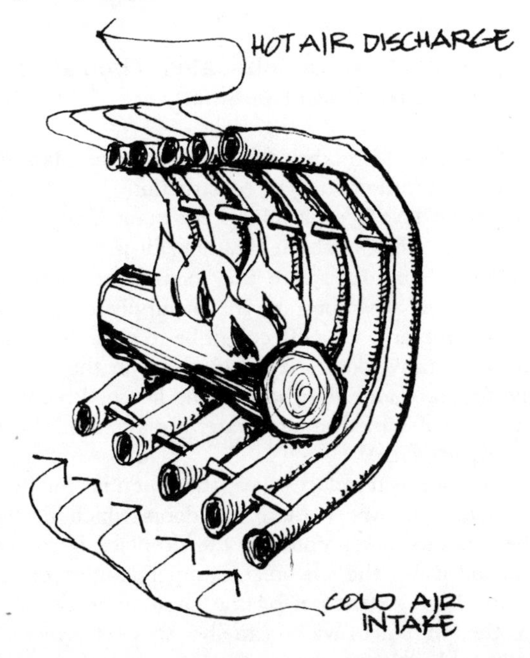

Figure 11-3. A tube grate for a fireplace. Fresh room air enters the lower ends of the tubes and is heated as it circulates under and around the back of the fire. Streams of hot air then emerge into the room from the upper ends of the tubes.

I am aware of no careful testing of these devices to measure their effectiveness quantitatively. There is little doubt that they work — they do extract additional heat from a fireplace. The emerging air does have to buck a general inward flow of room air into the fireplace, but this is not difficult even for the natural-convection units as long as the upper portions of the pipes are long and high enough. Many different size units are available, to fit the great variety of fireplace shapes and sizes. How much these units increase the heating efficiency of fireplaces cannot be determined without careful measurements.

It is not true that the hotter the air which emerges from the pipes, the more efficient the device is. In general, the temperature is high when the flow is low because heat is being added to a smaller quantity of air. When the flow is high, the air's temperature is less, but the total amount of heat in it is usually more. Also, determination of the net effect of tube grates requires measuring not only the heat output of the grate, but also any changes in the fireplace's normal heat flows due to the grate's presence. (See Chapter 10 for a similar discussion of stovepipe heat extractors.)

Glass doors for fireplaces are more ambiguous in their effects on energy efficiency. The doors are usually fitted into frames which seal against the fireplace opening. When the doors are closed, very little air can enter the fireplace except through air inlet dampers at the bottom of the unit.

Closing the doors on an operating fireplace has three principle effects: much less room air is carried up the chimney with the combustion products, much less direct radiation from the fire passes into the room, and somewhat more heat will be given off by the chimney walls. Thus there are opposing effects with respect to net energy efficiency.

Most of the heating from an ordinary fireplace is due to radiation received from the flames, coals, and warmed surfaces inside the fireplace. Virtually all kinds of glass which are practical for use in fireplace doors absorb much of the radiation from the fire. The glass itself, whén warmed, reemits some radiation, but the net effect is a large reduction in the direct heat output of the fireplace.

The decreased air flow into the fireplace reduces the heating load of the house. The amount of this reduction depends on particular details for each installation. One always important factor is the outdoor temperature, for this determines how much heat from some source will have been added to each pound of air by the time it reaches the fireplace to warm it to room temperature. In a particular case, the outdoor air might have a temperature of 30°F, and room temperature might be 70°F. If the air flow with the doors closed is cut by 2/3, from 1500 to 500 pounds per hour (this is a guess), then the savings in heat is 9600 Btu per hour.[7] Actual savings may range from 2,000 to 20,000 Btu per hour. This is a significant reduction in the house's heating needs.

Use of glass doors can cut back on excess air entering the fireplace without decreasing the wood combustion rate. Thus with the same amount of heat being generated, but less air diluting the combustion products, the flue gases are hotter. (This effect is enhanced by the decrease in radiant energy transferred into the room.) Thus the masonry around the fireplace and the entire chimney are hotter. Consequently more heat will enter the house by conduction through the fireplace and chimney walls. The amount depends critically on the type of chimney and its location. The effect is greatest with interior masonry chimneys with their walls directly exposed to the living spaces. The increased heat transfer is smaller for prefabricated, insulated, metal chimneys. The increased heat transfer from an exposed interior masonry chimney is probably significant, perhaps 50 percent more than the chimney-heat contribution without glass doors, or on the order of a few thousand Btu per hour.

Combining these three effects (of decreased direct radiation, decreased warm air losses, and increased heat given off from the chimney) to yield an overall assessment of the net effect of the use of glass doors in fireplaces is not yet possible. Too many effects depend critically on details of each situation, and too few measurements have been made. The net effect is probably positive if the doors are used mostly at night as the fire is dying down, and it is probably negative if the doors are closed most of the time a fire is burning. In any case, use of glass doors contributes to safety by preventing sparks etc. from getting out of the fireplace.

A related use of glass doors is to shut down the fireplace completely at night. Without such a device to cut off the air supply, (heated) room air escapes through the chimney all night long, presuming the chimney damper had to be left open to vent the fumes from the remains of the evening's fire. With a little more inconvenience but much less cost, the same objective can be met by placing a non-combustible material such as a sheet of asbestos or metal tightly up against the fireplace opening. The better the seal, the less warm air will be lost. It is not necessary in most cases to let in any air to help the smoke go up the chimney,[8] and any wood or charcoal which may not burn due to lack of oxygen, can be used in the next fire.

Some people place only a standard spark screen over the fireplace, but shut the flue damper before retiring at night as long as there is only a small amount of glowing charcoal left in the fire. Charcoal burns relatively cleanly, with no unpleasant odor or visible smoke. The energy efficiency of such a flueless charcoal fire is very high — possibly over 90 percent, since there is no heat loss up the chimney.[9] Carbon monoxide is usually produced in a charcoal fire, but in such very small quantities compared to the volume of air in a house and its natural ventilation rate, that only very rarely could it constitute a health hazard. Oxygen depletion is also insignificant.

Fireplace heat extractors are also available. In a unit called the Thriftchanger, the hot flue gases are sent through a set of tubes around which room air flows. Retrofitting such a unit to most existing fireplaces is a major undertaking unless the fireplace has an exposed metal chimney. In new systems, installation is not difficult. Since fireplaces are so inefficient, this type of unit is capable of extracting a large amount of heat. As is true with all heat extractors, the better they work, the cooler the flue gases are and hence the poorer is the draft. Good heat extractors should be added only to systems with some excess chimney capacity.

PREFABRICATED FIREPLACES

The most effective (and most expensive) fireplace accessory is virtually a whole new metal, or metal and glass, fireplace inserted into the old one. Most units consist basically of a double metal wall construction with a space in between through which air can circulate (Figure 11-4). As with the tube grates, some units come with blowers, and others rely on natural convection to circulate the air. Glass doors are included in some models; metal doors are also employed. Since masonry fireplaces come in so many shapes and sizes custom manufacturing and fitting are often necessary when retrofitting an existing fireplace.

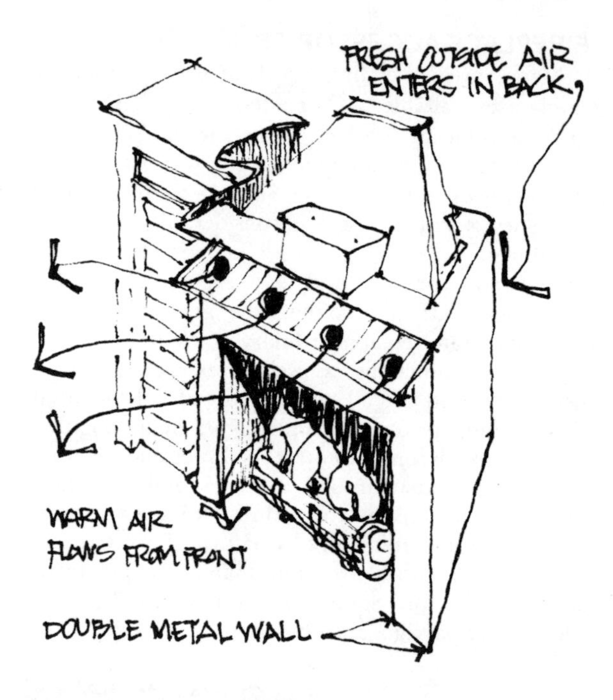

Figure 11-4. A prefabricated circulating fireplace (The Northern Heatliner).

This type of prefabricated circulating fireplace design is more common in new construction. The Heatilator fireplaces have been available for many years. In some units (e.g., the Northern Heatliner) it is heated *outdoor* air which circulates into the room, and some of this same air may then become the air supply for combustion in the fireplace. One effect can be to virtually eliminate uncomfortable floor-level drafts both in the room with the fireplace and elsewhere in the house. Prefabricated fireplaces designed to heat water are also available (e.g., the Hydroplace).

Circulating fireplaces have the potential for being as efficient as stoves. If there are doors which fit tightly, the amount of air entering the fireplace is completely controlled by the air inlet dampers under or in the doors. Heat extraction by the circulating air or water in the fireplace walls can be very effective. With reasonably careful design, the efficiency of this type of unit could approach the efficiency of a free standing stove. I am aware of no measurements which indicate actual efficiencies.

TERMINOLOGY

There is not a clear distinction between fireplaces and stove, at least the way the terms are often used. At the extremes, the differences are clear — the traditional open masonry fireplace, and the closed metal "parlor" stove. But in between, there seems to be no characteristic which clearly delineates the two. There are openable stoves (e.g., Franklins) and closable fireplaces (with glass door). There are also non-metal stoves (stone and ceramic) and metal fireplaces. Free-standing fireplaces perform much like an open stove, and some come equipped with glass doors making them closable. Some stoves have glass or mica windows making the fire visible just as in fireplaces with glass doors. Some free-standing fireplaces are just as portable as most stoves. The National Fire Protection Association (NFPA) uses the term "fireplace" to mean "A hearth, fire chamber, or similarly prepared place with a chimney."[10] The NFPA also uses the term "fireplace stove" to mean "a chimney-connected, solid fuel burning stove having part of its fire chamber open to the room." NFPA does not define the term "stove" or "wood stove" at least not in the publication wherein related terms are defined.

The basis for the confusion appears to be that there is in fact such a tremendous variety of wood-burning devices. Most of the time, "fireplace" indicates a device intended to be usable in an open configuration, whether or not it can also be closed and operated closed. But there are exceptions. Franklin fireplaces are also called Franklin stoves. There is also a stove, Better 'n Bens, designed to sit on the hearth in front of a fireplace which is aptly described as a fireplace stove. It normally can be operated only closed, but an accessory is available permitting it to be operated open. The confusion is partly a sign of the great variety of wood heaters available.

SUMMARY

Masonry fireplace net energy efficiencies range from a high of about 15 to 35 percent for continuous use to about zero percent for occasional use (no net heat gain or loss); efficiencies between zero and 15 percent are probably most common. Fireplaces will typically burn 3 to 10 times as much wood as a closed stove to generate the same amount of useful heat.

Accurate information on efficiencies of factory built fireplaces, and fireplaces equipped with heat-extracting or heat-saving accessories, is not available. The most efficient kind of fireplace is probably the type that heats air (or water) circulated through the fireplace's walls. A well-designed unit operated with doors closed could be as efficient as many closed stoves. Glass or metal doors alone are of uncertain value when used during a hot fire, but should improve net efficiencies when used at the end of a fire (such as overnight). Covering a fireplace opening with a metal or asbestos cover at night as a fire is dying down can accomplish the same thing. A problem with glass doors is that they need frequent cleaning. Tube grates probably improve efficiencies, especially if equipped with a blower.

Freestanding metal fireplaces are probably more efficient than traditional masonry fireplaces. As with any wood burner, it is important to have a flue gas damper, or doors, to stop interior air from constantly flowing up the chimney when there is no fire. Openable stoves (like the Franklin) have efficiencies in the range of 30 to 50 percent.[11]

No fireplace or open stove has the capability of heating for many hours without refueling, as some closed stoves can. However open wood burners do not generate as much creosote as do closed stoves.

Chapter Twelve Creosote and Chimney Fires

Wood combustion is never perfectly complete. Wood smoke almost always contains some unburned gases and a fog of unburned tar-like liquids. Some of these materials will condense out of the flue gases onto any surface which is not too hot. The condensate is usually dark brown or black, and has an unpleasant acrid odor. It is called creosote. If condensed on a relatively cool surface (such as an exterior stovepipe chimney), the creosote will contain a large amount of water along with the organic compounds, and will thus be very fluid. Water is usually absent if the condensation occurs on surfaces hotter than 150°F. The condensate may then be thick and sticky, like tacky paint or tar. Creosote may be found almost anywhere in a wood-heating system, from the top of a chimney to the insides of the stove itself.

Creosote's chemical composition is not well known, partly because it is such a complex mixture of compounds that its analysis is very difficult. Its composition is also variable, depending on the conditions of its formation, such as the type of wood and its moisture content, the conditions of combustion, and the temperatures at which condensation occurs. Creosote doubtless includes many of the compounds listed in Table A3-1, particularly acetic and formic acids, creosol and cresol.[1] All of these similar-sounding words are from a Greek word meaning to preserve, because they are good wood preservatives. The creosote used as a wood preservative is made by pyrolyzing wood tar (or coal tar) which is made by pyrolyzing wood (or coal). Its chemical composition is similar to, but not identical with, the fluid chimney deposits from wood stoves. Because of its acidity chimney creosote is corrosive to many materials, including steel and mortar (common chimney materials). Good chimney tile liners and stainless steel are relatively immune to its corrosion.

Creosote which remains in a chimney after its initial formation may later be significantly modified both in physical form and chemical content. The water and the more volatile organic compounds tend to evaporate, leaving the more tar-like substances behind. If these are subsequently heated by the flue gases from a hotter fire (this usually happens), they themselves are further pyrolyzed to the same final solid product that wood is — carbon. The physical form is usually flaky, and often shiny on one side. Partially pyrolyzed deposits can have a bubbly appearance. The flakes do not adhere strongly to a stove pipe and thus are easy to brush off; some of the other forms will not budge even under the action of a stiff wire brush.

The amount of creosote deposited depends mostly on two factors — the density of the smoke and fumes from the fire, and the temperature of the surface on which it is condensing. Highest smoke densities occur when a large amount of wood in relatively small pieces is added to a hot bed of coals and the air inlet damper is closed. Here, there is considerable pyrolysis of wood, but little combustion, and little air to dilute the smoke. In practice, creosote generation is highest during low-power, overnight, smoldering burns. Smoke densities are least when combustion is relatively complete, which tends to be the case when the amount of excess air admitted to the wood-burner is high. Leaky stoves, open stoves, and fireplaces typically have the least severe creosote problems.

One way to lower the average smoke density in an air-tight stove is to use less wood each time fuel is added, and/or to use larger pieces of wood; in either case, the air supply need not be turned down so much in order to limit the heat output and combustion is likely to be more complete. Of course, if less wood is added, stokings must be more frequent. A related procedure to limit creosote is to leave the air inlet moderately open after adding wood until the wood is mostly reduced to charcoal, and then close the inlet as much as desired. This will promote complete combustion during pyrolysis, when the creosote compounds are being formed, but there will be a significant heat surge while the gases are burning.

Extra air can also be added to the flue gases in the stove pipe; this is what the Ashley creosote inhibitor accessory does. But the net effect of adding dilution air is not obvious or necessarily beneficial. Dilution air will decrease the smoke density, but it will also decrease its temperature. These effects have opposing influences on creosote formation. The National Fire Prevention Association states that dilution air increases chimney deposits.[2] In any case, the cooling effect of dilution air does decrease the heat transfer through the stovepipe and chimney, thus decreasing the system's energy efficiency.

Creosote formation may also depend on the type of wood burned and on its moisture content. Dry hardwoods have a reputation for generating the least creosote, but the quantity can still be very large. No kind of wood eliminates creosote formation.

For a given smoke density near a surface, the cooler the surface, the more creosote will condense on it. The phenomenon is very similar to water vapor condensing on the outside of a glass of ice water on a humid day, except for an inversion — condensation occurs on the inside of a chimney, especially when cold air outside makes the inner chimney surface relatively cool. A stovepipe chimney outside a house on a cold day will be wet on the inside with creosote (including a lot of water) virtually all the time. A well-insulated, prefabricated metal chimney has the least serious creosote problems; its insulation helps maintain higher temperatures on its inner surface, and its low heat capacity allows it to warm up very quickly after a fire is started. Masonry chimneys frequently accumulate deposits at the beginnings of fires — their interior surfaces take a longer time to warm because the construction is so massive. Any type of chimney which runs up the outside of a house is more susceptible to creosote problems than the same type of chimney rising in the house's interior, due to the cooling effect of the colder outdoor air on the exterior chimney.

Average flue gas temperatures can be increased by minimizing the length of stove pipe connecting the stove to the chimney. This, of course, will also decrease the energy efficiency of the system, and it is often true that measures which decrease creosote formation also decrease heating efficiency. For instance, stoves which have high energy efficiencies due to their relatively good heat transfer (e.g., the Sevca, Lange 6303 and double-barrel stoves) are more likely to have chimney creosote problems precisely because they do such a good job extracting heat from the flue gases.

Generally, creosote is inevitable and must be lived with. Any kind of chimney deposit decreases the system's heating efficiency. Soot and dried creosote accumulations have a significant insulating effect; less of the heat in the flue gases is transferred into a house through dirty stovepipe and chimneys. The most annoying problem can be creosote dripping from a stovepipe or chimney, and the most dangerous problem is chimney fires, during which the creosote, or its pyrolyzed residue, burns.

Creosote dripping can usually be eliminated. Joints in vertical segments of stovepipe will not leak if, at the joints, the smaller, crimped ends always stick *down* into the receiving end. (Smoke will not leak out of the joints due to this direction of overlap.) Since this is not the usual orientation for stovepipe, a double-male fitting may be necessary at some point to connect the stovepipe to the stove, a prefabricated chimney, or a rain cap. Special drip-proof adaptors are available for connecting some sizes of stovepipe to Metalbestos brand prefabricated chimneys.[3] Common types of stovepipe elbows can leak creosote due to their swivel joints; rigid and accordion-type leakproof elbows are available. Horizontal or gently sloping sections of stovepipe should be oriented so their seams are on top. Joints between horizontal pipes and/or fittings are the most difficult to seal against dripping. A good high-temperature sealant can sometimes help, but is no guarantee. The joint must also be snug, and be well secured with sheet-metal screws. If all joints are made leak-proof, then the creosote will generally drip into the stove, where, when the fire is hot, it will be burned.

Chimney fires occur when the combustible deposits on the inside of a chimney burn. The deposits may be "raw" creosote, pyrolyzed creosote, or soot. Ignition requires adequate oxygen, which is usually available, and sufficiently high temperatures — the same conditions as for the ignition and combustion of any fuel. Chimney fires are most likely to occur during a very hot fire, as when cardboard or Christmas-tree branches are burned, or even when a stove burns normal wood, but at a higher than normal rate. A crackling sound can often be heard at the beginning of a chimney fire. As the intensity of the fire rises, the stovepipe will sometimes shake violently, air will be very forcefully drawn in through the stove, and the stovepipe may glow red hot. A tall plume of flame and sparks can be seen rising from the top of uncapped chimneys.

The most effective way to suppress a chimney fire is to limit its air supply although both water and salt are sometimes suggested. If a relatively air-tight stove is the connected appliance, this is easily done by closing the stove's air-inlet dampers, if all the stovepipe and/or chimney joints are tight, and if no other appliance is connected to the same flue. If the fire is in the chimney of a leaky stove or fireplace, one generally has to wait for it to burn out. Attempts can be made to limit the air supply entering through a fireplace by using a wet blanket or metal sheet to seal the fireplace opening; but the very strong draft can make this difficult and even dangerous, and only non-combustible materials should be used or an additional problem may be created.

In a properly designed and maintained chimney, the only potential hazard related to chimney fires is ignition of the building's roof or surroundings due to sparks and burning embers coming out of the top of the chimney. A spark arresting screen can decrease, but not eliminate, this possibility, but spark screens themselves are often not suitable for use with wood fuel because they can become clogged. The chimney itself and the stovepipe, when properly installed, are intended to withstand an occasional chimney fire without danger of ignition of their surroundings. During a chimney fire, one ought to check the roof and surroundings, and possibly wet down critical areas. If the chimney may not be up to safety standards, one should also keep a close watch on all surfaces near the chimney.

Some people start chimney fires fairly frequently, as a means of chimney cleaning. This deters very intense chimney fires and the small ones which do happen are always under a watchful eye. Under some circumstances, this practice may be reasonable, but generally it is a risky method to keep a chimney clean. There is always some danger of a house fire, but in addition, any chimney fire is wearing on a chimney; the high temperatures increase the corrosion rate of metals and the thermal expansion of masonry materials encourage crack formation and growth.

Chemical chimney cleaners are available. Opinions on their effectiveness vary, but apprently when used regularly, and as directed, they work, and do not damage chimneys. The usual chimney-cleaning method is the oldest — human energy and some kind of mechanical tool. A stiff wire brush, a heavy chain (perhaps in a bag) hung with a rope and worked up and down from the top of the chimney, and very small brushes have all been used. Professional chimney sweeps are also reappearing.

Some people clean yearly, others after every few cords of wood burned, but there are so many factors influencing creosote build-up that such generalizations are not appropriate in most particular cases. In new installations, or when changes occur (such as a different stove) the chimney should be checked frequently (after 2 weeks, then after a month, then after another 2 months, etc.) until it is clear how frequently cleaning is usually needed.

Chapter Thirteen Economics

A common reason people give for being interested in wood heating is to save money. In cold climates the annual cost of fuel or electricity to heat a house is typically many hundreds of dollars. Since a single wood stove can contribute a significant portion of the needed heat, there is the potential for saving hundreds of dollars a year.

Few people heat with wood for purely financial reasons, but awareness of the full economic implications should aid sound decision making. However, valid generalizations about the economics of wood heating cannot be made because of the many unknown variables in each individual case. This discussion will stress underlying issues and principles.

The two costs associated with wood heating systems are the initial purchase and installation cost, and the continuing fuel cost. The resulting savings are in lower conventional heating bills for natural gas, oil, electricity or LP gas. Wood heating systems rarely replace conventional systems, but are usually supplemental sources of heat.

Suppose a $300 stove is purchased and the installed cost of a prefabricated metal chimney is an additional $300, for a total initial cost of $600. Assuming the system will last 20 years, how much would have to be saved in fuel costs each year in order to break even? (If the conventional system burns oil, the net annual fuel saving is the money saved on oil minus the money spent on wood.) For simplicity let us also assume there is no inflation.

Spreading the $600 initial cost over 20 years would suggest $30 per year as the minimum annual savings to break even. But in fact, about $50 per year would have to be saved. The reason is that if instead of buying a stove and chimney the money were deposited in an insured savings account earning a guaranteed interest of say 5 percent, then one would have $1592 after 20 years. To be fair, the net annual fuel savings must also be thought of as being deposited in a similar account. The result in this example is that if $48.15 is saved each year, it will amount to $1592, including interest, after 20 years. Thus unless the net savings in fuel costs exceeded $48.15 per year, one would have been better off, in strict economic terms, to put the $600 in the savings bank rather than into the wood heating system. In general, the annual net savings in fuel costs must exceed about 8 to 12 percent of the initial cost of the equipment in order that the venture not lose money.[1]

The actual annual net savings in fuel costs depends on some factors which are usually not known precisely. For example, suppose a wood stove is in a house with an oil-fired central heating system. In order to be able to save money on fuel by using the stove, wood must be cheaper than oil *for the same amount of useful heat delivered to the living space of the house.* No. 2 heating oil has an energy content of about 139,000 Btu per gallon. If the oil heater has a net energy efficiency of 65 percent, then 90,000 Btu (0.65 × 139,000) of useful heat is obtained for each gallon of oil burned. If oil costs 40 cents per gallon, then $4.43 [($0.40 ÷ 90,000 Btu) × 1,000,000 is the fuel cost per million Btu of delivered heat. This figure is easily adjusted for the prevailing price of oil, but the net heating efficiency of one's own furnace or boiler is not easily determined, and can vary from 30 to 80 percent!

The fuel cost per million Btu of wood-stove heat can be estimated similarly, but with an additional uncertainty - the energy content of the wood. As indicated on page 23 in Table 3-2, the energy per cord varies over a large range, from less than 15 million to over 30 million Btu depending mostly on wood density. Unfortunately, the density of a given type of wood is variable over a range of about 10 percent, and the amount of solid wood per cord has a larger variation. Thus the actual energy in a cord of a given wood type has an inherent uncertainty of about 20 percent, and the difference between different types of wood can be 100 percent.

As an example, if wood costs $50 per cord and contains 22.5 million Btu per cord, and is used in a stove with a net efficiency of 50 percent, wood heat costs $4.44 per million Btu [50 ÷ (22.5 x 0.5)].

Figures 13-1 to 13-4 summarize the fuel cost comparisons of wood to oil, natural gas, electricity, and liquified petroleum (LP) gas (propane and butane are both forms of LP gas). The solid line in each figure represents the break-even fuel prices in the average case of a wood stove with an efficiency of 50 percent, a furnace or boiler efficiency of 65 percent (or 100 percent for electric heat) and wood with 22.5 million Btu per cord (high heat value). Under these assumptions, wood heat is cheaper (in terms of fuel costs) if the point on the graph corresponding to the actual prevailing fuel prices falls to the left of the solid line. For example, if wood costs $60 per cord and oil costs 40 cents per gallon, the corresponding point on the graph is at the intersection of the vertical $60-per-cord line and the horizontal

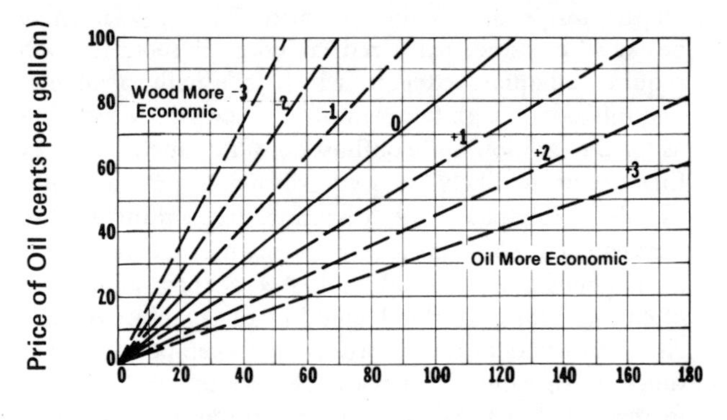

Figure 13-1. Fuel-cost comparison for heat from oil and from wood. The different lines correspond to different assumptions concerning the oil-furnace and wood-heater efficiencies, and the type of wood. Each line represents the break-even prices for wood and oil under particular assumptions (see text). If the point on the graph corresponding to the actual prices of oil and wood falls to the left of the appropriate break-even line, the fuel cost of wood heat is less than the fuel cost of oil heat.

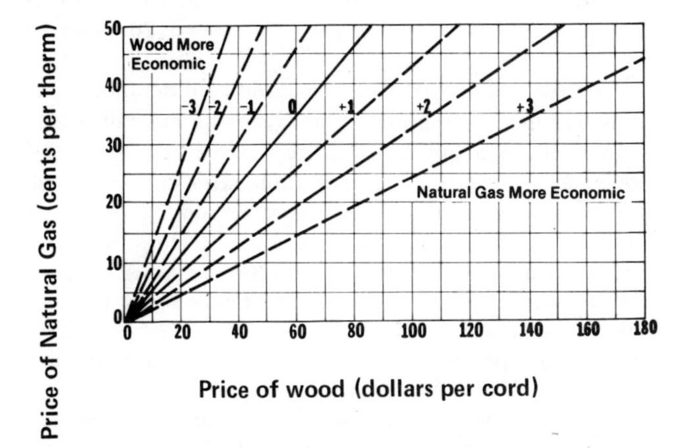

Figure 13-2. Fuel-cost comparison for heat from natural gas and from wood. The different lines correspond to different assumptions concerning the gas-heater's and wood-heater's efficiencies, and the type of wood. Each line represents the break-even prices for wood and natural gas under particular assumptions (see text). If the point on the graph corresponding to the actual prices of wood and natural gas falls to the left of the appropriate break-even line, the fuel cost of wood heat is less than the fuel cost of oil heat. (A "therm" is 100,000 Btu or about 100 cubic feet of gas.)

40-cents-per-gallon line. This point is to the right of the solid line, indicating oil heat is less expensive than wood heat in this case. If oil were more than 50 cents per gallon, wood would be the better buy, since the intersection of the two fuel-cost lines would then lie to the left of the solid line.

The solid line corresponds to the case of average wood, and average efficiencies for the stove and for the furnace.

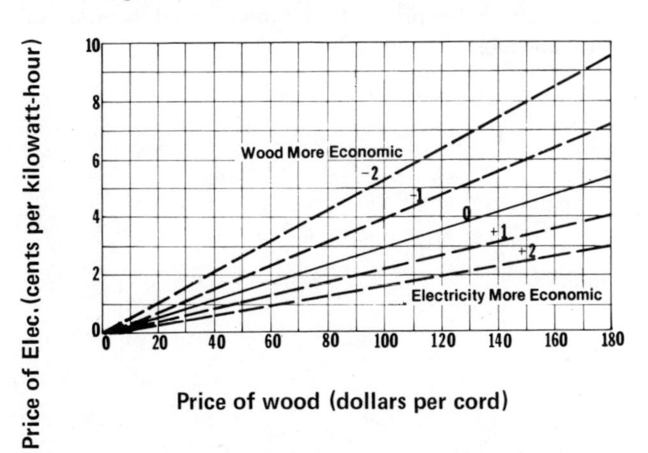

Figure 13-3. Energy cost comparison for electric heat and wood heat. The different lines correspond to different assumptions concerning the wood heater's efficiency and the type of wood. Each line represents the break-even prices for electricity and wood under particular assumptions (see text). If the point on the graph corresponding to the actual prices of electricity and wood falls to the left of the appropriate break-even line, the fuel cost of wood heat is less than the energy cost for electrical heat.

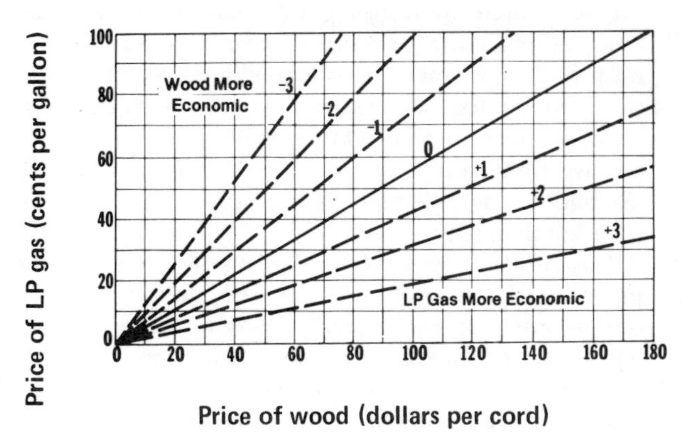

Figure 13-4. Fuel-cost comparison for heat from LP gas and from wood. The different lines correspond to different assumptions concerning the LP gas heater's and wood heater's efficiencies, and the type of wood. Each line represents the break-even prices for wood and LP gas under particular assumptions (see text). If the point on the graph corresponding to the actual prices of wood and LP gas falls to the left of the appropriate break-even line, the fuel cost of wood heat is less than the fuel cost of heat from LP gas.[2]

In practice, most people's situations are not average in all three of these respects. The dashed lines represent the break-even fuel prices for a variety of other conditions. To estimate which line is appropriate in any particular case, assess whether each of the three critical characteristics is high, average, or low, assign the appropriate number to each, and total the three numbers (see below). The appropriate line to use on the graphs is the one labeled with this total. For instance, if the wood stove has a low energy efficiency (e.g., a Franklin stove operated mostly with its doors open), the corresponding number for this category is -1. If the available wood has a high energy content (such as hickory or oak), + 1 is the score for this category. If the conventional heating system is

Stove net efficiency	high[4]	+ 1
	average	0
	low	− 1__
Wood energy content per cord	high[5]	+ 1
	average	0
	low	− 1__
Conventional heating-system net efficiency[3] (use 0 for electric heat)	high[6]	− 1
	average	0
	low	+ 1__
		____
	Total	___

an old, poorly maintained oil-fired boiler with a low energy efficiency, the number for this category is + 1. The sum of these three numbers, − 1, + 1, is + 1, and hence in this case the appropriate break-even line on the oil-wood graph is the one labeled "+ 1". Wood is the more economic fuel in this case if the point on the graph corresponding to the actual prices of oil and wood falls to the left of this "+ 1" line.

It is apparent from these graphs that, in terms of fuel costs, purchased wood is often competitive with electric heat. For the average case (the solid line), wood at $132 per cord (or less) is a better buy than electricity at 4 cents per kilowatt-hour. On the other hand, to be competitive with natural gas at 20 cents per therm in the average case would require the cost of a cord of wood to be less than $35.[7]

It is critical in these economic considerations that the price per *full standard cord* of wood be used. Wood dealers often do not carefully measure the wood they sell because it would take too much time to do so; in practice, wood is more often sold by "truckloads" than by cords. The only way to be sure what price is being paid *per cord* is to stack and measure the delivered wood to determine the number of cords (or fractions of a cord) it really is. Dealer's estimates can be significantly inaccurate.

As discussed at the beginning of this chapter, in order for a wood system to be economic overall, the annual savings in fuel costs must be enough to pay for the system's initial cost. This typically requires the annual fuel-cost savings to be about 10 percent of the initial investment. If wood is the more economic fuel the annual savings will be larger the more wood is used. If the price of wood is very near the break-even price, the savings will never be large. If wood is available at half of the break-even cost, and wood heat supplies half the house's seasonal heating needs, one fourth of the (previous) annual heating costs will be saved by using wood.[8] Wood furnaces can supply 100 percent of a house's heat needs. Many houses get all their heat from a single stove, although a 50 percent contribution may be more typical. The intensity of a stove's use is a critical parameter, and one which usually cannot be accurately predicted.

In practice, supplementing a conventional heating system with wood can result in a lower usage of total heat. If the occupants choose to spend much of their time around the stove, the rest of the house can be cooler than normal with little decrease in comfort; the resulting savings in needed heat is similar to that of a thermostat set-back, and can amount to 20 percent or more.

If fuelwood is free, the economic considerations are much simpler - using wood is always economic. Fuelwood can rarely be acquired at no cost, even if one considers only out-of-pocket costs. For those who cut and haul their own wood, there is at least the cost of gasoline and oil for the chainsaw and truck. If the chainsaw is used principally for personal fuelwood cutting, its cost is an indirect cost of the wood. The cost of cutting one's own wood is mostly in terms of time. However, for many people, wood cutting and splitting has recreational value and thus does not represent a cost.

Considerable time can also be involved in tending a wood heater - carrying in wood, refueling the burner, adjusting the controls and carrying out the ash. If one had to pay somebody a reasonable wage to do these chores, wood heating would be considerably more costly, but for a person who does these tasks for him or herself the cost of the time is zero. The value of these activities will vary from person to person.

Other hidden costs of heating with wood may include the need for a space to store the wood, protection of the wood from rain and snow, chimney cleaning, stovepipe replacement, stove refinishing (blackening).

Many people heat with wood without being aware of or caring whether or not it is economical. Whether or not purely economic arguments favor wood heating there can be considerable enjoyment in heating with wood — from watching the fire, from the physical exercise, from the beauty of a wood stove, from the feeling of independence from the outside world for heating energy, and because a renewable energy source is being used. In addition to these benefits, wood heating often does save money, especially if the wood is gathered rather than purchased, or if the wood heat replaces electric heat. If the prices of fossil fuels and electricity increase more than the price of fuelwood, wood heating will become more economical.

Footnotes

1.1 E. P. Eckholm, *Losing Ground,* (NY: W.W. Norton, 1976), Chapter 6.

1.2 H. C. Hottel and J. B. Howard, *New Energy Technology: Facts and Assessments,* (Cambridge: MIT Press, 1971), p. 4.

1.3 See, for example, *The Feasibility of Generating Electricity in the State of Vermont using Wood as a Fuel: A Study,* submitted by J.P.R. Associates, Stowe, Vermont, to the State of Vermont, Agency of Environmental Conservation, Dept. of Forests and Parks, 1975.

1.4 I. S. Goldstein, "Potential for Converting Wood into Plastics," *Science* 189 (1975), p. 847.

1.5 I. S. Goldstein, "Potential for Converting Wood into Plastics," *Science* 189 (1975), p. 847.

1.6 These figures are from H. C. Hottel and J. B. Howard, *New Energy technology - Some Facts and Assessments,* (Cambridge: MIT Press, 1971).

1.7 E.P. Eckholm, *Losing Ground,* (New York: W.W. Norton, 1976), Chapter 6.

1.8 The energy in wind and falling water comes from the sun

1.9 Report of the Governor's Task Force on Wood as a Source of Energy, State of Vermont (1975).

1.10 Actually, almost the entire effect of greenhouses in holding heat is not due to the infrared-radiation properties of glass, but due to the physical barrier the glass provides which prevents the warmer air inside from leaking out and colder outside air from getting in. The effect of carbon dioxide in the atmosphere should more properly be called the "atmosphere effect." See S.D. Silverstein, "Effect of Infrared Transparency on the Heat Transfer Through Windows; A Clarification of the Greenhouse Effect," *Science* 193 (1976), p. 229.

1.11 R.A. Bryson, "A Perspective on Climatic Change," *Science* 184 (1974), p. 753. See also W.W. Kellogg, and S.H. Schneider, "Climate Stabilization: For Better or Worse?," *Science* 186 (1974), p. 1163.

1.12 This assumes (and it is true) that potential fossil fuels are not being layed down as undecayed plant matter at the rate at which we are burning fossil fuels.

2.1 An object with mass *m* moving (but not rotating) with speed *v,* has an amount of kinetic energy equal to $\frac{1}{2}mv^2$.

2.2 Technically heat is the form of energy which is transferred by virtue of a temperature difference, and via molecular collisions.

2.3 Electrons also contribute in some solids and liquids.

2.4 In general, contributions to skin energy changes from conduction and evaporation should also be taken into account. See page 98.

2.5 $.12 \frac{Btu}{lb^\circ F} \times 100\ lb. \times (470^\circ - 70^\circ F) = 4800\ Btu.$

2.6 Evaporation at room temperature requires about 1050 Btu per pound; boiling at 212°F requires about 970 Btu per pound.

2.7 The surface is assumed to be flat or convex, to be spectrally black, and to be surrounded by a 70°F environment. All of these assumptions are approximately true for stoves in homes. Natural convection is computed as $.19(\Delta T)^{4/3}$ Btu/hr. sq. ft, where ΔT is the temperature difference (in °F) between the surface and its surroundings. Radiation is computed as 1.71×10^{-9} ($T^4 - 530^4$) Btu/hr. sq. ft. where T is the surface's temperature in degrees Rankine (= degrees Fahrenheit plus 460).

3.1 For example, David Havens, *The Woodburner's Handbook,* (Harpswell Press, 1973).

3.2 E. T. Howard, "Heat of Combustion of Various Southern Pine Materials," *Wood Science* 5 (1973), p. 194-197; and S.W. Parr and C.N. Davidson, "The Calorific Valve of American Woods," *J. of Indus. and Eng. Chem.* 14 (1922), p. 935; and Gottlieb, *J. Pratit. Chem.* 28 (1883), p. 414.

3.3 $.95 \times 8600 + .05 \times 17{,}400 = 9040$

3.4 Because wood shrinks as it drys, and because the amount of shrinkage is different for different species, the best parameter for assessing energy per cord is the density calculated as ovendry weight divided by the volume at typical moisture contents. However, the relative densities of woods are not *much* affected by the particular definition of density that is used.

3.5 Every pound of hydrogen in ovendry wood becomes 9 pounds of water when completely burned (9 is the ratio of the molecular weights of water to hydrogen). Thus if ovendry wood is 6 percent hydrogen, 0.54 pounds of water vapor are produced for each ovendry pound of wood burned. The approximate elemental composition of dry wood is: carbon, 49 percent; hydrogen, 6 percent; oxygen 44 percent; and ash, 1 percent.

3.6 0.79 pound $\times$ 1050 Btu/pound = 830 Btu

3.7 In fact, the loss is a little more, for an interesting reason. Wood actively holds onto water molecules — it is hygroscopic. If ovendry wood is moistened with liquid water, it actually heats up a little; the evolved heat is called the heat of wetting. When water is driven from wood in an oven or in a fire, not only does the evaporation energy (1050 Btu per pound) have to be supplied, but a little extra energy, just to break the gentle water-to-wood bond. For wood with a moisture content of 25 percent, this extra energy needed is about 100 Btu per pound of water. (The dryer the wood, the more energy it takes to remove one pound of moisture; at a moisture content of 1 percent, about 450 Btu is required to remove 1 pound of water.) See F.F.P. Kollman and W.A. Cote, Jr., *Principles of Wood Technology I: Solid Wood,* (Berlin: Springer-Verlag, 1968), p. 203. Thus, in the example in the text, 25 Btu is necessary to sever the bonds of 0.25 pounds of water. Thus the low heat value of wood with a moisture content of 25 percent is 7770 − 25 = 7745 Btu per ovendry pound. (See summary below.) [This assumes the high heat value as measured in a bomb calorimeter is determined using ovendry wood, or that appropriate corrections are made, so that this effect of the heat of wetting is not already included in the high heat value.]

1.25 pounds of 25 percent moisture wood [1 pound of ovendry wood and .25 pound of water].

High heat value	8600 Btu
minus latent heat in combustion-product water vapor	-567
minus energy to break hydroscopic bonds between the .25 pound of water and the wood	- 25
minus latent heat in evaporated moisture	-263
Low heat value	7745 Btu

3.8 L. Metz, "Fire Protection of Wood," *(Z. Ver. deut. Ing.* 80, (1963), pp. 660-667; L. Metz, *Holzschutz gegen Feuer,* (Berlin: VDI-Verlag, 2nd ed., 1942.) Species identification is no more specific in the above references than in Figures 3-5 to 3-8. Apparent inconsistencies in densities are probably due to variability among the particular samples used. A given tropical wood type may vary in density by as much as 50 percent depending on growing conditions.

3.9 G.E. Likens and F. H. Bormann, *Chemical Analyses of Plant Tissue from the Hubbard Brook Ecosystem in New Hampshire,* Bulletin No. 79, Yale

University: School of Forestry, (1970).

3.10 *Wood Handbook,* U.S. Department of Agriculture Handbook No. 72, (1974), pp. 2-5.

3.11 *Webster's Third New International Dictionary,* (Springfield: C. & G. Merrimam, 1961).

3.12 There is only a slight temperature dependence in the range of typical ambient temperatures; equilibrium moisture contents are only 5 to 10 percent (not percentage points) lower at 90°F than at 30°F for the same relative humidity.

4.1 Since exhaled air is somewhat depleted in oxygen one might expect human breath could have an extinguishing effect on fire by depriving it of the more oxygen-rich air around it. But in fact exhaled air still has at least 75 percent of its original oxygen; so this is not likely to be an important factor. (Mouth-to-mouth artificial respiration would be ineffective if exhaled air did not still have significant oxygen content.)

4.2 It is in this process of new bond formation that energy is released. For instance, a hydrogen atom and an oxygen atom are strongly attracted when they are near each other. The force pulls them together violently, so when they arrive at their normal bonded separation, they have a large amount of excess speed and hence, energy. If this energy remains in the form of motional (kinetic) energy, it will eventually be shared with the other molecules in the gas, and that means that the gas's temperature will have been increased (flames are hot). Some of the energy may also be emitted as radiation (visible, ultraviolet or infrared). These are the major types of energy released in flames — sensible heat (which is the motional energy of molecules) and radiation. A little energy is also in the form of latent heat since water vapor is one of the products of combustion.

4.3 The ranges widen a little at both ends if the fuel-air mixture is hotter, for then a little less heat need be supplied from the chemical reactions in order to propagate the flame.

4.4 Gas ranges, clothes, dryers, furnaces and water heaters have what are called premixed flames - the fuel and air are mixed together before they reach the combustion zone.

4.5 This condition is approximate since the velocity of the wind will not be uniform or constant due to turbulence, and since the flame velocity in a flammable gas depends on the oxygen concentration, which also is not uniform or constant.

4.6 There are "trick" birthday cake candles which are very hard to blow out. I suspect they have imbedded in their wicks small grains of a substance which, like explosives, has the oxygen needed for its combustion built into the material. Thus, when the ignition temperature of the substance is reached, it all burns at once producing intense (high temperature) heat for a short time. After the candle flame is blown out, the wick is consumed downward by smoldering until one of the special grains is encountered; the temperature of the glowing wick (burning wick "charcoal") ignites the grain which then ignites the unburned gases (smoke), and the candle flame is then back again. (For a discussion of why the glowing wick itself in a blown out candle does not ignite the smoke, see page 29.)

4.7 In the commercial manufacture of charcoal, great care must be exercised to avoid it spontaneous combustion. After pyrolysis is complete, th charcoal must be cooled and kept in an oxygen-free atmosphere for a few days before it is safe for shipment.

4.8 The intensity of surface energy absorption is roughly the same for any size piece of flaming wood; if there is a difference, the intensity is larger for larger pieces of wood because their flames are usually thicker. Typical flames are optically thin, so doubling the flames depth over a surface doubles the radiation intensity, other things being equal.

4.9 It is often said that large pieces of (isolated) wood cannot sustain a flame because the heat *loss* is too large. But in fact, the fraction of the combustion heat which escapes from a small piece of wood is, if anything, larger than that from a large piece.

4.10 J. Bryan, "The Fire Hazard of Wood," *Wood* 8 (1943), pp. 260-262..

4.11 It is not true that evaporating the water takes more energy than the wood has. The wettest green woods are about half water by weight; the water's evaporation requires only about 12 percent of the wood's energy in this worst case. See Figure 3-4.

4.12 Special chemical additions can produce various flame colors, but here we are speaking of pure organic fuels.

4.13 The spectrum of the carbon-particle emission is not quite that of a black body radiator at the same temperature, but is shifted a little towards the blue — the color temperature is about 100°C higher than the true temperature. This is because the particles are much smaller than visible wavelengths. See *Flames,* by A. G. Faydon and H. G. Wolfhard, (London: Chapman and Hall, 1960), p. 231.

4.14 The color, in this case, is due mostly to light scattering, the same effect that makes the sky blue.

4.15 Smoke compositions and the physics of smoke formation are discussed in "Air Quality and Smoke from Urban and Forest Fires," (Washington: National Research Council, NAS, 1976).

5.1 A more technical discussion of chimney design may be found in the following three publications: A. Kinkead, "Gravity Flow Capacity Equations for Designing Vent and Chimney Systems," *Proceedings of the Pacific Coast Gas Association* 53 (1962); R.L. Stone, "Fireplace Operation Depends upon Good Chimney Design," *ASHRAE Journal* (February 1969); *ASHRAE Handbook and Product Directory, 1975 Equipment Volume* (New York: American Society of Heating, Refrigerating and Air-Conditioning Engineers, 1975), Chapter 26.

5.2 A distinction is sometimes made between "static" and "total" pressure inside chimneys and ducts; total pressure includes the "velocity pressure" which is what one's hand would feel when facing into the flow as when you stick it out the window of a moving car. It is the static pressure which is meant in the above context of "draft."

5.3 This is often called "theoretical" draft. I prefer the word "static" because it is more accurately descriptive. There is nothing inherently more theoretical about draft when gases are stationary (static) than when they are moving.

5.4 The average molecular weight of the gas mixture also affects draft, but the effect is small (no more than a few percent).

5.5 The flue-gas flow (dimensions of mass/time) is calculated as

$$A\sqrt{2gh\, d_f (d_a - d_f) / k}$$

where A = area of flue

g = gravitational acceleration

h = chimney height

k = system resistance coefficient (see Table 5-1)

d_f = density of flue gas (depends on temperature)

d_a = density of outdoor air

5.6 Chimney flow is proportional to the square root of its height, other things being equal.

5.7 Taller chimneys also have a higher resistance to flow, but as indicated in the previous example, the resistance of even a very long straight section of flue passage is usually negligible compared to the whole-system resistance.

5.8 Industrial chimneys are often very high. The principle reason is pollutant dispersal. It is also true that additional height increases the flow capacity more for large diameter industrial stacks than for domestic chimneys because there is less cooling of the flue gases when the chimney is used at its full capacity.

5.9 The worst case is probably a one-story flat roofed house. With a chimney extending 4 feet above the roof line, the total vertical rise of the vent, from the stovepipe collar, can hardly be less than 10 feet.

5.10 The reasons are (1) that a masonry chimney has a larger exterior surface area (for the same interior area), and (2) that the resistance to heat transfer is dominated by the surface phenomena of radiation and convection, not by conduction through the material.

5.11 W.S. Harris and R. J. Matin, "Heat Transmitted to the I=B=R Research Home from the Inside Chimney," *ASHVE Transactions* Vol. 59, (1953), pp. 97-112.

5.12 Highly reflective surfaces can also contribute to keeping the outside cool.

5.13 For instance, by the Underwriters Laboratory.

5.14 A more quantitative account of some of the subjects discussed in this section can be found in A.G. Wilson, "Influence of the House on Chimney Draft," *ASHRAE Journal,* (December, 1960).

5.15 The size of a house's stack effect can be as large as 0.3 inches of water in terms of pressure difference or draft.

5.16 This discussion presumes the interior of the house does not have significantly air-tight partitions, e.g., that bedroom doors are not air-tight. If either the room the stove is in, or the room with the open window, can be effectively sealed off pressure-tight from the other room, then the effects are modified. It is also assumed that the house and the chimney have about the same height.

5.17 If the component of the wind velocity into the chimney is on the order of, or larger than, the flue-gas velocity.

5.18 This is approximately described by Bernoulli's equation. Specifically, the pressure decrease in moving air compared to still air is $\frac{1}{2}dv^2$ where d is the air's mass density and v is its speed.

5.19 These estimates are from R. L. Stone, "The Role of Chimneys in Heat Losses from Buildings," a paper given at The Annual Meeting of ASHRAE, Boston, Mass., June, 1975.

5.20 *ASHRAE Handbook and Product Directory, 1975 Equipment Volume,* (New York: American Society of Heating, Refrigerating and Air-Conditioning Engineers, 1975), p. 26-13.

5.21 Freely paraphrased from *Using Coal and Wood Stoves Safely!,* National Fire Protection Association, publication No. HS-8, 1974.

6.1 C. G. Segeler, Editor, *Gas Engineers Handbook,* (New York: Industrial Press Inc., 1969), pp. 12/156 and 12/342.

6.2 Some of the fuel's high heat value is converted to latent heat (potential energy of water vapor). See Chapter 2.

6.3 Some kitchen exhaust fans do not contribute to air exchange; they filter the air and recirculate it into the kitchen.

6.4 *ASHRAE Handbook of Fundamentals,* (New York: American Society of Heating, Refrigerating and Air-conditioning Engineers, 1972), p. 337, Table 1.

6.5 Not quite all of the high heat content of the fuel is recovered even in this ideal situation since not quite all of the water vapor will have condensed, releasing its heat of condensation. If the European system (based on low heat value) for reporting energy efficiencies were used, efficiencies in this ideal case would exceed 100 percent by a few percentage points.

6.6 C. B. Prakash and F. E. Murray, "Studies on Air Emissions from the Combustion of Wood-waste," *Combustion Science and Technology* 6 (1972), pp. 81-88.

6.7 The high reflectivity and low emissivity of shiny metal finishes (e.g., chrome plate) might also be useful for keeping combustion temperatures high. A chrome plated stove interior would reflect most of the fire's emitted radiation right back into the fire, but the surface would very quickly become dirty, losing its high reflectivity. Chrome on the outside of a stove would suppress infrared radiation from the stove into the room, decreasing the stove's heat output substantially, and it would also raise the temperature inside the stove. Shiny surfaces on one or both of the inner surfaces of a double-wall construction with an air gap in between would have the same kinds of effects.

6.8 The resistance to heat transfer is dominated by the surface phenomena of radiation and convection, not by conduction through the metal.

6.9 Based on measurements by the author.

6.10 B. A. Landry and R. A. Sherman, "The Development of a Design of Smokeless Stove for Bituminous Coal," *Transactions of the ASME* (January, 1950), p. 9.

6.11 Coal stoves have been tested, and an optimum design, including a secondary air inlet, was developed. See B. A. Landry and R. A. Sherman, "The Development of a Design of Smokeless Stove for Bituminous Coal," *Transactions of the ASME,* 72 (1950), p. 9.

6.12 There are some high temperature sealants which are purported to be much better, manufactured by Industrial Gasket and Shim Co., Inc., P. O. box 368, Meadow Lands, Penn.

6.13 J. L. Threlkeld, *Thermal Environmental Engineering,* 2nd ed., (Englewood Cliffs: Prentice-Hall, Inc., 1970), p. 24.

6.14 W. L. Wolf, Editor, *Handbook of Military Infrared Technology,* (Washington, D.C.: Office of Naval Research, Department of the Navy, 1965), pp.76-84.

6.15 Why is it then that steam radiators in houses are often painted silver? I know no good answer. Any other (non-metallic) color would make them more efficient radiators.

6.16 Marcus Bull, "Experiments to determine the comparative quantities of Heat evolved in the combustion of the principal varieties of Wood and Coal Used in the United States for fuel; and, also, to determine the comparative quantities of Heat lost by the ordinary apparatus made use of for their combustion," *Transactions of the American Philosophical Society* III, New Series, (1830), pp. 1-63.

6.17 L. E. Seeley and F. W. Keator, "Wood-Burning Space Heaters," *Mechanical Engineering* 62 (1940), p. 864.

6.18 F. B. Rowley and J. R. Allen, "Tests to Determine the Efficiency of Coal Stoves," *Transactions of the ASHVE* 26 (1920), p. 115.

6.19 B. A. Landry and R. A. Sherman, "The Development of a Design of Smokeless Stove for Bituminous Coal," *Transactions of the ASME,* Vol. 72 (1950), p. 9.

6.20 G. A. W. van Doornum, A. C. Bonapace and H. T. Surgnit, *Smokeless Combustion of Bituminous Coal in Domestic Appliances,* Information Circular No. 17, (Pretoria: Fuel Research Institute of South Africa, 1972).

6.21 R. W. Rowse and W. C. Moss, "Domestic Boilers and Stoves Using Solid Fuel," *Journal of the Institution of Heating and Ventilating Engineers* 18 (1950), pp. 32-78.

6.22 L. L. Fox, "Efficiency of Domestic Space-Heating Appliances Using Solid Fuel," *Journal of the Institute of Fuel,* (November 1952), p. 267.

6.23 J.W. Shelton, "An Analysis of Woostove Performance," *Blair and Ketchum's Country Journal* 3 No. 10 (October, 1976), pp. 47-50.

6.24 These efficiencies are computed using the American convention of assessing the wood's energy content as its high heat value, rather than the European convention of using the low heat value. Use of the European convention would result in efficiencies 4 to 7 percentage points higher. This is one reason the efficiencies of Jotul stoves appear to be so high in Jotul promotional literature. (Another reason may be that many of the tests are conducted using liquid fuels, not wood).

6.25 The heat given off by the stovepipe also tends to bring energy efficiencies closer together; the less effective heat transfer is in a store itself, the more use will be made of the stovepipe for heat transfer. (Measured stove efficiencies usually include the heat given off by a few feet of stovepipe.)

6.26 J. Dollgust, *Alte and Neve Bavernstube,* (Munich: F. Bruckmann, 1962). Jaan Jaakson, "Wood Burning Brick Stove," Wood Burning Quarterly, 1 No. 2 (Fall, 1976), p. 39.

7.1 This presumes that adding very small amounts of wood every ten minutes or so is not a practical way to achieve low powers. In non-airtight stoves, minimum power is limited by the amount of air leakage.

7.2 In some stoves, the thermostat's temperature is influenced also by the stove's surface temperature. This can have some advantages in evening out room temperature oscillations by providing some anticipation of the stove's output.

7.3 These heat output records were measured by the author in a calorimeter room with a volume of about 500 cubic feet and very low heat capacity. Blowers force about 700 cubic feet of air per minute through the room carrying away the stove's heat output. The temperature difference between the air entering and leaving the room is proportion to the stove's heat output. (See Appendix 4 and J. W. Shelton, " An Analysis of Woodstove Performance," *Blair and Ketchum's Country Journal* 3 No. 10 (October, 1976), pp. 47-50.

7.4 There can be a difference in the velocity of the combustion air. If the draft is reduced, the air inlet damper must be opened further to achieve the same flow. This will generally result in a lower air velocity over a larger inlet area. Velocities can affect the efficiency of air use in combustion.

7.5 The mass of a heavy stove (e.g., Fisher Papa Bear) is about 400 lbs. Most of this is steel or cast iron. Iron and steel have a heat capacity of .12 Btu per pound. If the stove's average temperature is 400°F at a typical power of, say, 30,000 Btu/hr, the heat stored in the stove is 400 lbs x .12 Btu per pound x (400-70)°F = 16,000 Btu, which is the equivalent of ½ hour of output. For a light stove weighing say 50 lbs, the stored heat is roughly four minutes of 30,000 Btu/hr output. Thus the initial lag, or in general, the moderating time, of a heavy stove is significant.

7.6 Based on the author's observations.

7.7 Landry and Sherman found this to be the case in coal burners; B.A. Landry and R.A. Sherman, "The Development of a Design of Smokeless Stove for Bituminous Coal," *Transactions of ASME* 72 (1950), p. 9.

8.1 $.20 \frac{Btu}{lb°F} \times 4000\ lb. \times 20°F = 16000\ Btu$

8.2 $.20 \frac{Btu}{lb°F} \times 4000\ lb \times (85° - 45°F) = 32000\ Btu$

8.3 If there is thermal insulation over the masonry on the living-space side, that masonry material will not contribute to temperature leveling. The material must be in good thermal contact with the inside of the house.

8.4 The air absorbs small amounts of the radiation directly - up to about 15 percent. The exact amount depends on water vapor and carbon dioxide concentrations in the air and on the average path length traveled by the radiation through the air.

8.5 Based upon the author's measurements.

8.6 This estimate is based on data in W.S. Harris and J.R. Martin, "Heat Transmitted to the I=B=R Research Home from the Inside Chimney," *Transactions of ASHVE* Vol. 59, (1953), pp. 97-112.

8.7 This is due more to a decrease in heat transfer efficiency than a decrease in completeness of combustion. A 20 degree decrease in the temperature of the combustion air results, approximately, in a 20 degree decrease in flue-gas temperature. Thus the amount of heat transferred to the room is less. The loss in heat output probably exceeds the amount of energy it would have taken to preheat the air up to room temperature.

9.1 The pamphlet, *Using Coal and Wood Stoves Safely!* NFPA No. HS-8, 1974, is the best single source on safety. It is available (in 1976) for $2.00 from NFPA, 470 Atlantic Avenue, Boston, Massachusetts 02210. Also relevant are NFPA Standard No. 211 *Chimneys, Fireplaces and Vents* and NFPA Standard No. 89M *Heat Producing Appliance Clearances,* each of which also costs $2.00.

9.2 G.C. McNaughton, "Ignition and Charring Temperatures of Wood," *Wood Products* 50 Forest Products Laboratory Report No. 1464. pp. 21-22. "Wood, Ignition of, at Low Temperatures," Card C60, Card Data Service, Underwriters' Laboratories, Inc., (Chicago: 1945).

9.3 The average must take into account the properties of emitted radiation by giving extra weight to high temperature regions.

9.6 This difference between the required clearances for interior versus exterior chimneys may be based on the higher heat losses expected from the three exposed sides of an exterior chimney.

9.4 Equivalent heat and corrosion resistant materials may be used, but seldom are.

9.5 This gauge is required in pipe with a 10 inch diameter or less. In larger diameter pipe, which is very rarely used in any residential application, thicker steel is required.

9.7 Also see "Smoke Detectors," Consumer Reports 41 (Oct., 1976), pp. 555-559.

10.1 A heat pipe is a sealed tube which transfers heat very effectively from one end to the other by boiling a fluid at one end and condensing it at the other. A wick inside the tube returns the condensate to complete the cycle.

11.1 S. Konzo and W.S. Harris, *Fuel Savings Resulting From Closing of Rooms and From Use of a Fireplace,* Bulletin 41, No. 13 (November 16, 1943); University of Illinois Engineering Experiment Station Bulletin Series No. 348.

11.2 L.L. Fox, "Efficiency of Domestic Space-Heating Appliances Using Solid Fuel," *Journal of the Institute of Fuel* 25 (1952), pp. 267-276. (British); R. H. Rouse and W.C. Moss "Domestic oilers and stoves using solid fuel," *Journal of the Institution of Heating and Ventilating Engineers* 18, British (1950), pp. 32-78, unpublished measurements by Richard L. Stone; unpublished measurements by the author.

11.3 T. Kollock, "Efficiency of Wood in Stove and Open Fireplaces," *Forest, Fish and Game,* 3 (1911), pp. 95-97.

11.4 R.L. Stone, "Fireplace Operation Depends on Good Chimney Design," *ASHRAE Journal,* (February, 1969). Also *ASHRAE Guide and Data Book, 1975 Applications Volume,* (New York: American Society of Heating, Air-Condition and Refrigerating Engineers), pp. 26.24-26.26.

11.5 Count Rumford, "Of Chimney Fireplaces," *The Complete Works of Count Rumford,* (Boston: The American Academy of Arts and Sciences, 1875), pp. 484-557.

11.6 Paul R. Achenbach, "Physics of Chimneys," *Physics Today* (December 1949), pp. 18-23; Vrest Orton, *The Forgotton Art of Building a Good Fireplace,* Yankee Inc. (1909), p. 44; David Havens *The Wood Burner's Handbook, (New York: Harpswell Press, 1973), p. 93.*

11.7 $.24 \frac{Btu}{lb°F} \times 1000\ lb \times (70° - 30°F) = 9600\ Btu$ *(per hr.)*

11.8 An exception is installations where chimney flow reversal can occur when the chimney cools down, or, equivalently, chimneys which are not self-starting.

11.9 There is still some chemical energy loss, and latent heat loss too, unless humidification is a desired objective; some heat may also conduct out of the fireplace through the back if it is in an exterior wall.

11.10 *Glossary of Terms Relating to Heat Producing Appliances,* NFPA No. 97M, (Boston: National Fire Protection Association, 1972).

11.11 Based on measurements of British coal stove's burning anthracite, (R.H. Rowse and W.C. Moss, "Domestic Boilers and Stoves Using Solid Fuel," *Journal of the Institution of Heating and Ventilating Engineers* 18 (1950), pp. 32-78, and on measurements by the author.

12.1 Creosol and cresol are two distinct compounds.

12.2 *Fire Protection Handbook,* 13th ed. (Boston: National Fire Protection Association, 1969), p. 967.

12.3 Manufactured by Thompson and Anderson, Inc. 446 Stroudwater Street, Westbrook, Maine 04092.

13.1 The required annual saving in order to break even, expressed as a fraction of the initial investment, and expressed in terms of current dollars, is

$(1 + i)^n i/((1 + i)^n - 1)$

where n is the life expectancy of the system in years and i is the annual interest rate obtainable in quaranteed savings accounts. Some values of this expression are as follows:

i	5 years	10 years	20 years	30 years	50 years
2%	.21	.11	.061	.045	.032
5%	.23	.13	.080	.065	.056
8%	.25	.15	.10	.089	.082
11%	.27	.17	.13	.115	.111

E.G., if the stove and chimney are expected to last 20 years, and the annual fuel savings were invested in a guaranteed account earning 8% interest, then annual savings must exceed .10 (from the above table) of the initial cost of the system, in order to break even economically.

13.2 The energy content of LP gas varies from about 91,500 to 102,000 Btu per gallon, depending on its composition. In Figure 13-4, the value of 97,000 was assumed.

13.3 These furnace or boiler efficiencies should take into account all heat losses from the furnace or boiler and from plenums, ducts and pipes which do not contribute to the desired heating of the house. Most efficiency ratings of central heaters do not include these losses.

13.4 High, average and low efficiencies for stoves are assumed here to be about 65, 50, and 40 percent respectively.

13.5 High, average and low energy contents for wood are assumed here to be about 30, 22.5 and 17 million Btu (high heat value) per cord respectively.

13.6 High, average and low conventional system efficiencies; are assumed here to be 85, 65, and 50 percent respectively.

13.7 Inflation is likely to result in price increases for both conventional fuels and wood. If both prices rise at the same rate, the relative economics is unchanges. However, the dollar amount of the minimum break-even annual savings will have to increase each year to keep pace with inflation, as in this discussion all quantities are expressed in terms of the value of dollars at the time of the stove purchase. If the costs of fossil fuels and electricity increase faster than the cost of wood, wood will become more economic.

13.8 In general, the annual fuel-cost savings is

$(1 - R) f C$

where f is the fraction of the total heat which is supplied by the wood system, R is the ratio of the actual cost of wood (per cord) to the break-even cost of wood, and C is the conventional annual heating cost (if no wood heat were used). For instance, if without wood heat one now pays $700 per year for electric house heating (C = $700), if 75 percent of the heating needs would be met by a wood stove (a small house with a stove in use much of the time) (f = 0.75), if the break-even line in Figure 13-3 is the line marked +1 and electricity costs 4 cents per kilowatthour (then the break-even cost of wood is $175 per cord), and if wood is available for $60 per cord (then R = 60/175 = 0.34), then the annual fuel-cost savings would be

$(1 - 0.34) \times 0.75 \times 700 = \$347$.

Glossary

AIRTIGHT STOVES

A stove which is sufficiently airtight that its performance would be indistinguishable from a stove of the same overall design which was literally airtight. In a literally airtight stove, all the air which enters it goes in through the air inlet(s). Generally, non-airtight stoves are less energy efficient and their level of heat output is less controllable than airtight stoves.

BAFFLE PLATE

A partition inside a stove to control the direction of flow of combustion air, flames, or flue gases.

Btu or BRITISH THERMAL UNIT

A unit for measuring energy, equal to the amount of energy necessary to increase the temperature of 1 pound of water by 1 degree Fahrenheit.

CHIMNEY CONNECTOR

That portion of a stove's venting system between the stove and its chimney. Chimney connectors are usually made of stovepipe.

CIRCULATING STOVES

A stove with an outer jacket, beyond the main structure, with openings at the bottom and top so that air can circulate. Circulating stoves transfer more than half of their heat output as heated air. (See Radiant Stove.)

COMBUSTION PRODUCTS (OR PRODUCTS OF COMBUSTION)

The products of the chemical reactions which constitute combustion, typically consisting of carbon dioxide, water vapor and small amounts of some incompletely burned organic compounds.

CORD

A common measure of firewood and pulpwood, equal to the amount of wood in a carefully stacked (parallel) pile of wood which is 4 feet high, 8 feet wide and 4 feet deep. The amount of solid wood in this 128 cubic foot pile is usually estimated to be between 80 and 90 cubic feet.

CREOSOTE

Chimney deposits originating as condensed organic vapors or condensed tar fog. Creosote is often initially liquid, but may dry and/or pyrolyze to a solid or flaky form.

DAMPER

A valve (usually a movable or rotatable plate) for controlling the flow of gases and/or the draft in stoves and in stovepipe.

DIFFUSION FLAME

A flame into which the needed oxygen diffuses, as opposed to a premixed flame, which is the burning of an already thoroughly mixed combination of fuel gas and oxygen.

DRAFT

The difference in air pressure inside and outside a chimney or stove. Draft also is used to mean air flow.

DRAFT REGULATOR

A device designed to limit excessive draft in a fuel-burning appliance by letting air into the venting system. Draft regulators are usually installed in chimney connectors.

DRUM OR BARREL STOVE

A stove made of a steel drum or barrel which was intended as a container. The conversion to a stove usually involves cutting holes and bolting on a door, a flue collar and legs.

EFFICIENCY OR ENERGY EFFICIENCY

As applied to a wood stove, the fraction (or percentage) of the chemical energy in the wood which is converted to useful heat by the stove, including the heat from an average amount of exposed stovepipe (about 6 feet).

EXCESS AIR

Air admitted to a burner which is in excess of the amount theoretically needed for complete combustion.

FIREBRICK

Brick capable of withstanding high temperatures, such as in furnaces and kilns. Firebrick is often used to mean only "hard" or "dense" firebrick as distinguished from "soft" or 'insulating" firebrick.

FLAME VELOCITY

The velocity at which a flame (or flame front) would propagate through a flammable mixture of a fuel gas and air; or the rate of propagation of the combustion reaction zone.

FLUE COLLAR

The part of a stove to which the chimney connector or chimney attaches.

FLUE GASES

The gaseous combustion products from fuel-burning appliances, plus whatever air is mixed with them - i.e., the gases in an operating flue.

GREEN WOOD

Undried freshly cut wood from a live tree.

HEARTWOOD

The wood in the center of a tree extending out to the sapwood. The heartwood no longer participates in the tree's life processes. It is usually darker in color and more resistant to decay. Young trees have no heartwood.

HIGH HEAT VALUE

The chemical energy per unit mass of wood. The high heat value represents the amount of chemical energy released if one mass unit of wood is completely burned. Some of the energy will be in the form of latent heat (a form of potential energy) unless the water vapor in the combustion products condenses.

IGNITION TEMPERATURE

The minimum temperature of a flammable mixture of gases at which it can spontaneously ignite.

INFRARED RADIATION

The invisible (and harmless) radiation emitted by any hot object, which is converted into heat when it

is absorbed.

INSULATING BRICK

Low density (high porosity), low thermal conductivity firebrick intended for use in kilns and furnaces to insulate them, reducing heat losses. Its conductivity and its heat storage capacity are both 1/5 to 1/3 that of hard firebrick.

LATENT HEAT

The potential energy in water vapor which is converted into (sensible) heat when the vapor condenses. A pound of water vapor at room temperature has about 1050 Btu of latent heat.

LINER OR STOVE LINER

A layer of metal or brick placed immediately adjacent to a side or bottom of a stove, intended either to protect the main stove structure from getting too hot, or to insulate the combustion chamber, making it hotter and thus promoting more complete combustion. Liners are usually designed for easy replacement.

LOW HEAT VALUE

The chemical energy per unit mass of wood, minus the latent heat in the water vapor which would result from the complete combustion of the wood. The low heat value is the amount of (sensible) heat produced when a unit mass of wood is burned completely and no resulting water vapor condenses.

NET EFFICIENCY OR NET ENERGY EFFICIENCY

As applied to a wood stove, the net useful heat added to a house, divided by the chemical energy in the wood burned. Net efficiencies differ from efficiencies if the stovepipe is extra long or extra short, if useful heat is given off by the chimney, or if use of the stove causes extra outdoor air to be drawn into the house.

OPEN OR OPENABLE STOVE

A stove designed to be operable with open doors, exposing the fire to direct view. An example is the Franklin stove.

OVENDRY WOOD

Wood which has been dried to constant weight at about 215 degrees Fahrenheit and low humidity. Ovendry wood is defined to have zero moisture content.

PRIMARY COMBUSTION

The burning of solid wood and some of the combustible gases, which takes place in that portion of the stove where the wood is. The distinction between primary and secondary combustion is somewhat artificial and hence not always clear.

PYROLIGNEOUS ACID

The acidic brown aqueous liquid obtained by condensing the gaseous products of pyrolysis of wood. Pyroligneous acid is the same as creosote in its wettest form.

PYROLYSIS

The chemical destruction of wood by the action of heat alone, in the absence of oxygen and hence without burning. The products of pyrolysis are gases, tar fog and charcoal.

RADIANT STOVE

A stove without the outer jacket that circulating stoves have. Radiant stoves transfer more than half their energy output in the form of radiation.

SAPWOOD

The wooa extending from the heartwood out to the bark in a tree. Sapwood participates in transporting sap up and down a tree. Sapwood usually is lighter in color than heartwood and is usually more susceptible to decay.

SEASONED WOOD

Wood which has lost a significant amount of its original (green) moisture. The term has no more specific (and universally accepted) meaning.

SECONDARY COMBUSTION

The burning of the combustible gases and smoke which are not burned in primary combustion.

SECONDARY COMBUSTION CHAMBER

The place where secondary combustion occurs.

SENSIBLE HEAT

Energy in the form of random motions of molecules, atoms and electrons. Sensible heat is the form of energy that is transferred from a warm object to a cooler object when they are in direct physical contact. Sensible heat can be sensed or felt directly by human skin. Infrared radiation (sometimes called radiant heat) is a completely different energy form which is transformed into sensible heat when it is absorbed by a surface such as skin. Latent heat is a form of potential heat which also cannot be sensed by humans, but can be converted into sensible heat.

SMOKEPIPE DAMPER

A damper installed in a smokepipe to regulate flow and draft.

STACK EFFECT

The effects resulting from the warm air in buildings on a cold day being relatively buoyant, just as are the flue gases in a chimney or stack. Effects include pressure differences between inside and outside the building, airflow into the building in the lower stories and airflow out of the building in the upper portions.

STOVEPIPE

Single-walled metal pipe and fittings intended primarily to be used for chimney connectors but also sometimes used for chimneys.

SUSTAINED YIELD

The rate at which wood can be harvested from an area forever, without decreasing the area's productivity. Sustained yield harvesting involves taking wood at a rate no larger than the rate at which new wood is growing.

THIMBLE

A device to be installed in combustible walls, through which stovepipe passes, intended to help protect the walls from igniting due to stovepipe heat. A thimble by itself is not usually adequate (see text). The simplest thimbles are simply metal or fire-clay sleeves or cylinders. The common thimbles are double-walled air-ventilated metal cylinders.

Appendix I

HUMAN TEMPERATURE SENSATION

If one grasps a pot or rack in a hot oven, it hurts. Is this because it is so hot? This reason cannot be sufficient because there are hot things which, when touched, do not burn. Aluminum foil in a hot oven can be touched with no accompanying feeling of hotness. Red hot coals such as might accidentally fall out of a stove, can be picked up and juggled back into the fire with bare hands with no ill effects (but be careful!). The foil is as hot as the pot, and a glowing coal is much hotter — yet they do not *feel* hot, and they do not necessarily burn.

Skin evidentally is not a very good thermometer. This is principally due to the fact that there is a lot of flesh around the very few cells in skin which are the body's temperature sensors. Each sensor cell actually does a fairly good job of sensing its own temperature in the skin; the problem is the temperature of the sensor cells is not necessarily the same as that of the surface being touched.

A hot, heavy metal pan is capable of heating your skin to a painful temperature. Because of its heaviness, it has a lot of heat stored in it, and because it is a metal, that heat can conduct quite readily into your skin, making it nearly as hot as the pan.

Hot aluminum foil does not burn when touched; it is so thin that despite the fact that aluminum is an excellent conductor of heat, the total amount of heat in the thin foil is so small that when distributed in your skin, your skin is only warm, not hot. A hot glowing coal can be touched momentarily without burning because 1) its capacity to store heat is not great, 2) it is a poor conductor of heat, and 3) a thin layer of ash and/or cooler charcoal is often between your finger and the coal while you're touching it.

THERMAL COMFORT

Air temperature is, of course, the most important aspect of a person's environment affecting thermal comfort. But three other factors can also influence one's feeling of adequate warmth: the air's humidity, its motion, and the amount of radiant energy traveling through surrounding space.

Most people have experienced the additional discomfort of high humidity on a warm day; the air feels hotter when it is humid, even though its temperature (as measured more objectively with a thermometer) is the same. At room temperature, the effect is less often noticed, but is there. Many experiments have been done using a set of rooms with very carefully controlled temperature and humidity. As human subjects move from room to room, they state whether each new room feels warmer, the same, or cooler than the previous room. From these studies, it is found that the following three sets of conditions all give the same feeling of warmth - they all feel like the same temperature:

Actual Temperature (°F)	Relative Humidity (%)	Sensation of* Temperature
71.5	0	70
70.0	50	70
68.5	100	70

In other words, a person who is comfortable at 70°F and 50 percent relative humidity will be just as comfortable at a slightly cooler temperature if the humidity is higher. He or she will also be just as comfortable at a higher temperature, if the air is made appropriately drier.

Alternatively, if the actual temperature in the test rooms is the same and the humidity is varied, then a person's subjective temperature sensation varies as follows:

Actual Temperature (°F)	Relative Humidity (%)	Sensation of* Temperature
70	0	68.5
70	50	70.0
70	100	71.5

*This subjective or "effective" temperature scale is made to agree with the actual temperature when the relative humidity is 50%.

In other words, humid air feels warmer than dry air when the actual temperatures are identical.

The explanation for this effect is related to evaporation of water from the skin. Evaporation is a cooling process - molecules of water need extra energy to break away from their neighbors and become water vapor. This energy comes partly from the air over your skin, but mostly from the skin itself. Having lost energy, it is cooler (unless the energy is replaced from some other source). When the air is dry, evaporation is enhanced and your skin is losing more energy; when the air is more humid, evaporation rates become less and your skin is losing less energy.

The decreased energy loss from your skin in humid air can be compensated by decreasing the temperature of the air. As indicated previously, in the vicinity of 70°F, an increase in humidity by 50 percentage points can be balanced by a 1.5°F decrease in temperature.

It does not follow that humidification saves energy. In a house with a humidifier, the thermostat can be set somewhat lower than in a house without humidification and the feeling of air warmth in both houses will be the same. However, the energy necessary to humidify the air is usually far greater than the decrease in the sensible heat loss from the house.[1] This is true regardless of the type of humidification including such simple humidifiers as house plants. The net effect is almost always an increase in total energy consumption (usually of the furnace).

Movement of air also influences thermal comfort. Increased air speed enhances both conduction of heat out of the skin and evaporation. At typical room temperatures and relative humidities, an air velocity of 3 feet per second makes the air feel about 4°F cooler than stationary air at the same actual temperature.[2]

The effect of infrared radiation on thermal comfort is especially important since so much of the heat output of stoves is radiant. Thermal comfort depends on skin temperature. One can be perfectly comfortable in cool dry air despite the skin's loss of energy by conduction (to the cool air) and evaporation (into the dry air) as long as enough infrared radiation is being absorbed by the skin. Very few home heating systems are designed this way, with cooler air temperatures being compensated for by higher infrared radiation intensity. One reason is the difficulty of having the radiation coming from all directions, so that an occupant will be equally warm on all sides. In a house heated only with an open fireplace, one's back, away from the fire, can be much cooler than one's front. This is due both to unequal amounts of radiation, coming from the two directions and to drafts. In a typically snug house heated with a closed stove, there is little if any discomfort from such temperature contrasts.

In winter, when exterior walls, and windows are cooler, the infrared environment is less intense than in warmer weather. The air is also dryer. Thus one's skin loses more energy by both radiation and evaporation. Hence, air temperatures need to be a little higher (particularly in rooms with exterior walls) to produce the same sensation of human comfort. Since thermostats regulate air temperature, they need to be set a little higher in the colder weather to provide the same degree of thermal comfort, other things being equal (e.g., clothing).

1. J. W. Shelton, "The Energy Cost of Humidification," ASHRAE Journal, January (1976), p. 52-55.

2. The effect is not linear. See *ASHRAE Handbook of Fundamentals,* (1972), p. 142.

Appendix 2

MOISTURE CONTENT SCALES

There are two common ways of reporting moisture content in wood. In this book, and in most technical writings, moisture content is always based upon the ovendry weight of the wood:

$$\text{Moisture content (ovendry wood basis)} = \frac{\text{weight of moisture removed in oven drying}}{\text{weight of ovendry wood}}$$

Using this scale, wood which is half water by weight has a moisture content of 100 percent.

A second way to report moisture contents is based on the weight of the moist wood:

$$\text{Moisture content (moist wood basis)} = \frac{\text{weight of moisture removed in oven drying}}{\text{initial weight of wood, including its moisture}}$$

Using this scale, wood which is half water by weight has a moisture content of 50 percent.

These different scales for reporting moisture contents are another possible cause for discrepancies among lists of energy contents. 20 percent moisture content on an ovendry wood basis is the same as 25 percent moisture content on a moist wood basis. To facilitate comparisons between writings using the two conventions, Table A2-1 gives conversions.

MOISTURE CONTENT ON AN OVENDRY-WOOD BASIS PERCENT	MOISTURE CONTENT IN EITHER SCALE PERCENT	MOISTURE CONTENT ON A MOIST-WOOD BASIS PERCENT
0%	0%	0%
5.3	5	4.8
11.1	10	9.1
17.6	15	13.0
25.0	20	16.7
33.3	25	20.0
42.9	30	23.1
53.8	35	25.9
66.7	40	28.6
100.0	50	33.3
150.0	60	37.5
233.0	70	41.2
Infinite	100	50.0
——	150	60.0
——	200	66.7
——	250	71.4

TABLE A2-1. Conversions between moisture contents as expressed in the moist wood and oven-dry wood scales. To use the table for either conversion, find the value to be converted in the center column. Then to convert from dry to moist basis read to adjacent number in the right column. To convert from moist to dry, read the adjacent number in the left column. If m and d represent the moisture contents on the moist-wood and dry-wood bases respectively, then $m = d/1+d)$, and $d = m/(1-m)$.

Appendix 3

PYROLYSIS[1]

A thin slab of wood which is heated slowly and gradually, such as in an oven, so that all of the piece is at the same temperature, will undergo the following changes (it is presumed for the moment that combustion does not occur): For temperatures up to about 212°F (100°C) the major effect of heat is to drive out moisture which was present in the wood. This consumes heat energy (such processes are called endothermic) since evaporation of water requires energy. Each pound of water driven off requires about 1150 Btu of heat energy —

about 1050 Btu for the heat of evaporation of water at room temperature, plus about 100 Btu to break the hygroscopic bond between water and wood.

As the temperature increases from 212°F to about 540°F (about 280°C) a small amount of additional water is driven out, and carbon dioxide, carbon monoxide, formic acid, acetic acid, glyoxal and probably many other compounds are evolved out of the wood. The processes are still endothermic — heat energy is consumed, not generated. All of the compounds except water and carbon dioxide are combustible, but the actual mixture, including the water vapor and carbon dioxide, may not be flammable (ignitable) due to excessive dilution with the non-combustible gases.

The bulk of pyrolysis occurs between 540°F (about 280°C) and about 900°F (about 500°C). The reactions evolve heat — they are exothermic. Hence, the temperature of the wood starts to rise spontaneously even in the absence of oxygen. Large amounts of gases are generated. The most abundant ones are carbon monoxide (a component of automobile exhaust), methane (the principle ingredient of natural gas), methonal (wood alcohol), formaldehyde, and hydrogen, as well as formic and acetic acids, water vapor and carbon dioxide (Table A3-1). Wood tar is another product of heating wood, and although the tar molecules do not exist as a vapor, some tar in the form of tiny droplets is carried out of the wood by the moving gases. The mixture is sufficiently rich in combustibles that it is ignitable and serves as the fuel source for the large wood flames. By the time the temperature of the wood reaches about 900°F, pyrolysis is essentially complete and the final solid product is charcoal. If denied oxygen so that it cannot burn, charcoal is stable to very high temperatures — charcoal does not melt until its temperature reaches about 6300°F (3480°C) and its boiling point is about 7600°F (4200°C), temperatures much much higher than the highest temperature achievable in a wood fire.

In a normal wood fire, the wood is not heated so slowly and uniformly that it is all the same temperature, as was assumed above. The inside of a piece of wood may not even be warm when the outside is already fully charred and burning. Additional "secondary" pyrolysis reactions take place. Gases and tar generated below the wood's surface must pass through the surface layers on their way out of the wood. Water vapor can react with carbon in charred layers and become carbon monoxide and hydrogen gases. Carbon dioxide can be similarly transformed into carbon monoxide. These reactions consume heat — they are endothermic. The tar droplets from inside the wood also can react with charcoal, and this reaction evolves heat (it is exothermic).

1. Most of the following information is from F. L. Browne, *Theories of the Combustion of Wood and its Control,* Report No.2136, Forest Products Laboratory, Madison, Wisconsin (1963).

ACIDS
- Formic
- Acetic
- Propionic
- Butyric
- Aconitic ?
- Tricarballylic ?
- Ketoglutaric ?

ALCOHOLS
- Methanol
- Ethanol

CARBONYLS
- Formaldehyde
- Acetaldehyde
- Acetone
- Diacetyl
- Furfural
- Methyl Furfural

HYDROCARBONS
- 3,4-Benzpyrene
- 1,2,5,6-Dibenzanthracene
- 20-Methylcholanthrene

PHENOLS
- Cresols
- Creosol
- Guaiacol
- Guaiacol derivatives
  - 4-Ethyl
  - 4-Propyl
  - 6-Methyl
  - 6-Ethyl
  - 6-Propyl
- Pyrogallol ethers
  - 1-0-Methyl
- 1,3-Dimethyl

PHENOLS (continued)
- 1,3-Dimethyl Pyrogallol derivatives
  - 5-Methyl
  - 5-Ethyl
  - 5-Propyl
- 1-0-Methyl-5-Methyl Pyrogallol
- Veratrole
- Xylenols

OTHERS
- Ammonia
- Carbon dioxide
- Resins
- Water
- Waxes

TABLE A3-1. Some of the compounds which have been reported to be present in hardwood smoke. Cigarette smoke and smoked meats contain many of these compounds. [Data quoted in V. Jahnsen, "The Chemical Composition of Hardwood Smoke," Ph.D. Thesis, Purdue University (1961).]

Appendix 4

EFFICIENCY MEASURING METHODS

Three general methods can be used to measure the energy efficiency of stoves. All require careful weighing of the wood to be burned, and determining its moisture content (by drying to constant weight at about 215 degrees Fahrenheit) in order to know the chemical energy in the wood. The methods differ in how they assess the amount of heat given off in the living spaces of the house.

The most direct method is to measure this useful heat output itself. The experimental problem is that the radiant heat moves out in all directions in the room and natural convection and drafts move the sensible heat in unpredictable ways. The difficult experimental task is to measure all this dispersed energy.[1] One approach is to gather it all together in one pipe or duct where measuring its quantity is easier. This can not easily be done in a house. It requires a specially built room which is very well insulated and air tight.[2]

Air is circulated through the room, entering and leaving through ducts, and moved by blowers. (Water run through panels or "radiators" used to collect the heat could also be used.) The heat output from the stove in the room then has virtually no place to go except into this moving air. Measuring the flow of this air and how much warmer it gets while passing through the room then determines the heat output of the stove. The major disadvantage of this efficiency measuring method is that the stove's environment in the test room is different from that in a house. Air velocity in the test room is higher and the temperature may be higher. Both of these conditions make the test room efficiencies higher than home efficiencies. Correction factors can be determined, but only with considerable effort.

The energy efficiencies of oil and natural gas domestic furnaces are often determined by measuring the amount of energy which goes up the chimney. Since energy cannot be destroyed, any of the total energy in the fuel burned which did not go up the chimney must have been left in the house. The three important kinds of energy in the stack are sensible heat, latent heat and chemical energy (smoke, etc.). Ideally each of these contributions should be independently and carefully measured, which is rather difficult.

It is interesting that, in general practice, when this method is used none of the above energy measurements is made directly. Instead, temperature and the concentration of carbon dioxide (and sometimes carbon monoxide) are measured. From these measurements alone, one can deduce the energy loss up the stack *if* one also knows the (elemental) composition and energy content of the fuel which is burning, and the fuel's burning rate (e.g., pounds per hour). (This type of test is sometimes called a Bacharach test.)

The most serious problem in applying this method to wood stoves is knowing the elemental composition and energy content of the part of the fuel which is actually undergoing combustion at the time of the test. Initial gentle warming of wood can in principle result in only evaporation of some of its moisture content, a process which actually consumes energy; thus the energy content, or energy release rate, is actually negative at this stage, and the elemental composition of the mass which is being released is the same as water, H_2O, since that is what it is. Towards the end of most burning cycles, the composition of the "wood" remaining approaches that of charcoal, which is mostly carbon, and has an energy value nearly twice that of wood (about 13,000 Btu per pound). Between these two extremes, the energy content and elemental composition of wood have values somewhere in between; exactly where is impossible to know without making very detailed measurements which essentially entail a completely different approach to measuring efficiences. Also, temperatures must be constant, and combustion must be nearly complete. (If only CO_2 and CO are measured, there should be no other carbon containing molecules or particles in the chimney gases.) For oil and natural gas furnaces, these conditions can be closely met. With wood in a domestic-sized heater, they can only be roughly approximated.[3]

The third efficiency-measuring technique is to place and operate a stove in an actual, ordinary, lived-in house with a conventional heating system as well, and to measure how much less energy the conventional system uses during periods when the stove is in use, compared to when it is not. The advantage of this method is that it directly indicates how much conventional energy is saved, including such effects as possible increased heating needs of the house due to the operation of the wood heater (this is especially applicable to fireplaces and open-type stoves; see Chapter 6). Disadvantages are that the result depends on the efficiency of the conventional heating system, on how much the wood heater is used, as well as on the wood heater's efficiency. The result is not just a characterization of the wood heater alone and thus would usually be of little direct value to another user in another installation.

Great care must be taken that the actual heating needs of the house are the same during the test periods when the wood heater is in use, and when it is not; or, if the needs are not the same, which is usually the case, compensation must be made. Ideally this requires monitoring the weather (temperature, wind and sunshine) and being able to compute the heating needs of the house under all conditions. ("Degree-day" data are a good approximation, but neglect wind and sun effects.) If the amount of activity in the house (number of people, and use of appliances) is not the same, additional compensation must be made for these internal heat sources. Because of all these complications, reasonably accurate results require each test period to last for at least a week, and the whole test should be repeated two or three times to check for consistency.[4]

There is not a simple practical way to measure a stove's energy efficiency. The only really simple measurement is temperature. Low flue gas temperatures can indicate either efficient heat transfer, or a lot of excess air, and in any case, the completeness of combustion is unknown.

1. It has been done, in a British study of coal and coke fueled stoves and fireplaces. See R. W. Rowse and W. C. Moss, "Domestic Boilers and Stoves Using Solid Fuel," *Journal of the Institute of Heating and Ventilating Engineers,* Vol. 18, (1950), pp. 32-78.

2. J. W. Shelton, "An Analysis of Wood Stove Performance, *Blair and Ketchum's Country Journal 3* No. 10 (Oct., 1976).

3. The only example of this method, applied to wood stoves, that I am aware of is reported in L. E. Seeley and F. W. Keator, 'Wood Burning Space Heaters," *Mechanical Engineering,* Vol. 62, (1940), p. 864.

4. This method has been used to test a fireplace burning coal. See page 80 and S. Konzo and W.S. Harris, *Fuel Savings Resulting from Closing of Rooms and from Use of A Fireplace,* University of Illinois, Engineering Experiment Station, Bulletin Series No. 348 (1943).

Appendix 5

HEAT LOSSES FROM HOUSES AND ANNUAL WOOD CONSUMPTION

There are two principle modes of heat loss from houses—heat is lost by being conducted through the exterior boundries of the house, and heat is carried out of the house by natural convection of air via air "infiltration" or "exchange." Insulation and use of thermopane (or storm or combination) windows reduce conduction losses. Careful construction, weather stripping and vapor barriers reduce air infiltration losses.

The colder it is outside (relative to inside), the more heat is lost by both mechanisms. The standard method for estimating heat losses is to first compute the heat loss per degree of temperature difference between the indoors and outdoors; then the needed heating rate at any outdoor temperature is obtained by multiplying by the appropriate temperature difference.

Table A5-1 gives "U-factors" for various types of construction. The U-factor is the amount of heat (in Btu) which would conduct through 1 square foot of each type of construction in one hour, when the air temperature is 1°F colder on the outside than on the inside. If the U-factor for each kind of construction in a house is multiplied by the number of square feet of that kind of construction, the sum of these products is the conduction heat loss for the whole house for a 1°F temperature difference between the inside and the outside.

For example, suppose a single story house has overall exterior dimensions of 24 ft × 30 ft, 8 ft high walls with 3½ inches of insulation, 6 inches of insulation in the ceiling with an attic above, 10 thermopane windows, each of which is 3 × 5 ft, and a floor with 3½ inches of insulation over an unheated basement. The total conduction heat loss per °F is computed as follows:

Area (square feet)	U-Factor	Product
Floor 24' × 30' = 720	0.04	29
Ceiling 24' × 30' = 720	0.05	36
Windows 10 × 3' × 5' = 150	0.61	92
Walls (excluding windows) 2 × (24' × 8') + 2 × (30' × 8') − 150 = 714	0.07	50
		207 Btu per hour per °F

Air infiltration losses cannot be predicted as accurately, since they depend critically on the care with which each particular house is built, particularly around the doors and windows. In relatively tight houses (e.g., houses with insulation, storm or combination windows, a vapor barrier [e.g., a plastic sheet under the wall board], and careful tight construction in general), infiltration is equivalent to about one air change per hour. This is typical of most new houses. In drafty houses, the air infiltration rate is about two air changes per hour. In specially designed and constructed energy-conservative houses, one half air change per hour is possible. (The rates actually depend on wind and temperature; these numbers are rough averages for winter conditions.) The resulting heat loss per °F is obtained by multiplying the volume of the house times 0.018[1]. Continuing the example above, the volume of the house is 24' × 30' × 8' = 5760 cubic feet. If its infiltration rate is 1 air change per hour, the coresponding heat loss factor is

5760 cubic feet × 1 × 0.018 Btu per hour per cubic foot ≃ [2] 104 Btu per hour per cubic foot.

This is 50 percent of the conduction heat loss factor, and the total is

207 + 104 = 311 Btu per hour per °F

For other houses, the infiltration loss may range from 30 to 100 percent of the conduction loss. This range is large, and there is no simple way to determine the actual amount of air-infiltration heat loss. This results in substantial uncertainties (roughly 30 percent) in heat loss estimates of houses.

Once this heat-loss coefficient for the house has been obtained, it need only be multiplied by the expected maximum temperature difference to obtain the needed furnace or heater output capacity. Local weather bureaus keep records of the average annual minimum temperature. If this temperature were −10°F and the house were to be maintained at 65°F, the maximum temperature difference would be 65° − (−10°) = 75°F. Thus in the example used previously, the maximum needed heating rate would be

311 Btu per hour per °F × 75°F = 23,000 Btu per hour.

Since estimates such as this have inherent uncertainties of roughly 30 percent they should not be taken too literally. Typically a heater with a capacity slightly larger than the predicted need is selected in order to be on the safe side. However if the estimate is being made to size a wood heater to be used as a supplemental (as opposed to exclusive) heat source, purchasing an oversize unit is unnecessary. If the occupants do not mind if the house's temperature is 5 or 10 degrees below the normal indoor temperature during the coldest winter days, the heater can also be undersized.

If a house is to be heated 100 percent by wood, the total annual heat energy needed is obtained by multiplying the house's heat loss coefficient (Btu per hour per °F) times the number of degree days[3] per year times 24 (hours per day). For example, if the number of degree days per year is 4400, then continuing the previous example, the needed heat energy per year is

311 Btu per hour per °F × 4400°F days × 24 hours per day ≃ 33,000,000 Btu.

If the wood heater has an energy efficiency of 50 percent, then twice this amount of energy is required in the form of wood, or 66 million Btu. If the wood used has an energy content of 22 million Btu per cord (see Table 3-1) then 66 ÷ 22 = 3 cords of wood would be required. In general, the number of cords needed per year is

(annual heat needed) ÷ (stove's energy efficiency × wood's high heat value per cord).

where the efficiency is expressed as a fraction (55 percent is 0.55 as a fraction.)

[1] 0.018 is the density of air times its specific heat:(0.75 pounds per cubic foot) × (.24 Btu per pound per °F) = 0.018 Btu per hour per cubic foot.

[2] The symbol "≃" means "approximately equal to".

[3] Degree-day information can be obtained from local weather

stations, newspapers or fuel-oil dealers. Degree days are a measure of the cummulative heat needed to keep a house comfortably warm (65° to 70°F). As defined in the ASHRAE Handbook of Fundamentals, "For any one day, when the mean temperature is less than 65°F, there exists as many degree days as there are Fahrenheit degrees difference in temperature between the mean-temperature for the day and 65°F". Thus the degree days for a day over which the means temperature is 45°F is 65 − 45 = 20 degree days. Annual total degree days are the sum of the daily figures.

WALLS ON OUTSIDE OF BUILDING	
Wood siding, studs, and interior wall board or plaster with no sheathing or insulation	0.33
Wood siding, building paper, sheathing, studs, and interior wall board or plaster with no insulation	0.27
Wood siding, building paper, sheathing, studs, insulating board with interior wall board or plaster	0.19
Wood siding, paper, sheathing, 3 5/8" fiberglass or rockwool, studs, and interior wall board or plaster	0.07
Brick 8" thick, with plaster on one side (no insulation)	0.47
Brick 12" thick, with plaster on one side (no insulation)	0.33
Brick 8" thick, with air space, insulatin board and plaster on one side	0.22
Brick 12" thick with air space, insulating board and plaster on one side	0.20
Brick 4" thick, paper, wood sheathing, studs, insulating board and plaster	0.21
Brick 4" thick, paper, sheathing, 3 5/8" fiberglass or rockwool studs, interior wall board or plaster	0.08
Concrete block 8" thick with no insulation or interior finish	0.53
Concrete block 8" thick with air space, insulating board and plaster	0.23
Masonry wall of any type and thickness below grade	0.06
CEILINGS WITH ATTIC SPACE ABOVE	
Plaster with no floor above or in attic	0.30
Plaster with a tight floor above or in attic	0.18
Plaster with 3 5/8" fiberglass or rockwool	0.07
Plaster with 6" of fiberglass or rockwool	0.05
CEILINGS WHICH ARE THE UNDERSIDE OF A SHINGLED ROOF	
Plaster, rafters, sheathing, building paper and shingles	0.29
Plaster, rafters, 3 5/8" fiberglass or rockwool or equivalent, sheathing, paper and shingles	0.06
Plaster, rafters, 6" fiberglass or rockwool or equivalent, sheathing, paper, and shingles	0.04
FLOORS OVER ENCLOSED, UNHEATED SPACE, OR BASEMENT	
Finish flooring over sub-floor on joists	0.15
Finish flooring over sub-floor on joists with ½" insulating board on bottom of joists	0.09
Finish flooring over sub-floor on joists with 3 5/8" fiberglass or rockwool or equivalent between joists	0.04
FLOORS OVER EXPOSED SPACE	
Finish flooring over sub-floor on joists with ½" insulating board on bottom of joists	0.20
Finish flooring over sub-floor on joists with 3 5/8" fiberglass or rockwool or equivalent between joists	0.06
CONCRETE FLOORS	
Concrete floor directly on ground	0.10
Concrete floor on ground below grade	0.04
WINDOWS	
Single glass (no Storm window)	1.13
Single glass with well fitting storm window or combination	0.56
"Thermopane" or insulating glass	0.61
DOORS	
Single door	0.50
Ordinary door plus storm door	0.30

TABLE A5-1. U-factors for various types of construction. [Adapted from IBR (Institute of Boiler and Radiator Manufacturers) Heat Loss Calculation Guide, First Edition, (Aug. 1965)].

Appendix 6

ENERGY CONSERVATION MYTH: LIMITED THERMOSTAT SETBACK AT NIGHT

There seems to be disagreement over the energy-conserving potential of short-term or temporary thermostat setback, such as overnight. Many published lists of energy-saving tips seem to suggest that there is some optimum amount of setback, which, if exceeded will result in greater energy consumption. For instance, Western Mass. Electric Company recently distributed a pamphlet which includes the suggestion, "Set thermostat to provide a comfortable temperature and avoid constant readjustments. Don't lower the thermostat more than 6 to 8 degrees at night."

The suggestion that larger setbacks than the suggested 6-8 degrees may result in larger energy consumption seems to be based on the notion that if the structure itself (walls, floors, ceilings) and its contents get too cold during the setback period, more energy may have to be used to reheat all this mass than was saved during the setback period, thus resulting in a net increase in energy consumption.

In fact this notion is nonsense. The energy necessary for warmup or pickup after a setback never exceeds the savings during the setback. The reason is based on the fact that temperature losses depend on temperature differences. However, when temperatures are changing rapidly, as they are likely to with frequent thermostat adjustments, the heat capacity of the structure may have to be taken into consideration in calculating conductive losses. Heat conducted out of a building is finally lost only after penetrating the walls or roof, since a sudden drop of temperature inside the structure can permit some of the heat stored in the walls to return to the heated space. Thus we really need to know the temperature difference (or gradient) at the outermost layer of the building's exterior. Examination of the fundamental heat-conduction equation shows that after a thermostat setback this temperature gradient at the outer part of the walls and roof can never exceed its steady-state value before the setback (assuming constant outdoor conditions). And in fact, it is always less, indicating that energy is always conserved, regardless of the duration or amount of setback. (Not all heat is lost by conduction; in typical homes, half of the total heat loss is due to air exchange. This loss is essentially proportional to the indoor-outdoor air-temperature difference and thus is clearly reduced anytime the indoor air temperature is reduced.)

The amount of energy saved by night setback of course depends on climate, the heat-loss coefficient of the building, the normal set point, the heat capacity of building, and the heating-plant capacity. Typical residential savings for 5 to 10° setbacks have been determined by Mr. Nelson of Honeywell, and are reported in an article in the August 1973 issue of the ASHRAE Journal. A 5 to 15% savings can be expected under typical circumstances for these modest setbacks. The savings are not directly proportional to amount of setback, but can easily exceed 50% for large setbacks in mild climates. And in fact, the energy savings are often larger than just the decreased heat loss. The utilization efficiency of most heating systems increases with the intensity of use. During the pickup time after a setback the heating system is likely to be "on" for a longer time than usual, during which its utilization efficiency is higher. In addition, refrigerators, freezers and water chillers will consume less energy during setbacks because of their cooler environments.

Larger setbacks *can* result in *larger* energy consumption in systems with proportional control, such as a system in which the high demand during the pickup period might call into service an additional or auxiliary heating system which was less efficient than the basic unit. However such a circumstance is unnecessary and would most likely be the result of poor design or inadequate maintenance.

The usual practical limit to the amount of night setback is comfort, perhaps at night, but more usually in the morning if one does not have an automatic timed thermostat. The fast response of most hot-air systems may allow larger setbacks. In those circumstances where possible discomfort associated with morning pickup time does *not* provide a limit to the amount of setback, the health of house plants may, or ultimately, the prevention of freezing damage to water pipes.

In engineering new houses, there are some difficult choices. Although temporary thermostat setback always saves energy, the amount of setback tolerable from a comfort standpoint may depend on the pickup time, i.e., on the capacity of the heating system. The larger the capacity of the heating system, the shorter the warm-up time and hence the larger the possible setback. However, oversized heating systems (except electric resistive heating) have lower efficiencies. The optimum choice thus depends on the style of the inhabitants. Also, massive construction, as for instance in passive south-window solar heating systems, does not permit as much savings from letting the inside temperature fall at night because it will not fall very much. That is a beauty and liability of such systems; they respond slowly to any thermal perturbations.

In summary, the *limited* aspect of night setback does not belong in an energy conservation list but perhaps in a discussion of comfort. From the point of view of energy conservation alone, the larger the setback the better, regardless of its duration.

Index

Section Two Manufacturers and Importers

Manufacturers and Exclusive Importers Alphabetical Address Index Including Descriptive Statement and Product Identification by Specification Chart Divisions And Product Illustrations

EDITOR'S NOTE

The following information has been furnished by manufacturers and exclusive importers. A listing by address only indicates that no further information was provided as requested or the firm had not been contacted prior to the Encyclopedia's printing deadline.

Firms contacted prior to the Encyclopedia's first printing were offered the option on a fee basis to have products illustrated.

Companies desiring changes and/or additions to this listing section may contact the publisher and the additional information will be included in our next printing.

Division Indications: EXAMPLE - II (Heating Stoves) refers to the appropriate specification section where the companies' products will be found if appropriate information has been provided prior to the first printing of the Encyclopedia.

Information contained in this section has been taken from the importer's or manufacturer's literature. The Woodburners Encyclopedia does not guarantee or testify to the accuracy of the information supplied herein.

* indicates that firm is EXCLUSIVE IMPORTER

Abundant Life Farm
PO Box 63
Lochmere, NH 03252
(603) 524-0891
II (Heating Stoves)

Aglow Heat-X-Changer
PO Box 10427
Eugene, OR 97401
(503) 343-7605
VII (Fireplace Accessories)

With the Aglow Heat-X-Changer, your fireplace becomes a heating system. This durable unit helps you save on your monthly heating by increasing the effectiveness of your existing fireplace. Cold air from the room is drawn into the unit. The air is heated by the fire and is then recirculated. With the Heat-X-Changer, your fireplace becomes more than just an attractive fixture: it becomes a forced-air heating system. The Heat-X-Changer is adaptable to most masonry fireplaces.

Albright Welding Co.
Rte. 15 Box 108
Jeffersonville, VT 05464
(802) 644-2987
XI (Woodsplitters)

L. O. Balls Woodsplitters - kits, components, complete units. Motorized or tractor-hydraulics powered. 3-point hitch or road tires. 26" or 48" capacity. Plans available. We have manufactured splitters for over eight years. Many sold in U.S. and Canada. Guaranteed. Write or call for more information and prices.

American Stovalator
Rte. 7
Arlington, VT 05250
VII (Fireplace Accessories)

The American Stovalator turns your fireplace into a safe, efficient heating system. An attractive unit that fits into, and against, your existing fireplace, the Stovalator heats your home using the same fire you would normally build in your fireplace.

American Way
Dept. 10-0G
190 Range Road
Wilton, CT 06897
VII (Fireplace Accessories)

Amherst Welding, Inc.
330 Harkness Rd.
Amherst, MA 01002
VII (Fireplace Accessories)

Acquappliances, Inc.
135-H Sunshine Lane
San Marcos, CA 92069
(714) 744-1610
VII (Fireplace Accessories)

Ashley Automatic Heater Co.
1604 17th Ave. S.W.
PO Box 730
Sheffield, ALA 35660
II (Heating Stoves) III (Circulating Heaters)

Ashley Spark Distributors, Inc.
710 N.W. 14th Ave.
Portland, OR 97209

Atlanta Stove Works, Inc.
PO Box 5254
Atlanta, GA 30307
Or Contact
Whole Earth Access Co.
I (Cookstoves; II (Heating Stoves);
III (Circulating Heaters; V (Free-Standing Fireplace);
VI (Fireplace Accessories)

The Atlanta Stove Works is an 85 year old company and has been making cast iron stoves since its existence. It has two manufacturing facilities—one in Atlanta and one in Birmingham, Alabama. The Birmingham Plant is one of the most automated, grey iron, foundries in the country.

Atlanta Stove produces a variety of products, including

Franklin fireplaces, coal and wood stoves, cast iron cookware, barbecue grills and cast iron grates.

High quality castings are produced in the ultra-modern foundry where iron analysis and consistency is assured. Years of experience have proven Atlanta Stove to be a reliable and consistently high quality producer of woodburning equipment.

Cookstove, #15-36

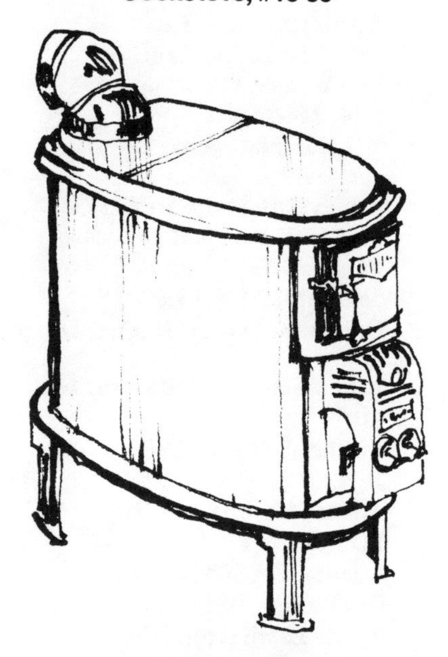

Heating Stove, #2502

Atlantic Clarion Stove Co.
Brewer, ME 04412

Autocrat Corporation
New Athens, Ill. 62264
(618) 475-2121
I (Cookstoves); II (Heating Stoves);
III (Circulating Heaters)

Bahia Bar-B-Que
PO Box 1806
St. Paul, MN 55111

Bellway Manufacturing
Grafton, VT 05146
(802) 843-2432
IV (Furnaces & Boilers)

The new Bellway Hi-Temp woodburning heating units are of an improved design, tested in Vermont winters for nearly 20 years. All heavy steel welded smoke tight construction, the fuel being stored in top section, feeding down as needed to the high temperature burner of charcoal in lower section. Burns wood green or dry, garbage and catalogs. A slow burning hot fire up to 1800 degrees plus 1000 degrees in afterburner.

All units are rated like oil burners in btu and operate 10 hours in 0 degree weather without attention. Complete with casings for gravity or forced warm air with all electric controls, grate shaker handle, long ash shovel and poker.

Besta Heater Ovens
Box 887
Charlestown, NH 03603

Birmingham Stove & Range
PO Box 2647
Birmingham, ALA 35202
(205) 322-0371

Black's
58 Maine St.
Brunswick, ME 04011

Blazing Showers
PO Box 327
Point Arena, CA 95468
OR CONTACT
WHOLE EARTH ACCESS CO.
(707) 882-9956
VIII (Hot Water Heaters)

Blazing Showers enables you to obtain your hot water needs from your woodburning stove. Our basic stove pipe model uses the best sizing in the stove pipe to fill your storage tank with hot water. Our water heating system is designed for easy installation and maximum efficiency while maintaining the normal operation of your wood stove. The products require no gas or electricity and help make you more self-sufficient. Thus, Blazing Showers is perfect for use above or in conjunction with solar or even regular fossil fuel water heating.

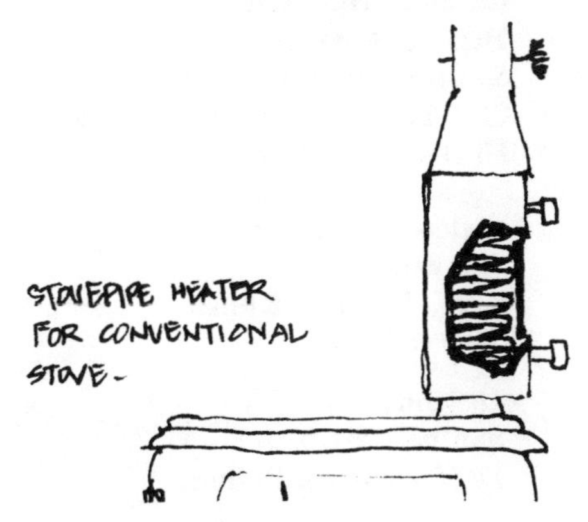

Hot Water Heaters, Firebox and Stovepipe

Boston Stove Co.
Dept. CJ1
155 John St.
Reading, MA 01867
VII (Fireplace Accessories)

***Bow & Arrow Stove Co.**
14 Arrow St.
Cambridge, MA 02138
(617) 492-1411
I (Cookstoves); II (Heating Stoves);
III (Circulating Heaters)

Started in early 1975 by three architects and a graphic artist in their office, the Bow & Arrow Stove Company retails a unique selection of efficient heating and cooking stoves, as well as other energy-saving products for the home. The company also imports and distributes a variety of wood and coal stoves: Parra sauna stove, Petit Godin wood/coal stove, Preporod STG-5 cookstove, Preporod DP-5 coal heater, Supra 401 wood heater and the Rayburn No. 4 Room Heater. Bow & Arrow also distributes the Drip/Pruf universal stovepipe system and the Taba cooking grill for use with combination stove/fireplaces. Dealer inquiries are invited. The Bow & Arrow retail catalog is available for fifty cents.

Cookstove, Preporod #STG-5

Heating Stove, Fyrtønden

Heating Stove, Parra Sauna

Heating Stove, Petit Godin #3720

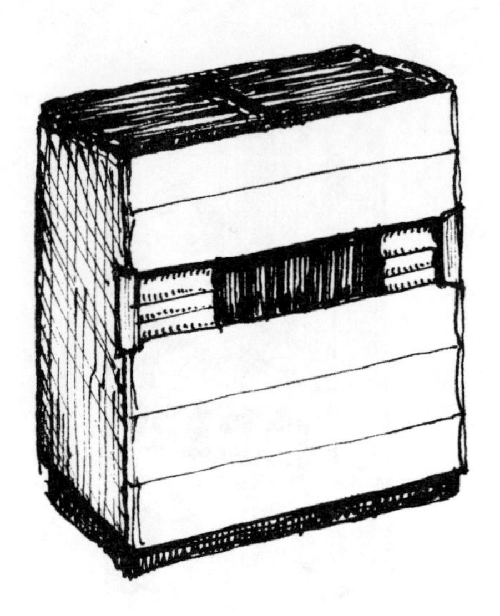

Circulating Heater, Supra #401

Canaqua Co.
Box 6
High Falls, NY 12440
(914) 687-7457

1908 Baker St.
San Francisco, CA 94115
(415) 346-0752

Canaqua Co. is a small R&D group which designs and manufactures sealing systems for industry, specializing in high-temperature sealants and gaskets. Our products and services have won wide acceptance in heavy industry, where performance requirements are far more stringent than any every encountered in woodstoves. Woodstove manufacturers are encouraged to call Canaqua for help with their sealing problems. In most cases we will be able to recommend a product already on the market.

Carlson Mechanical Contr's, Inc.
Box 242
Prentice, WI 54556
(715) 564-2481 or (715) 428-3481
IV (Furnaces & Boilers)

Many installed in northern Wisconsin. Beat the high cost of home heating with one of these units. Safe, dependable, well constructed, of welded steel plate and tubing. Connects to your existing hot water heating system. No expensive automatic controls to buy. Utilizes the pump and controls on your present boiler.

C & D Distributors, Inc.
Box 766
Old Saybrook, CT 06475
(203) 388-5665
VII (Fireplace Accessories); XI (Wood Splitters)

Chim-A-Lator
8824 Wentworth Ave., So.
Minneapolis, MN 55420
(612) 884-7274
VII (Fireplace Accessories)

Chimney Heat-Reclaimer Corp.
Dept. Y 53 Railroad Ave.
Southington, CT 06489
(203) 628-4738 (Ext. Y)
IX (Heat Reclaimers)

Waldo G. Cumings
Fall Road
East Lebanon, ME 04027
(207) 457-1219

20 Schuler St.
Sanford, ME 04073
IV (Furnaces & Boilers)

The Central Heating Wood Burning Furnace was developed for 1200 square feet single story and split entry type ranch style homes. The furnace is designed to be used as an auxiliary heating system, that can be tied into existing hot air duct system. It also was intended as a back-up separate system for the use in electric heated homes and circulated hot water heated homes.

The concept was to keep simplicity and safety in mind, along with the cost, so the furnace could be utilized in many situations. the basic reason, was to "Help People Help Themselves" to save energy and dollars, also creating jobs and boosting our economy. The wood is a solar energy renewable natural resource, that saves energy and keeps our dollars in the Northeast.

Dampney Company
85 Paris St.
Everett, MA 02149
(617) 389-2805

THURMALOX 270 heat resistant coating silicone for stove manufacturers. THURMALOX 270 is a silicone based coating having unimpaired film properties at temperatures up to 1200 degrees F. THURMALOX 270 also is free of zinc metal. It therefore does not contribute to any weldment embrittlement. This is true whether or not the coating is removed from steel surfaces prior to welding. Prevents oxidation-corrosion of the exteriors of non-weather exposed hot surfaces.

Damsite Dynamite Stove So.
RD 3
Montpelier, VT 05602
(802) 223-7139
II (Heating Stoves); IV (Furnaces & Boilers)

We are craftsmen who produce (individually, by hand) a line of sophisticated stoves designed for maximum heating efficiency and safety at prices far below those of stoves of equivalent quality. We can do this because of very low advertising budget and direct selling from our factory. Our stoves all hold their fires overnight and usually have parts of logs still burning in the morning, not just coals. In addition two of our stoves are designed with special chambers for drying green wood before burning. They are all unconditionally guaranteed against faulty workmanship.

Sam Daniels Co.
Box 868
Montpelier, VT 05602
(802) 223-2801
IV (Furnaces & Boilers)

Dawson Mfg. Co.
Box 2024
Enfield, CT 06082

Charles Dedrick, Inc.
Stone Ridge, NY 12484

Deforge Industries
PO Box 216
Winooski, VT 05404
(802) 863-2653
VII (Fireplace Accessories)

Manufacturers of the famous Vermont Fireplace Heater. Converts your fireplace or free-standing stove to an efficient forced hot air furnace. Extremely durable with its unique double wall grate construction. Custom built models available. Contact factory for prices and information.

Didier Mfg. Co.
1652 Phillips Ave.
Racine, WI 53403
(414) 634-6633

Double Star
c/o Whole Earth Access Co.
2466 Shattuck Ave.
Berkeley, CA 94704
(415) 848-0510
II (Heating Stoves); V (Free-Standing Fireplaces)

We have done considerable research into the various Franklins, Parlors, and Pot Bellys being brought in from Asia, and discovered substantial differences among them. The Double Star brand stoves in our judgement are in quality and value the best we've seen. In general the castings are heavier, better finished and better fitting. (When comparing Asian stoves, even from catalogs, always at least check net weight—not to be confused with shipping weight!) In our opinion these are also quite comparable to the domestic stoves which most of them are patterned after—at much better prices.

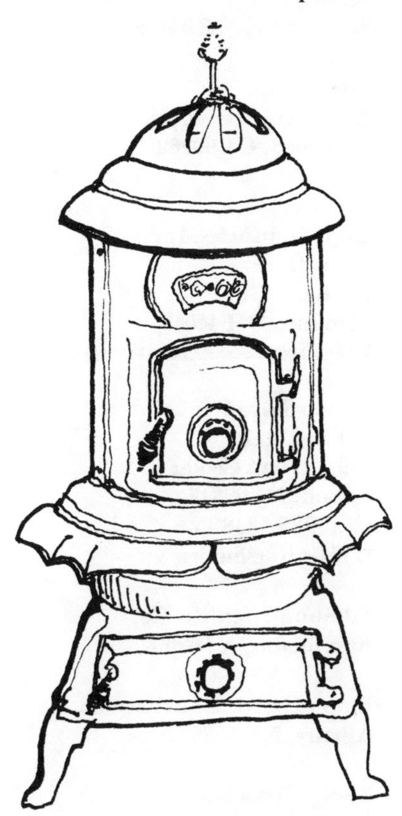

Heating Stove, Comfort Heater

Free-Standing Fireplace, Franklin #26

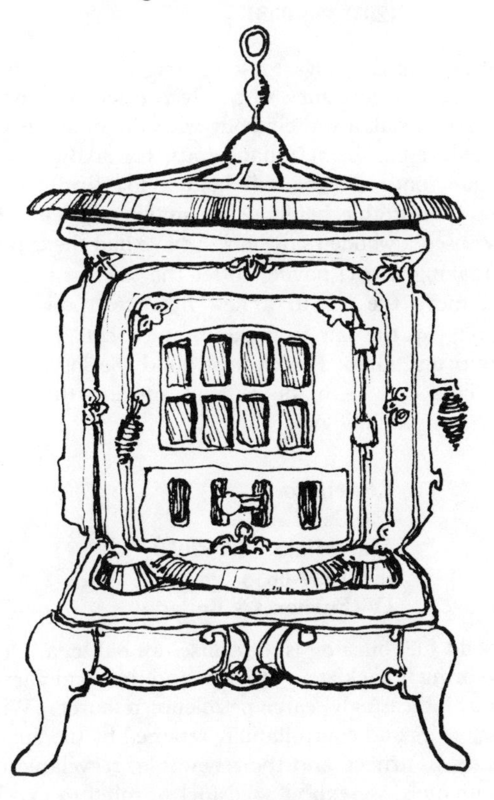

Heating Stove, Parlour Stove

Dover Corp.
Peerless Div.
PO Box 2015
Louisville, KY 40201

Dover Stove Company
Main St.
Sangerville, ME 04479

Dyna Corp.
2540 Industry Way
Lynwood, CA 90262

Eagle Industries, Inc.
Dept. D
PO Box 67
Madison, OH 44057
VII (Fireplace Accessories)

Edison Stove Works
PO Box 493
469 Raritan Center
Edison, NJ 08817
(201) 225-3848
II (Heating Stoves)

Edmund Scientific Co.
Barrington, NJ 08007

Empire Stove & Furnace Co.
Albany, NY 12207

Energy Associates
PO Box 524
Old Saybrook, CT 06475
(203) 388-0081

When cutting logs to stove length, many chainsaw users have a tendency to steady the log with one foot and try to balance their body with the other foot. This procedure can be very dangerous. Gator-Buck, with its unique tooth design, will hold a log firmly in place at a comfortable height while making all cuts. If you have used a wooden saw buck, you've had the experience of making a cut, having to set the saw on the ground and move the log to a new balanced position, then picking up the saw to make the next cut. This can be very tiring. Gator-Buck's teeth hold the log so that all the cuts can be made without moving the log with little saw-pinch, if any.

Enwell Corp.
750 Careswell St.
Marshfield, MA 02050
(617) 837-0638
IV (Furnaces & Boilers)

Solid fuel burning is, of course, an old idea. Our aim is to bring it back as an efficient and practical alternative to our increasingly scarce petroleum resources. With the efficiencies and controllability attained by the Spaulding Concept Furnace and the renewable/recyclable nature of our fuels, we expect solid fuel burning to experience a true rebirth.

The furnace burns fuels such as unsplit fireplace wood up to 30" long, chips, bark, sawdust, household trash, recycled municipal waste, rubber, and used engine oil interchangeably and efficiently. Converts your waste materials to energy. Output thermostatically controlled in three modes from 20,000 to 100,000 Btu's per hour. Fuel capacity sufficient for twelve hours burning at maximum output. Needs servicing only twice a day in the coldest weather—less often in milder weather.

Todd Evans, Inc.
110 Cooper St. Box X
Babylon, NY 11702

Fabsons Engineering
PO Box F-11
Leominster, MA 01453

Fireplaces (N.S. Limited)
Suite 215 Duke St. Tower
Halifax, Nova Scotia, B3J1N9

Fire-View Distributors
PO Box 370
Rogue River, OR 97537
(503) 582-3351

Fisher Stoves, Inc.
504 So. Main St.
Concord, NH 03301
(603) 224-5091
II (Heating Stoves)

Air tight, controlled burn stove with fire brick lining. Constructed with ¼" and 5/16" steel plate with a heavy cast iron door which has a superior seal that eliminates fire hazard from sparks. Two cooking surfaces with different temperatures and there is a portable oven available. The stoves are manufactured in Concord, NH so they can be ordered with right or left hand doors, back or side exhaust outlet and the length of the legs can be custom ordered. Unconditional 90 day guarantee plus 25 year guarantee on material and workmanship,

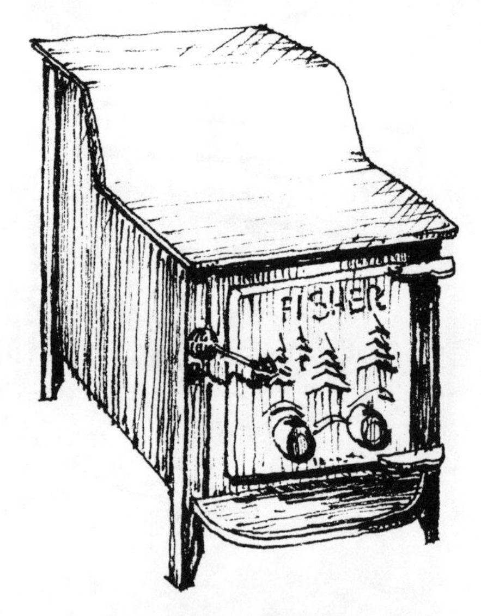

Heating Stove, Papa Bear

paint and fire brick excluded. Dealers throughout New England and Eastern Canada.

Heating Stove, Grandpa Bear

Fisk Stove
Tobey Farm
Box 935
Dennis, MA 02638
(617) 385-2171
X (Barrel Stoves)

Fisk Stove Plans—from the unavailability, expense, inefficiency and limitations of conventional woodburning stoves grew an air tight, thermostatically controlled, fire-brick lined, dual phase (downdraft/updraft), long burning, modular construction drum stove designed to be built by "handy" homeowners with normal home workshop tools. Modular design allows construction of any modular combination from basic small furnace ala Riteway to furnace cookstove complete with oven and running hot water; successfully operated over two years as sole heat for plumbed, uninsulated New England house, wide local publicity, including TV, caused writing of step-by-step plans with charts and diagrams.

Forest Fuels, Inc.
7 Main St.
Keene, NH 03431
(603) 357-3311

Forest Fuels, Inc. - Wood/Gas Burner System: A process which is based on chipped forest material—wood, bark, sawdust, shavings or wood pellets. The fuel is dried and fed into a reactor where it is heated to the point of distillation. This is done under starved air conditions when the distillation is completed, the charcoal remains and is burned as well producing carbon monoxide.

Franklin Fireplaces
1100 Waterway Blvd.
Indianapolis, IN 46202

Fuego Heating Systems
PO Box 666
Brewer, ME 04412
(207) 989-5757

Futura Stove Works
Main St. Box 14W
La Farge, WI 54639
XI (Wood Splitters)

Garden Way Research
PO Box 26 W
Charlotte, VT 05445
(802) 425-2137
II (Heating Stoves); VII (Fireplace Accessories); XI (Wood Splitters)

The Garden Way Box Stove is a simple and economical answer to heating with wood. The Shaker design is an extremely efficient use of materials as well as a high-efficiency heater. The 11 gage thick steel walls will last for many years of burning.

The nearly air-tight construction assures you of precise control of the fire.

L. W. Gay Stove Works, Inc.
Marlboro, VT 05344
(802) 257-0180
II (Heating Stoves)

We manufacture a Norwegian-like box stove. A copper coil can be installed in the firebox for heating water, and brackets and soapstone slabs can be supplied to convert it to a circulator. We also make a stovepipe/air heat exchanger and two kinds of stovepipe water heaters—in one the copper coil is exposed directly to the flue gases and in the other it isn't. Which to use depends on stove and chimney. We like to work on special problems, particularly those having to do with combined wood/solar systems. How about a laundry stove?

Gemco
404 Main St.
Marlborough, NH 03455

General Products Corp.
150 Ardale St.
West Haven, CT 06516
VII (Fireplace Accessories)

Glo-Fire
Spring & Sumner Streets
Lake Elsinore, CA 92330
(714) 674-3144

Golden Enterprise
PO Box 422
Windsor, VT 05089
VII (Fireplace Accessories)

The Thermal Grate is a tubular steel device designed to cradle a fire in the same manner as a conventional grate while generating warm air and discharging it into the room. As a fire is kindled in the bed of the grate, it heats the steel tubes and the air inside them. This warm air expands and is forced out the top of the tubes. Cool air is drawn in through the bottom and in turn is heated and discharged. As the fire gets hotter, the forced air action grows stronger, circulating warm air throughout the room.

Greenbriar Products, Inc.
Box 473G
Spring Green, WI 53588
V (Free-Standing Fireplaces)

HDI Importers
Schoolhouse Farm
Etna, NH 03750
(603) 643-3771

Heatilator
Box 409
Mount Pleasant, IA 52641
(319) 385-9211
VI (Pre-Fabricated Fireplaces)

Heat Reclamation Division
939 Chicopee St.
GPO Box 366
Chicopee, MA 01021
(413) 536-1311 Telex 95-5342
IX (Heat Reclaimers)

The Thermo$aver, all metal, is mounted in the flue pipe above your furnace. It encloses hollow tubes and has a fan/blower mounted on one end. The hot exhaust from the furnace circulates through the chamber heating the air in the tubes (sealed from the furnace air to avoid fumes). This heated air is blown into the room. The Thermo$aver turns off and on automatically—working only when the furnace is hot.

Heritage Fireplace Equipment Co.
1874 Englewood Ave.
Akron, OH 44312
(216) 798-9840

M. B. Hills, Inc.
Belfast, ME 04915
(207) 338-4120
IV (Furnaces)

Hinckley Foundry
13 Water St.
Newmarket, NH 03875
(603) 659-5804

150 years ago the Shakers designed the first true airtight stove. It incorporated many practical features not found in most modern woodburners. Air tightness was insured by casting the firebox cap in a single piece thereby eliminating eight leaky seams. Efficiency was almost doubled by passing the hot gasses through a secondary heat exchanger over the firebox. They knew that a large ash apron not only kept things neat, but made a great footrest or drying place and provided more heating surface. Safety was not overlooked either. The lean-to door stayed shut whether it was latched or not. True to Shaker beliefs, the stove remained simple, unencumbered with decoration and provided heat enough for any drafty home or meeting house.

***Home Fireplaces (Morso Canadian Importer)**
Markham, Ontario L3R1GE
(416) 495-1650
971 Powell Ave.
Winnipeg, Manitoba R3H OH4
(204) 774-3834
I (Cookstoves); II (Heating Stoves); III (Circulating Heaters); V (Free-Standing Fireplaces); VI (Pre-Fabricated Fireplace); VII (Fireplace Accessories)

A wholesale distributor of wood burning, factory built fireplaces, Franklins, wood burning stoves, pre-fabricated

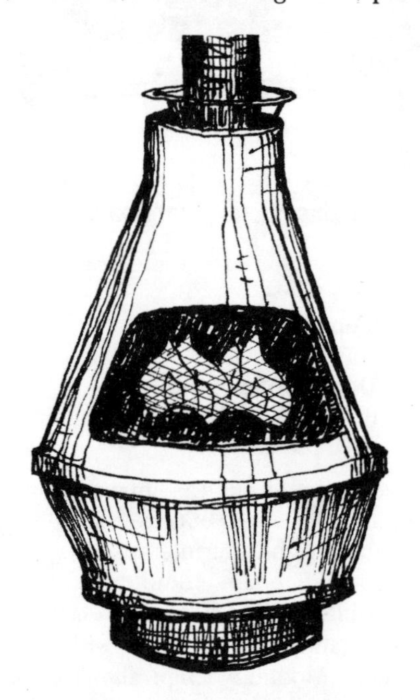

Free-Standing Fireplace, Sunfire

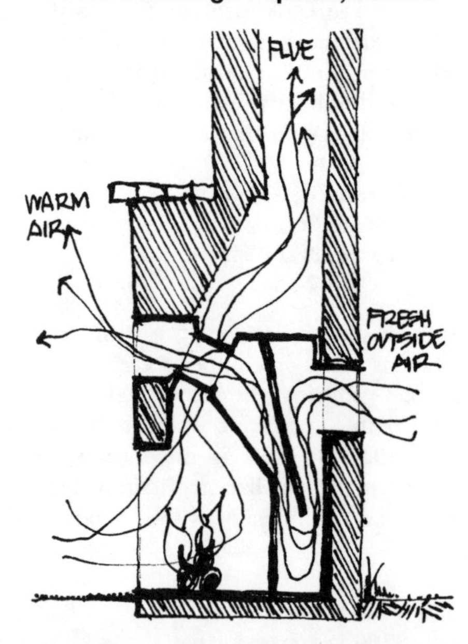

Pre-Fabricated Fireplace, Northern Heatliner #37

chimneys, mantels, fireside furnishings, made to measure screens, and Glassfyre doors. Home Fireplaces, through its affiliated manufacturers, supplies the market with the energy saving, super heating fireplaces, the Sunfire and the Northern Heatliner.

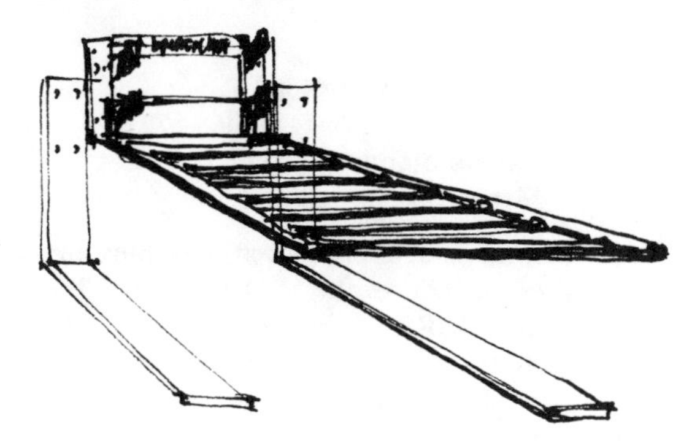

Fireplace Accessories, Emberchef Barbecue

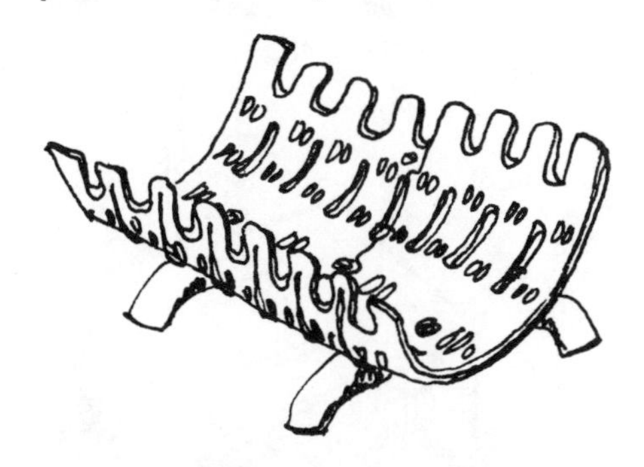

Fireplace Accessories, Fuel Miser Grate

Fireplace Accessories, Hammered Swedish Ensemble

Household Woodsplitters
PO Box 143
Jeffersonville, VT 05464
(802) 644-2253
XI (Wood Splitters)

Compact - can be easily stored and transported (even in the family car). **Rugged** - even though the Household is conveniently small, it is no toy. Its construction is of the finest steels and built to withstand almost anything. **Portable** - the Household can easily be towed from garage to woodpile using the convenient handle located behind the wedge. **Easy to use** - the Household Woodsplitter has one simple control. Move the valve handle forward to split - back to retract. **Dependable** - because the Household Woodsplitters ram is fully hydraulic out and back, there is nothing to catch, bind, break or come unsprung.

Hunter Enterprises Orillia Limited
PO Box 400
Orillia, Ontario, Canada
(705) 325-6111
III (Circulating Heaters); IV (Furnaces & Boilers)

The standard furnace comes equipped for natural draft operation, with a Valley comfort thermostat and damper control. This furnace is designed to function automatically without the use of electricity. The Valley Comfort Furnace, with an electrically controlled thermostat, blower and filter unit is an efficient, forced air heating system. These features help to increase the heat distribution capabilities of the Valley Comfort Furnace for evenly controlled home heating comfort and convenience.

Hydraform Products Corp.
PO Box 2409
Rochester, NH 03867
(603) 332-6128
II (Heating Stoves)

Hydraform Products Corp., a firm dedicated to the development of a highly engineered new stove designed to solve the problems of existing stoves, namely chimney clogging, wasted heat up chimney, costly short split logs, short burning times, long burning times but not practical heat output, difficult loading, insufficient Btu output to heat average house, firepot burnout and cracking, poor warranty, long delivery on replacement parts, and red hot stove for high Btu output.

Heating Stove, Larger Eagle

Inglewood Stove Company
Rte. 4
Woodstock, VT 05091
(802) 457-3238
II (Heating Stoves)

The first Tortoise Stove was made by Charles Portway

in 1820. The production of ornamental stoves ceased in 1900. The current stove is based on a model over 70 years old, rescued from a demolished chapel, with only minor changes in draughting and air control to meet present day standards. The decision to recommence production was influenced by the aesthetic qualities of the ornamental design, and the relevance of a slow burning stove in these days of fuel crisis.

Isothermics, Inc.
PO Box 86
Augusta, NJ 07822
(201) 383-3500
IX (Heat Reclaimers)

Manufacturer of residential waste heat reclaimers which increase heating system efficiency by providing extra heat at no additional cost. Used in stack temperatures from 400 degrees F. Tested proven safe and effective.

Jernlund Products, Inc.
1620 Terrace Dr.
St. Paul, MN 55113

Kenenatics
1140 No. Parker Dr.
Janesville, WI 53545

Kickapoo Stove Works, Ltd.
Rte. 1-A
LaFarge, WI 54639
(608) 625-4431
II (Heating Stoves)

At Kickapoo Stove Works we use craftsmanship and quality materials to manufacture the high efficiency Kickapoo wood heat stove. We also custom manufacture fireplace stoves and wood furnaces. We market our products directly from our Wisconsin factory outlets and nationally by mail order, as well as wholesale to retailers. Write for our free catalog of products and accessories. We are a small company with high standards of craftsmanship and personal attention to our customers.

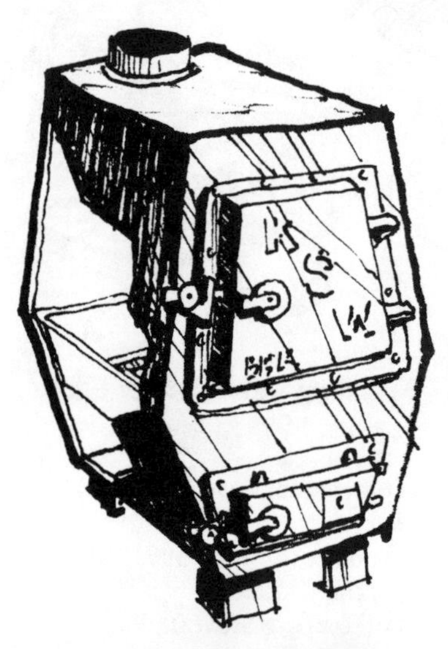

Heating Stove, BBR-1

Knotty Wood Splitters
Hebron, CT 06248
(203) 228-9122
XI (Wood Splitters)

KNT, Inc.
PO Box 25
Hayesville, OH 44838
(419) 368-3241 or (419) 368-8791
V (Free-Standing Fireplaces); VI (Pre-Fab. Fireplace)

KNT, Inc., manufacturer of wood / coal burning stoves and fireplaces, introduces the MARK V Dual Injection Fireplace. The MARK V is the latest of a series of advanced concept supplemental home heating units, developed by KNT. Combining the heating efficiency of a modern furnace with the grace and charm of a built-in fireplace, the MARK V is intended for today's discriminating homebuilder who seeks to combine heating efficiency with customer eye-appeal.

Free-Standing Fireplace, Impression #102

Pre-Fabricated Fireplace, Mark V

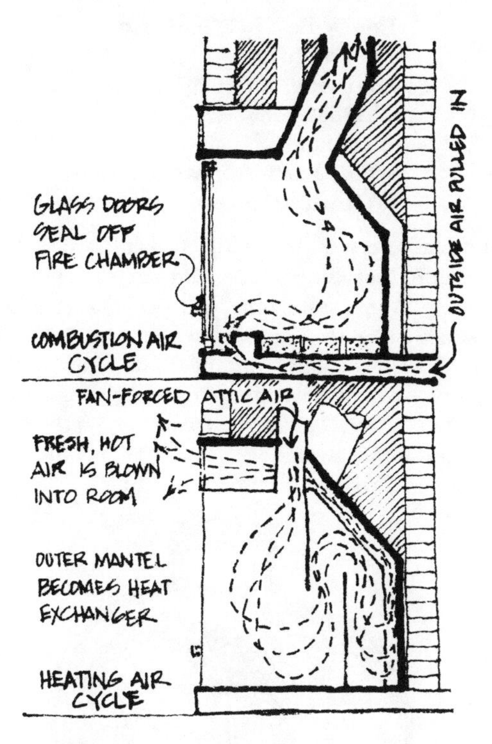

Pre-Fabricated Fireplace, Mark V (section)

***Kristia Associates (Jøtul Importers)**
PO Box 1118
Portland, ME 04104
(207) 772-2112
II (Heating Stoves); V (Free-Standing Fireplaces); VI (Pre-Fabricated Fireplaces)

A Jøtul stove is a lifetime investment. The stove is very rugged and handsome with traditional Norwegian design. A full inventory of replacement parts are also available from the factory. Jøtuls come in a variety of styles and sizes to fit almost any woodburner's needs.

Heating Stove, Jøtul #118

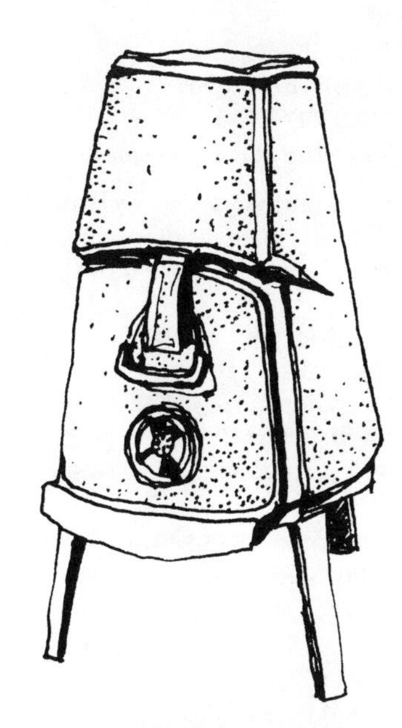

Heating Stove, Jøtul #4

Heating Stove, Jøtul #602

La Font Corp.
1319 Town St.
Prentice, WI 54556
(715) 428-2881
VII (Fireplace Accessories); XI (Wood Splitters)

Lance International
PO Box 562
1391 Blue Hills Ave.
Bloomfield, CT 06002
(203) 243-9700
VII (Fireplace Accessories); IX (Heat Reclaimers)

W. F. Landers Co.
PO Box 211
Springfield, MA 01101
(413) 786-5722

Lassy Tools, Inc.
Plainville, Ct 06062
(203) 747-2748
VII (Fireplace Accessories)

Heat-catchers are designed to reclaim heat from fireplaces. They have been widely sold from coast to coast with excellent results reported by the users.

Newton Lee
Rte. 1 Box 116
Worcester, VT 05682
(802) 223-3119
VII (Fireplace Accessories)

Leyden Energy Conservation Corp.
Brattleboro Rd.
Leyden, MA 01337

Locke Stove Co.
114 West 11th St.
Kansas City, MO 64105
(816) 421-1650
II (Heating Stoves); III (Circulating Heaters)

Warm Morning Model 701B - deluxe woodburning circulator with large furniture-styled cabinet; porcelain enamel cabinet finish and built-in thermostat.

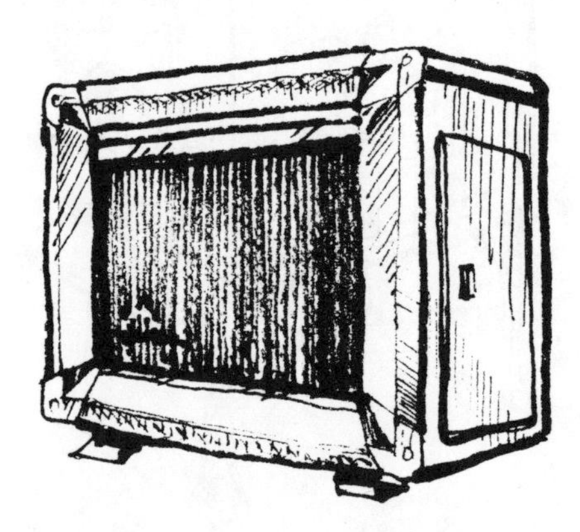

Circulating Heater, Warm Morning #701B

Log House Designs
Chatham, NH
(603) 694-3183
Fryeburg, ME 04058

Located in a Log House in the White Mountains of New Hampshire, we manufacture a Log Carrier, size 41" × 26"; 9.50 oz. coated Cordura (three times stronger than canvas) with 1" yellow nylon soft webbing handles. Colors bright red, California blue. We have found the log carrier to be a very practical item. We burn 7-8 cords of wood a year and use one all the time. It eliminates dirt and debris from collecting on clothing and makes the work of carrying in wood easier. No more kicking in the door, with both arms carrying wood piled to your face.

Longwood Furnace Co.
Gallatin, MO 64640
IV (Furnaces)

Louisville Tin & Stove Co.
PO Box 1079
Louisville, KY 40201
(502) 589-5380
or contact
Whole Earth Access Co.
II (Heating Stoves)

Manufacturers of blued sheet metal wood burning heating products since 1888. Products include stoves, stove pipe ovens, stove pipe, elbows, pokers and other accessories. The stovepipe oven and the Derby Drum Heater are particularly unique. The oven replaces a section of six inch stovepipe permitting cooking without requiring a range. The Derby Heater offers the economy of a basic sheet steel stove, but also includes a large convenient reinforced front door opening.

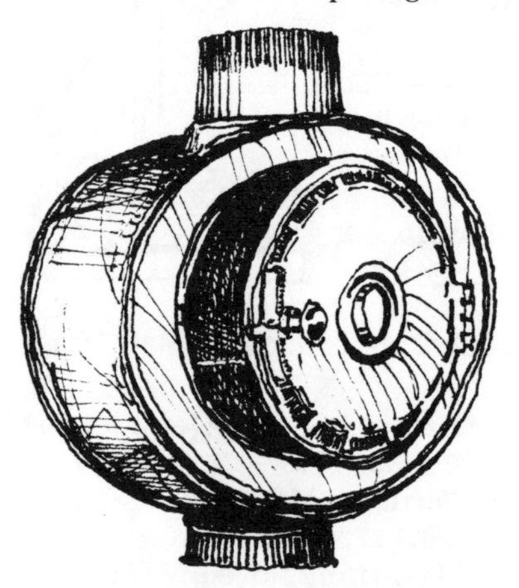

Flue Oven, Progress Drum Oven

Lynndale Manufacturing Company, Inc.
1309 North Hills Blvd.
Suite 207
North Little Rock, AK 72116
(501) 758-9602

PO Box 1154
Harrison, AK 72601
(501) 365-2378
IV (Furnaces & Boilers)

The old adage, necessity is the mother of invention,

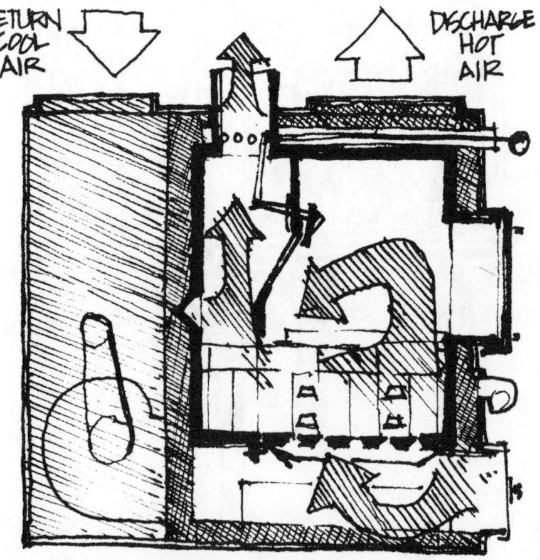

Furnace, "Wood Burner" #910

is still true today and is evidenced by the creation of the "wood burner". The "wood burner" was created because no central heating system existed which could give consumers advantages of a conventional furnace and economy of wood. Thus, 10 years of research and 5 months of design work gave birth to the "wood burner". Top priorities in construction are long life and low maintenance cost. Residential, Commercial, Industrial, Agricultural, Domestic Hot Water, combines with Solar.

Maine Wood Heat Co.
RD #1 Box 38
Norridgewock, ME 04957
(207) 696-5442

Majestic Company
Huntington, IN 46750
(219) 356-8000
V (Free-Standing Fireplaces); VI (Pre-Fab. Fireplaces); VII (Fireplace Accessories)

Malleable Iron Range Co.
715 N. Spring St.
Beaver Dam, WI 53916
(414) 887-8131

Malm Fireplaces, Inc.
368 Yolanda Ave.
Santa Rosa, CA 95404
(707) 546-8955
V (Free-Standing Fireplaces); VI (Pre-Fab. Fireplaces)

Marathon Heater Co.
Box 165 RD 2
Marathon, NY 13803

Marco Industries, Inc.
PO Box 6
Harrisonburg, VA 22801

Markade-Winnwood
4200 Birmingham Rd., NE
Kansas City, MO 64117
(816) 454-5260
II (Heating Stoves); IV (Furnaces & Boilers); V (Free-Standing Fireplaces); VII (Fireplace Access.); X (Barrel Stoves)

Martin Industries
PO Box 730
Sheffield, AL 35660
(205) 383-2421
I (Cooking Stove); II (Heating Stove); III (Circulating Heater); V (Free-Standing Fireplace)

Mechanical Product Development Corp.
Box 155
Swarthmore, PA 19081

***Merry Music Box (Styria Importers)**
20 McKown
Boothbay Harbor, ME 04538
(207) 633-2210
I (Cook Stoves); II (Heating Stoves)

We have used Austrian wood and coal burning heaters and ranges in our home for over 10 years now and our Styria products are built for us, according to our own specifications. We are your *only* source for these wonderful Austian ranges and heaters. We know them inside and out!

Metal Building Products, Inc.
35 Progress Ave.
Nashua, NH 03060
(603) 882-4271
III (Circulating Heaters); VI (Pre-Fab. Fireplaces)

Wood stove with 900 CFM Blower. Will heat area 3000 to 4000 square feet. Smaller models for homes. Contact manufacturer for prices.

Metal Concepts, Inc.
PO Box 25596
Seattle, WA 98125
(206) 365-3055
VII (Fireplace Accessories)

The Thermolux Fireplace Furnace is a supplemental heating device that takes some of your room air, forces it through a heat-exchanger under the coals and blows it back out into the room to keep you warmer. The Thermolux features a clean, straight-line design that really looks good. It enhances the charm of your fireplace while saving lots of precious heating dollars.

Modern-Aire
Modern Machine and Welding
Highway 2 West
Grand Rapids, Michigan, 55744

Modern Kit Sales
PO Box 12501
N. Kansas City, MO 64116

Mohawk Industries, Inc.
173 Howland Ave.
Adams, MA 01220
(413) 743-3648
II (Heating Stoves)

Using the down-draft principle, simple controls permit

Heating Stove, Tempwood

the Tempwood to work continuously for 12-14 hours without refueling. Air-tight and fully welded out of 13 gauge and ¼" steel, the Tempwood can produce 60,000 Btus an hour. Gives more heat for less wood and only has to be emptied at the very most five (5) times a year.

New England Fireplace Heaters, Inc.
372 Dorset St.
So. Burlington, VT 05401
(802) 658-4848
VII (Fireplace Accessories)

The New England Fireplace Heater converts an existing fireplace into an efficient heat source. Heats in excess of 1200 square feet. Fits most fireplaces, virtually no assembly or installation and has directional top flues and a stainless steel grate. Perfect for home or camp. A full one year guarantee and a complete functional and decorative accessory line available.

Newmac Mfg. Inc.
236 Norwich Ave.
Box 545
Woodstock, Ontario N4S 7W5, Canada
(519) 539-6147
IV (Furnaces & Boilers)

The Newmac Wood and Oil Combination Furnace burns wood and oil in separate combustion chambers; switches from wood to oil automatically; belt driven twin 10" blowers; burns wood completely to a fine powder; heavy stainless steel firebox liner; can be fired with wood when power is off; wood and oil fires thermostatically controlled.

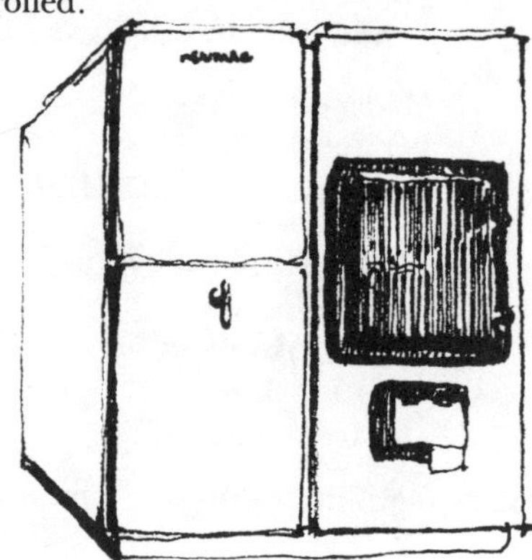

Furnace, Oil-Wood (front)

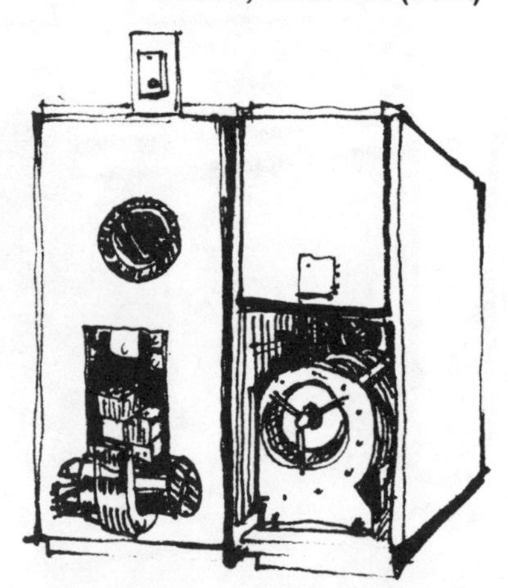

Furnace, Oil-Wood (rear)

Nichols Environmental
5 Apple Rd.
Beverly, MA 01915

Nortech Corporation
300 Greenwood Ave.
Midland Park, NJ 07432
(201) 445-6900
XI (Wood Splitters)

Manufacturer of the "Screw-Wedge" log splitters. Portable and economical. Split all wood safely, easily, and quickly.

***Old Country Appliances (Tirolia Importer)**
PO Box 330
Vacaville, CA 95688
or contact
Whole Earth Access Co.
(707) 448-8460
I (Cooking Stoves)

Tirolia cooks, bakes, and heats your house. This is Austria's finest range, famous in Europe since 1919, now available for immediate delivery in the USA. It is a modern, high quality, heavy duty, energy efficient, *easy* to use and clean kitchen range, soon to be a household word in American kitchens. It features air tight controlled draft, and full fire brick lining. High impact enamel finish available in colors; avocado green, coppertone brown, and white. Burns wood or coal. Hot water models available on special order. Many other features. Prices from $450 to $850. Send 35¢ for our illustrated brochure and technical information.

Cookstove, Innsbruck #D5N

Pioneer Lamps & Stoves
71A Yesler Way
Pioneer Sq. Station,
Seattle, WA 98104
I (Cook Stoves)

Portland Stove Foundry
57 Kennebec St.
Portland, ME 04104
(207) 773-0256
I (Cook Stoves); II (Heating Stoves); III (Circulating Heaters); V (Free-Standing Fireplace); VII (Fireplace Accessories); X (Barrel Stove)

***Preston Distributing Co. (Poele Importer)**
10 Whidden St.
Lowell, MA 01852
(617) 458-6303
III (Circulating Heaters)

It is natural to find safe, economical and efficient heating equipment being used in a country whose citizens are noted for their practical and frugal traits. In France people continued to depend upon coal and wood heaters long after the evolution of oil and gas units and these coal and wood heaters were constantly being updated and perfected. The results? Stoves that are outstanding in dependability, appearance and performance.

Preway, Inc.
Wisconsin Rapids, WI 54494
(715) 423-1100
V (Free-Standing Fireplaces); VI (Pre-Fab. Fireplace

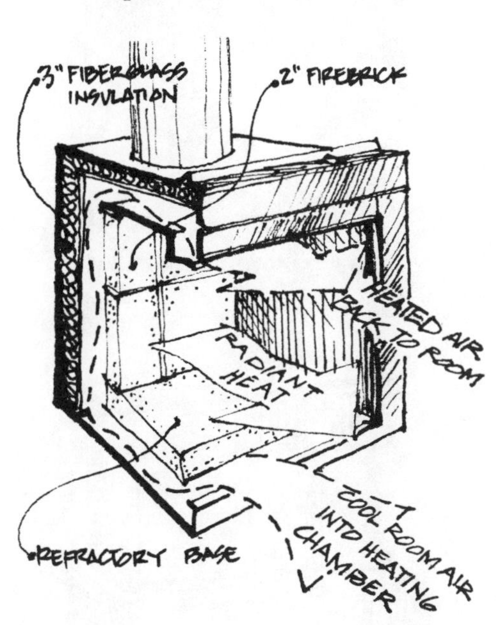

Pre-Fabricated Fireplace, Custom Built-In

Radiant Grate
31 Morgan Park
Clinton, CT 06413
(203) 669-6250
VII (Fireplace Accessories)

Ram & Forge
Brooks, ME 04921
(207) 722-3379
II (Heating Stoves); IV (Furnaces & Boilers)

REM Industries
408 C Simms Bldg.
Dayton, OH 45402

Ridgway Steel Fabricators, Inc.
Box 382
Bark St.
Ridgway, PA 15853
(814) 776-1323 or (814) 776-6156
VI (Pre-Fabricated Fireplace); VII (Fireplace Access.)

Hydroheat manufactures and distributes the most revolutionary and unique fireplace products available. The Hydroplace is a double-walled, water-circulating fireplace unit that captures fireplace heat and distributes it throughout the entire home. As water circulates through

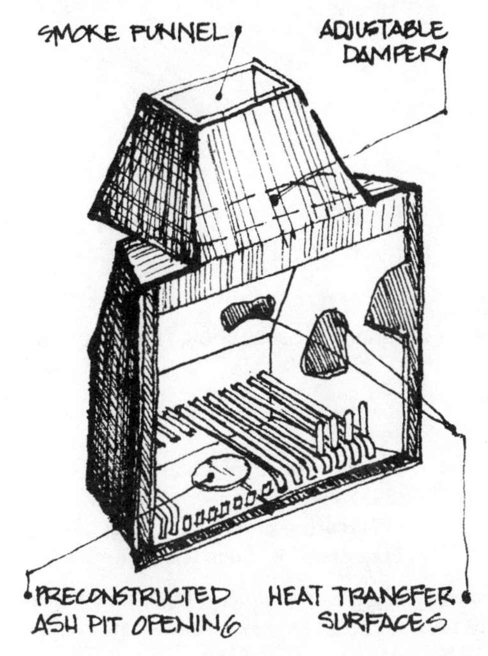

Pre-Fabricated Fireplace, Hydroplace

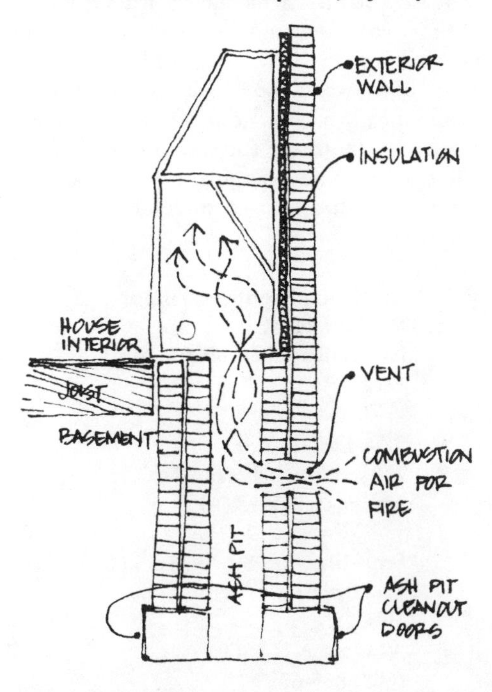

Pre-Fabricated Fireplace, Hydroplace (section)

the hollow steel jacket,it absorbs fireplace heat and then distributes it throughout the home via the existing central heating system. The Hydroplace resembles any conventional unit and is complete with floor, grates and adjustable damper. It is easily adaptable to homes with forced-air or electric heat.

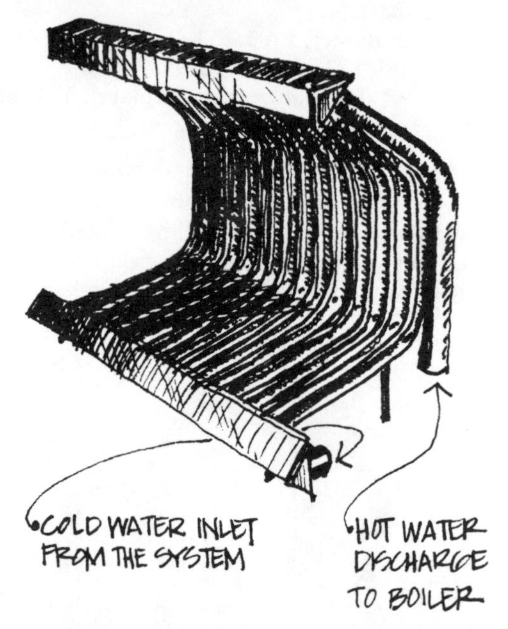

Fireplace Accessories, Hydrohearth

Riteway Manufacturing Co.
PO box 6
Harrisonburg, VA 22801
(703) 434-7090
II (Heating Stoves); III (Circulating Heaters); IV (Furnaces & Boilers)

Riteway pioneered the concept of complete combustion nearly four decades ago, and this unique design is still built into every Riteway heater, furnace, and boiler made. Riteway products achieve complete combustion by burning those gases that usually escape up the chimney when burning wood. Riteway's multi-fuel furnaces and boilers can burn wood, coal, gas or oil in any unit. They automatically switch from one fuel to another—no one needs to be there. Riteway offers a varied and proven line of heating equipment for those who are interested in a fresh approach to residential and commerical heating.

Ro Knich Products, Inc.
PO Box 311-E
No. Chicago, IL 60064

S/A Distributors
730 Midtown Plaza
Syracuse, NY 13210

***Scandinavian Stoves, Inc. (L. Lange & Co. Importer)**
Box 72
Alstead, NH 03602
(603) 835-6029
I (Cook Stoves); II (Heating Stoves)

Scandinavian Stoves, Inc. distributes in the United States stoves and fireplaces made by L. Lange & Co. of Denmark. The Lange Co. has been making fine cast iron wood and coal stoves for 125 years. The stoves are available in plain cast iron, or in enamel finish including red, blue, green, brown and black. They are designed for durability and efficiency. Most models will hold a fire many hours on one load of wood. Models include both coal and wood burners; a ship's stove; cook stoves; plus units suitable for heating anything from a bedroom to a whole house. Prices range from under $100 to over $700.

Heating Stove, Lange #6204

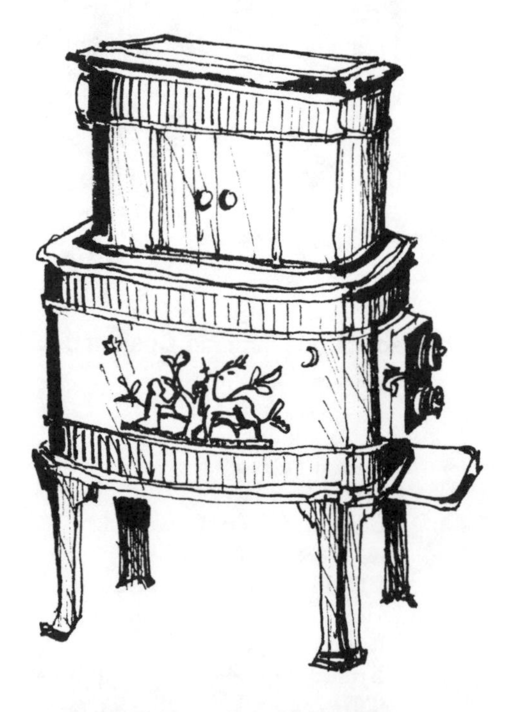

Heating Stove, Lange #6302K

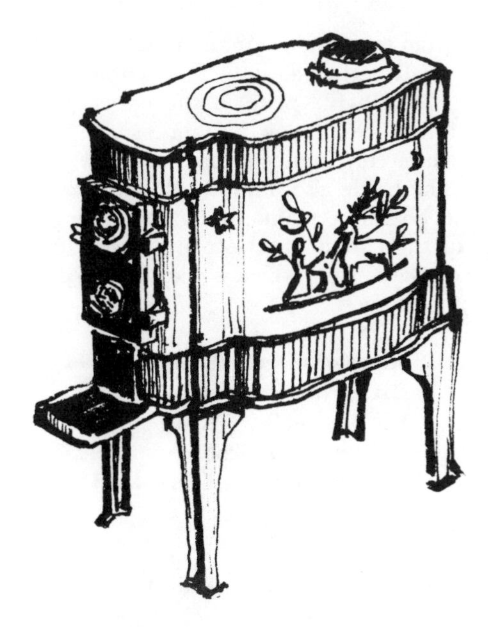

Heating Stove, Lange #6302A

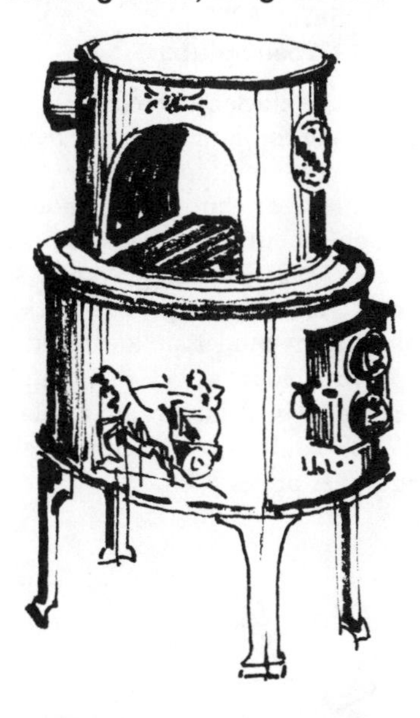

Heating Stove, Lange #6303

Scot's Stove Co.
11 Ells St.
Norwalk, CT 06850
II (Heating Stoves)

Self Sufficiency Products
1 Appletree Square
Minneapolis, MN 55420

Shenandoah Manufacturing Co., Inc.
PO Box 839
Harrisonburg, VA 22801
(703) 434-3838
II (Heating Stoves) III (Circulating Heaters); VII (Fireplace Accessories)

John P. Smith
174 Cedar St.
Branford, CT 06405
(203) 488-7225

Solar Sauna
Box 466
Hollis, NH 03049

Sotz Corp.
23797 Sprague Rd.
Columbia Station, OH 44028

Southeastern Vermont Community Action, Inc.
7-9 Westminster St.
Bellows Falls, VT 05101
(802) 463-4447
II (Heating Stoves)

A stove made from recycled propane tanks. Used mostly for heating, but has a cooking surface also. 237 lbs. $250 FOB Bellows Falls, Vermont. 6" flue.

***Southport Stoves (Morsø Importer)**
(Division of Howell Corporation)
248 Tolland St.
East Hartford, CT 06108
(203) 289-6079
II (Heating Stoves); V (Free-Standing Fireplaces

Southport stoves is the sole importer into the United States of the Danish MORSØ cast iron woodburning stoves and fireplaces. The controlled draft engineering features of these units, combined with the air-tight construction, results in a long-burning wood stove which will produce the maximum amount of heat with greater efficiency. The porcelainized enamel finish on the MORSØ stoves and fireplaces enhances the beauty and provides a permanent rustproof protection. The #2BO box stove illustrated is also available without the heat exchanger and both units are available in larger models.

Heating Stove, Morsø #2BO

Heating Stove, Morsø #1125

Sturges Heat Recovery, Inc.
PO Box 397
Stone Ridge, NY 12484
(914) 687-0281
VII (Fireplace Accessories); IX (Heat Reclaimer)

Suburban Manufacturing Company
4700 Forest Dr.
PO Box 6472
Columbia, SC 29206
(803) 782-2649
III (Circulating Heaters)

Suburban Manufacturing Co.
PO Box 399
Dayton, TN 37321

Sunshine Stove Works
RD 1 Box 38
Norridgewock, ME 04957
(914) 887-4580
II (Heating Stoves)

The design of our stove is based on the well known Scandinavian front-end combustion system, in which the primary draft enters the stove at the front rather than underneath; a baffle plate in the upper part of the firebox forces the air flow into an S-shape so that the wood burns evenly from front to back, slowly and completely. The baffle also improves the heating efficiency since the hot air remains in the stove longer before going up the chimney. A secondary airdraft helps to burn the volatile gases which is necessary for more complete combustion.

Heating Stove, Sunshine Stove

Superior Fireplace Co.
Div. of Mobex Corp.
PO Box 2066
Fullerton, CA 92633
VI (Pre-Fabricated Fireplace)

Taos Equipment Manufacturers, Inc.
Box 1565
Taos, NM 87571
(505) 758-8253
XI (Wood Splitters)

The Stickler is a vehicle-powered screw type log splitter. It offers the unique combination of extremely high power with very low cost and an absence of maintenance. Powered by the rear hub of a motor vehicle, the Stickler will split about one cord of hardwood per hour and retails for $199, including engine shut-off switch and replaceable alloy steel tip. The Stickler is available in four bolt-pattern models, allowing it to be used on any pickup, van, four wheel drive and on 90% of American car models. Adapters are available for $39.50. Garden tractor and farm tractor Sticklers are also available at higher prices.

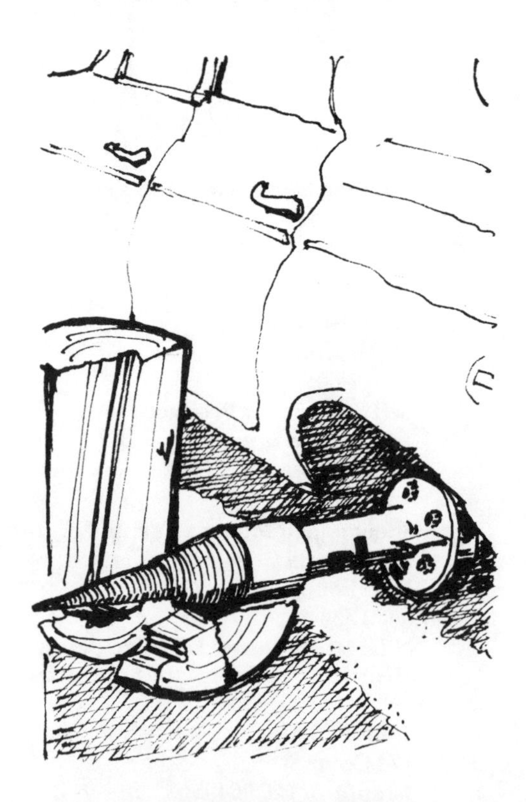

Wood Splitters, The Stickler

***Tekton Design Corp. (Tasso & Kedelfabric-Tarm importer)**
Conway, MA 01341
(413) 369-4685
IV (Furnaces & Boilers)

Tekton Design Corp. imports central heating boilers from Denmark, where traditional high energy costs have forced the development of very advanced wood burning equipment. Several models are all-in-one units that supply heat and domestic hot water as well. They burn wood or coal in one firebox and oil or gas in another. The oil or gas burner will take over automatically when the wood or coal fire dies out. Other models, some very large, have single fireboxes. They can be combined with an existing hot water system for automatic multi-fuel heating. Reasonably priced, they sell for $750 to $2,000.

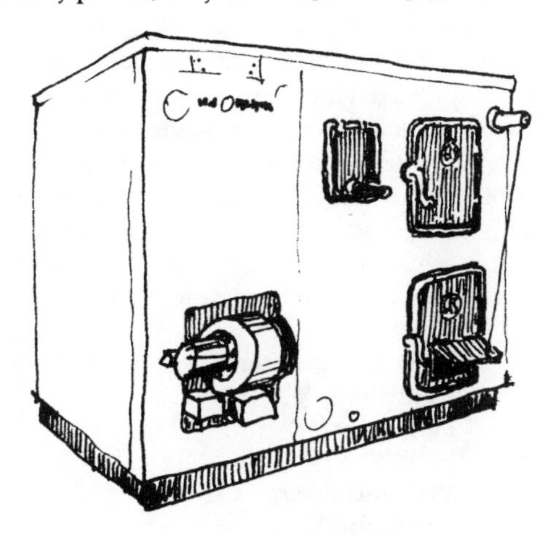

Furnace, HS#OT-70

Furnace, Tasso #A3

Temco
PO Box 1184
Nashville, NH 37202
(615) 297-7551
V (Free-Standing Fireplace); VI (Pre-Fabricated Fireplace

Thermalite Corp.
Dept. PS Box 69
Hanover, MA 02339

Thermo Control Wood Stoves
Central Bridge, NY 12035
II (Heating Stoves)

Thermo-Rite
The Fireplace House
1950 Wadsworth
Denver, CO 80215

Thermograte, Inc.
300 Atwater St.
St. Paul MN 55117
or contact
The Whole Earth Access Co.
(612) 489-8863
VII (Fireplace Accessories)

When you sit before a roaring fire, the heat you feel is radiant heat. You roast on the front side but your back side is cold. Most of the heat generated goes up the chimney. Thermograte captures heat that normally goes up the chimney, heats the air inside the tubes and delivers warm air back into the room.

Torrid Manufacturing Co., Inc.
1248 Poplar Place So.
Seattle, WA 98144
(206) 324-2754
V (Free-Standing Fireplace); IX (Heat Reclaimers)

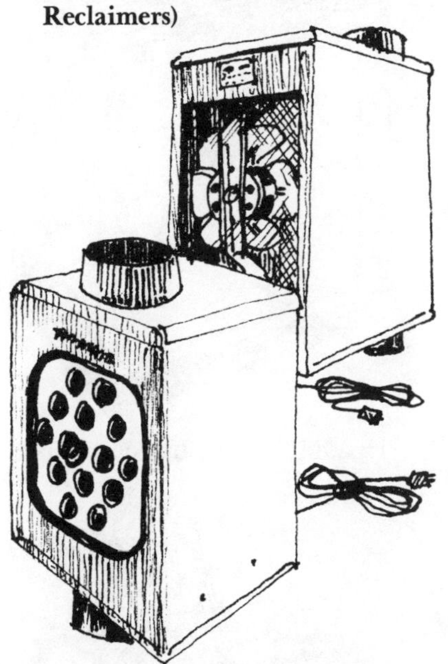

Heat Reclaimer, Air Heat Saver

Triway Mfg. Inc.
7819 Old Highway 99
Box 37
Marysville, WA 98270

United States Stove Company
PO Box 151
South Pittsburg, TN 37380
(615) 837-8631
II (Heating Stoves); III (Circulating Heaters); V (Free-Standing Fireplaces)

United States Stove Company has a history of over

100 years in the same location, located in the mid-South. Product lines consist of gas, oil, coal and wood heating equipment for the home. Shipments are made to all 50 states and foreign countries. A rapid expansion program is under way to take care of market requirements. Sizeable operations in both steel working and foundry divisions.

Circulating Heater, Wonderwood #726

Heating Stove, Boxwood #132-S

Free-Standing Fireplace, Franklin #301-ST

Vaporpack, Inc.
Box 428
Exeter, NH 03833
(603) 778-0509
IX (Heat Reclaimers)

Vermont Castings, Inc.
Box 126
Prince Street
Randolph, VT 05060
(802) 728-3355
II (Heating Stoves)

The Defiant is an all cast iron, airtight, thermostatically controlled, automatic, combination heater/fireplace/cook stove. It measures 33" high, 36" long, by 22" deep and weighs 330 pounds. It holds 70 pounds of 26" long wood, will provide up to 55,000 Btu's per hour of heat and will hold a fire 16 hours and longer. It can be operated either as an updraft or horizontal combustion stove and features preheated primary and secondary air, a secondary combustion chamber of over 1800 cubic inches and a 60" long internal flamepath. A flue collar accepts standard 8" pipe which may be reduced to 7" if desired. The thermostat control lever, door pins, finials and handles are nickel plated.

Vermont Counterflow Wood Furnace
Plainfield, VT 05667
IV (Furnaces)

Vermont Energy Products
100 Broad St.
Lyndonville, VT 05851
(802) 626-8842

Vermont Iron Stove Works, Inc.
The Bobbin Mill
Warren, VT 05674
(802) 496-2821
II (Heating Stoves)

Vermont Iron Stove Works, Inc. is engaged in the design and manufacture of high-quality efficient, economical, sturdy, and handsome cast iron wood stoves. We are seeking to preserve and enhance the nearly-

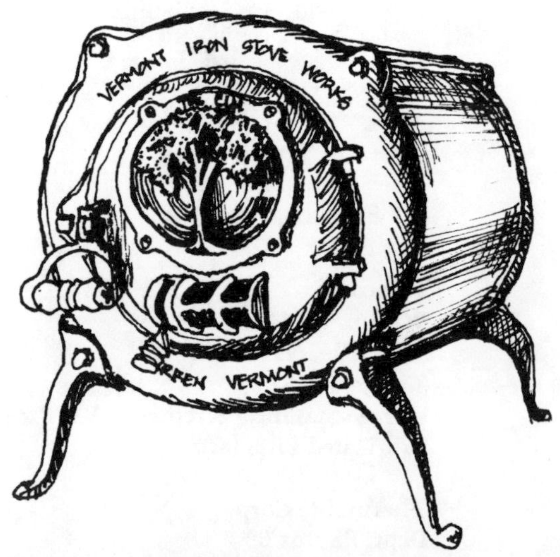

Heating Stove, The Elm

forgotten art of cast iron stove-making by providing a line of stoves whose designs are not merely copies or mimics of traditional stoves but are unique artforms in themselves. Our stoves are also designed with knowledge of the most up-to-date aspects of wood combustion technology. Our first stove, The Elm, is a cylindrical automatic parlor stove with a window in the door. This simple, yet unique, stove will be available by mid-summer, 1976. Three other stoves, The Maple, The Birch, and The Oak, will come out soon.

Vermont Soapstone Co.
Pekinsville, VT 05151
II (Heating Soves)

Vermont Techniques, Inc.
PO Box 107
Northfield, VT 05663
(802) 485-7905
VII (Fireplace Accessories)

Vermont Woodstove Company
307 Elm St.
Bennington, VT 05201
(802) 442-3985
II (Heating Stoves)

Vermont Woodstove Company is sole manufacturer and distributor of the unique "DownDrafter" woodstove. Two models are offered capable of heating either a few rooms or a house. Both are thermostatically controlled and both employ a true downdraft combustion principle for the utmost in safety, efficiency and convenience. The larger DDI has built-in chamber through which room air is blown. (It can also be coupled to an existing ductwork system.) Both models emphasize quality of materials, workmanship and design. Informative literature is available on a number of woodburning topics. Write or call direct for more information.

Volunteers in Technical Assistance
3706 Rhode Island Ave.
Mt. Rainier, MD 20822

E. G. Wasburne & Co.
83 Andover St.
Danvers, MA 01923

Washington Stove Works
PO Box 687
3402 Smith St.
Everett, WA 98201
or contact
Whole Earth Access Co.
(206) 252-2148
I (Cook Stoves); II (Heating Stoves); III (Circulating Heaters); V (Free-Standing Fireplace); VII (Fireplace Accessories); X (Barrel Stoves)

The production of cast-iron stoves of superb quality is a 100-year-old tradition at Washington Stove Works. Iron casting is an art which has stubbornly withstood the onslaught of modern technology. While there are machines which produce castings, the truly beautiful, big, handsomely-figured pieces are still the work of the skilled artisan using his hands. Most of Washington's stoves — at least, the free-standing types — are ageless designs from the past, such as the Franklin, the Parlor Stove and the Cannonball (potbelly). All of these are made up of parts cast in original molds to retain their authentic appearance.

Heating Stove, Arctic #25

Barrel Stove Kit, Oil Drum Conversion Door and Legs

Waterford Ironfounders Limited
Waterford, Ireland
Waterford 5911 Cable Iron Waterford; Telex 8763E1
I (Cooking Stoves)

Manufacturers of Waterford cooking ranges (traditional in Ireland since 1870). These ranges are constructed mainly of cast iron, with large hobs/hot plates for boiling and simmering. Large oven capacity for consistently good baking, roasting, casseroles, etc. Fireboxes burn timber, peat and soft coal efficiently and economically. The ranges operate on natural draft with adeaquate built-in draft controls and connect directly or by 5 inch diameter flue pipe to house flue. The finish is matte black.

White Mesa, Inc.
110 Laguna N.W.
Albuquerque, NM 87104
(505) 247-1066

The Hearth Heater is a fireplace surrounded by an air chamber. Cool floor level air is swept into the chamber, around the firebox, rapidly heated and forced into the room by the thermostatically controlled fan. This means more heat from a smaller fire than is possible with the typical "cone", yet allows the visual enjoyment. The firebox is 12 ga., the skin 16 ga. steel. Write for full details and cost.

Whitten Enterprises
Arlington, VT 05250

Whittier Steel & Mfg., Inc.
10725 S. Painter Ave.
Santa Fe Springs, CA 90670

Whole Earth Access Company
2466 Shattuck Ave.
Berkeley, CA 94704
(415) 848-0510
X (Barrel Stoves)

Whole Earth Access Company Oil Drum Conversion Kit: With the use of a screwdriver, drill, wrench, and hand metal-cutting saw (or jigsaw or cutting torch) plus this kit, and a little sand, you can convert a standard fifty-five gallon oil drum into a quite serviceable wood-burning stove (or trash burner). Kit inlcudes cast iron legs, collar, door, door frame, draft control, and secondary draft control cover, as well as nuts, bolts, washers, stove cement, and threaded steel rods to run between leg castings for added support. It even includes instructions. Castings are also available separately.

Wilson Industries
2296 Wycliff
St. Paul, MN 55114
(612) 646-7214

The Yukon combination furnace has a heat exchanger which puts comfort into the home, not out the chimney. Its Honeywell control system operates the unit at high efficiency—automatically. Cast iron grates enable furnace to withstand high temperatures of a wood fire. Firebrick lining retains precious heat. Heavy steel construction insures a long life. Secondary air fan controls burning of wood. Its 10 inch, belt driven blower is quiet, yet will even handle a 3½ ton air conditioner. Combustion chamber holds firewood up to 24 inches long and 10 inches in diameter with a separate combustion chamber for oil or gas. Thermostatically controlled gas or oil burner ignites if wood gets too low.

***Woodburning Specialties (Hunter Importer)**
PO Box 5
No. Marshfield, MA 02059
III (Circulating Heaters); IV (Furnaces)

Yankee Woodstoves
Cross St.
Bennington, NH 03442
(603) 588-6358
II (Heating Stoves)

Nearly all types of wood burning stoves are round in shape. Since metal expands and contracts with alternate heating and cooling, the round shape evenly takes these forces with minimum stress and no distortion. Yankee Woodstoves uses the round shape in its simplest, yet most rugged configuration: a heavy 18 guage steel drum. Reinforcing ribs in the upright models also give extra rigidity. Additionally, cast iron parts are used wherever extra ruggedness and dependability are required in the overall construction. The final result is a stove that performs as well as the most expensive types but costs considerably less.

Section Three Manufacturers Specification Charts

EDITOR'S NOTE

The following divisions are made up of specifications charts indicating the most significant categories of information for each type of product (division). Categories of information change from division to division and reflect the most important characteristics for each product type. The editor acknowledges that there are blank spaces under some products. We have attempted to furnish the maximum information possible, however in some instances information was not available from the manufacturer. We are hopeful that as addendums are published to the specification section of the Encyclopedia, manufacturers will provide the missing information for their products. Information contained in the specification section has been taken from manufacturers' literature and specification sheets. The Wood Burner's Encyclopedia does not guarantee or testify to the accuracy of the information supplied herein.

Sizing of Appliances: It is the editor's opinion that all manufacturers should rate the heating capacity of their products in Btu's (British Thermal Units per hour). Other indications of the capacity of an appliance can be vague, inaccurate and problematic to a consumer as an attempt is made to accurately choose the right size appliance. With this in mind the Encyclopedia will only rate those appliances stating a specific Btu capability. We hope that manufacturers will be encouraged to rate all appliances in this manner. When additional products are rated in Btu's per hour, this information will be included in future addendums.

Much careful consideration has been given to the question of publishing the *price* of products listed in this section. As a result of lengthy examination, prices have *not* been included. This position has been taken due to the continuing change in retail prices and the effect of shipping cost differentials from east coast to west coast. Therefore, we suggest you contact the manufacturer or importer to determine the current sales price.

SW = Shipping Weight

Division I — Cook Stoves

EDITOR'S NOTE

Cook Stoves are defined as any appliance that has both a cooking surface and a warming or baking oven.

DIVISION I — COOK STOVES							
MANUFACTURER	ATLANTA STOVE WORKS, INC.			AUTOCRAT CORP.			
PRODUCT	#15-36	#8316	#5115LB	HILLCREST 76-82VF-HS	RIDGETOP 566-82VC-HS	NOBLE 27-12A	KITCHEN HEATER 3666
OVERALL SIZE							36"H x 14½"W x 31½"
SIZE OF COOKTOP	35¼" x 21¼"	30¼" x 21½"	36" x 21½"	21¼" x 37½"	21¼ x 34½"	19½" x 24½"	18½" x 11"
NUMBER OF GRIDDLES	4/8" - 2/5½"	4/8"	6/8"	6/8"	6/8"	4/7"	1/8"
NUMBER OF OVENS	1	1	1	1	1	1	
SIZE OF OVEN(S)	15"x14"x11"	13½"x13"x10"	15"x14"x11"	17"x18"x11"	17"x18"x11"	12"x16"x9"	
HEIGHT OF COOKING SURFACE	29½"	28¾"	31"	31"	31"	26½"	36"
WEIGHT	285 LBS.	155 LBS.	362 LBS.	240 LBS.	215 LBS.	100 LBS.	160 S.W.
FIREBOX SIZE							8½"x17"x8½"
FLUE SIZE	7"	6"	7"				6"
FLUE LOCATION						TOP	
BODY FINISH	PLAIN BLACK	PLAIN BLACK	SEMI-ENAMELED	PORCELAIN ENAMEL	PORCELAIN ENAMEL		PORCELAIN ENAMEL
BODY MATERIALS	CAST IRON	CAST IRON	CAST IRON	20 GA. STEEL	20 GA. STEEL	22 GA. STEEL	
FIREBOX MATERIALS							CAST IRON LININGS
OVEN THERMOMETER			YES	YES			
GRATES				FIREBACK - DUPLEX	FIREBACK - DUPLEX	FIREBACK - 2 GRATES	DUPLEX
WATER HEATER			RESEVOIR	RESEVOIR - 22 QT.			
FEATURES	WARMING OVEN	17½" WOOD LENGTH	HIGH SHELF	HIGH SHELF	HIGH SHELF		
	15" WOOD LENGTH		15" WOOD LENGTH				

1 DIVISION

COOK STOVES

SPECIFICATIONS

MANUFACTURER	BOW AND ARROW STOVE CO.	HOME FIREPLACES, LTD	MARTIN INDUSTRIES		THE MERRY MUSIC BOX		
PRODUCT	PREPOROD STG-5	COOK STOVE CS-54	MARCO PRIDE RANGE	SUPREME RANGE	STYRIA 92	STYRIA 106	STYRIA 130
OVERALL SIZE	33 1/2"H x 28 1/2"W 20"D	35"W			34"H x 26"D x 36"L	34"H x 34"D x 42"L	34"H x 40"D x 51"L
SIZE OF COOKTOP	20" x 16 1/2"		35" x 23"	40" x 23 5/8"	23" x 16 1/2"	29" x 21"	36" x 24"
NUMBER OF GRIDDLES		6	6/8"	5			
NUMBER OF OVENS	1	1	1	1	1	1	1
SIZE OF OVEN(S)	10 3/4"H x 15 1/2"W 18"D		16" x 15 1/2" x 10 1/2"	17" x 16" x 11 3/4"	8 1/2" x 11" x 19"D	9" x 14" x 23"D	9" x 14" x 29"D
HEIGHT OF COOKING SURFACE			31"	33 3/4"	34"	34"	34"
WEIGHT	222 LBS.	285 S.W.	322 S.W.	466 LBS.	409 LBS.	611 LBS.	1040 LBS.
FIREBOX SIZE	5"H x 6 1/2"W x 16"D		18" x 8" x 8"	18" x 17 1/4" x 8"	5" x 4 1/2" x 15"D	5" x 7 1/2" x 18 1/2"D	5" x 7 1/2" x 20"D
FLUE SIZE	5"		7"	7"	5"		6"
FLUE LOCATION	LEFT; RIGHT				REAR	REAR	REAR
BODY FINISH	VITREOUS ENAMEL			PORCELAIN ENAMEL	ENAMEL	ENAMEL	ENAMEL
BODY MATERIALS		CAST IRON	CAST IRON				
FIREBOX MATERIALS	FIREBRICK LINING		LINING	FRONT-BACK LININGS	FIREBRICK LINING	FIREBRICK LINING	FIREBRICK LINING
OVEN THERMOMETER			YES	YES			
GRATES	YES		YES	YES			
WATER HEATER				RESEVOIR	RESEVOIR 2 1/2 GAL.	RESEVOIR 4 1/2 GAL.	RESEVOIR 6 1/2 GAL.
FEATURES	GLASS OVEN DOOR	WARMING OVEN	WARMING OVEN	HIGH SHELF	TOWEL RACK	TOWEL RACK	TOWEL RACK
					12" WOOD LENGTH		

MANUFACTURER	OLD COUNTRY APPLIANCES					PORTLAND STOVE FOUNDRY CO.	
PRODUCT	INNSBRUCK D5N	SALZBURG D7N	VIENNA D9N	ALPINE D6	THRIFTMASTER SD4	QUEEN ATLANTIC 408	QUEEN ATLANTIC 308C
OVERALL SIZE	33 1/2"H x 29"W 23 1/2"D	33 1/2"H x 35 1/2"W 23 1/2"D	33 1/2"H x 43 1/4"W 23 1/2"D	31 1/2"H x 33 1/2"W 21 1/2"D	28 1/4"H x 26 3/4"W 20 1/2"D	32 1/2"H x 30 3/4"D 52"L	32 1/4"H x 30 3/4"D 44 1/2"L
SIZE OF COOKTOP							
NUMBER OF GRIDDLES						6	6
NUMBER OF OVENS	1	1	1	1	1	1	1
SIZE OF OVEN(S)	8 1/4"H x 11 3/4"W 21 1/4"D	8 1/4"H x 11 3/4"W 21 1/4"D	9 1/2"H x 15 3/4"W 21 1/4"D	8 1/4"H x 10 1/4"W 19"D	8"H x 10"W x 17"D	20" x 20"	20" x 20"
HEIGHT OF COOKING SURFACE	33 1/2"	33 1/2"	33 1/2"	31 1/2"	28 1/4"	32 1/2"	32 1/4"
WEIGHT	355 LBS.	420 LBS.	540 LBS.	310 LBS.	245 LBS.		
FIREBOX SIZE	15 3/4"D	15 3/4"D	16 1/2"D	8"H x 13 3/4"D	6 1/4"H x 11 3/4"D	9"W x 23"L x 7"D	9"W x 23"L x 7"D
FLUE SIZE	5"	5"	6"	5"	5"		
FLUE LOCATION	SIDE; TOP; BACK	SIDE; TOP; BACK	SIDE; TOP; BACK	SIDE; TOP; BACK	SIDE; BACK		
BODY FINISH	ENAMEL	ENAMEL	ENAMEL	ENAMEL	ENAMEL	PORCELAIN ENAMEL	STOVE BLACKING
BODY MATERIALS						CAST IRON	CAST IRON
FIREBOX MATERIALS	FIRE CLAY LINING	FIRE CLAY LINING	FIRE CLAY LINING	FIRE CLAY LINING	FIRE CLAY LINING	LININGS	LININGS
OVEN THERMOMETER						YES	YES
GRATES	YES	YES	YES				
WATER HEATER		10 QT. RESEVOIR	19 QT. RESEVOIR	10 QT. RESEVOIR		6 GAL. RESEVOIR	6 GAL. RESEVOIR
FEATURES	BAKING SHELF	BAKING SHELF	BAKING SHELF	BAKING SHELF	BAKING SHELF	WARMING OVEN	WARMING OVEN
						HIGH SHELF	HIGH SHELF

1 DIVISION

COOK STOVES — SPECIFICATIONS

MANUFACTURER	PORTLAND STOVE FOUNDRY CO.				WASHINGTON STOVE WORKS		
PRODUCT	ATLANTIC #8	ATLANTIC KITCHEN HEATER 121B	TROLLA #354	TROLLA #325	OLYMPIC 18-W	OLYMPIC 8-15	OLYMPIC 7-14
OVERALL SIZE	28½"H × 38½"L	36"H × 15"W × 27½"D	24¾"H × 29"W × 18"D	15¾"H × 21½"W × 14"D			
SIZE OF COOKTOP	30" × 26½"				32" × 22½"	34" × 23½"	26½" × 19½"
NUMBER OF GRIDDLES	4/8"		2/12½"	2/9½"	2	2	2
NUMBER OF OVENS	1		1		1	1	1
SIZE OF OVEN(S)	20" × 22"		13" × 13" × 7"		13"H × 18"W × 18"D	10"H × 18"W × 15"D	8½"H × 16½"W × 13½"
HEIGHT OF COOKING SURFACE	28½	36"	24¾" (32")	15¾ (22¾")	32"	30¾"	30"
WEIGHT			220 LBS.	92½ LBS.	320 S.W.	245 S.W.	160 S.W.
FIREBOX SIZE	24"L	8½"W × 17½"L × 8"D			8½"H × 8"W × 16"D	8¾"H × 8"W × 19¾"D	7"H × 6½"W × 18"D
FLUE SIZE	6"	6"	5"	4"	7"	6"	6"
FLUE LOCATION	TOP	REAR	TOP	TOP			TOP
BODY FINISH	STOVE BLACKING	ENAMEL STEEL CASING	PORCELAIN	ENAMEL	PORCELAIN ENAMEL	MATTE BLACK	MATTE BLACK
BODY MATERIALS	CAST IRON	CAST IRON	CAST IRON	CAST IRON			
FIREBOX MATERIALS		SIDE LININGS			LINING	LINING	LINING
OVEN THERMOMETER					YES	YES	YES
GRATES					YES	YES	YES
WATER HEATER	6 GAL. RESERVOIR				COPPER COILS AVAILABLE		
FEATURES		HOT WATER COIL			HIGH SHELF	HIGH SHELF	

MANUFACTURER	WASHINGTON STOVE WORKS		WATERFORD IRONFOUNDRIES LTD.				
PRODUCT	FOC'SLE HEATER NEPTUNE 1A	FOC'SLE HEATER NEPTUNE 2A	WATERFORD	WATERFORD TRUBURN	WATERFORD STANLEY		
OVERALL SIZE	22"H	23"H					
SIZE OF COOKTOP	20¾" × 15"	24½" × 17"	29½" × 19½"	31" × 18½"	36" × 21¾"		
NUMBER OF GRIDDLES							
NUMBER OF OVENS	1	1	1	1	1		
SIZE OF OVEN(S)	9" × 9½" × 7"	11" × 11¼" × 8"	14" × 13½" × 9"	13¼" × 14½" × 13¼	15¾" × 15½" × 13"		
HEIGHT OF COOKING SURFACE			25¼"	30¼"	28"		
WEIGHT	143 S.W.	188 S.W.					
FIREBOX SIZE							
FLUE SIZE	5"	5"					
FLUE LOCATION	TOP	TOP					
BODY FINISH			MATTE BLACK	MATTE BLACK	MATTE BLACK		
BODY MATERIALS		CAST IRON	CAST IRON	CAST IRON	CAST IRON		
FIREBOX MATERIALS							
OVEN THERMOMETER							
GRATES		YES					
WATER HEATER							
FEATURES							

EDITOR'S NOTE

Heating Stoves are defined as appliances designed for space heating purposes. Not intended to be used with open doors.

*indicates optional at extra cost

**indicates hearth included

II DIVISION — HEATING STOVES SPECIFICATIONS

MANUFACTURER	ABUNDANT LIFE FARM	ASHLEY PRODUCTS DIVISION			ATLANTA STOVE WORKS, INC.		
PRODUCT	"COZY" PARLOUR STOVE	COLUMBIAN 25-HF REGULAR	CAROLINIAN 23-HF	CAROLINA 23-H	#2502	WOODBOX HEATER #23	WOODBOX HEATER #27
WOOD LENGTH	16"					20½"	24"
FIREDOOR SIZE		(TOP) 13¼ x 15" (FRONT) 10¾ x 12"	(TOP) 12" x 13" (FRONT) 10¾ x 12"	12" x 13"			
OVERALL SIZE		34"H x 20"W x 30"D	30"H x 18"W x 28½"	30"H x 18"W x 30½"D	35½"H		
WEIGHT		130 S.W.	110 S.W.	101 S.W.	147 S.W.	90 LBS.	115 LBS.
FIREBOX SIZE		23"H x 17½"W x 25"L	21"H x 16"W x 22¼"L	21"H x 16"W x 22¼"L	22" x 17" x 23½"		
FLUE SIZE					6"	6"	6"
FLUE LOCATION	REAR	TOP	TOP	TOP	TOP	TOP	TOP
BODY FINISH							
BODY MATERIALS	CAST IRON	STEEL	STEEL	STEEL	STEEL	CAST IRON	CAST IRON
THERMOSTATIC DRAFT CONTROL		YES	YES	YES	YES		
BTU RATING							
FEATURES	*	DELUXE MODEL AVAILABLE					

MANUFACTURER	ATLANTA STOVE WORK, INC.	AUTOCRAT CORPORATION					
PRODUCT	WOODBOX HEATER #32	AUBURN 2322	AUBURN 2326	KING DOWN-DRAFT 2722	KING DOWN-DRAFT 2726	WILDWOOD 2126	PINEWOOD 2018
WOOD LENGTH	27"						
FIREDOOR SIZE		10½"W x 9½"L	11⅜"W x 13⅜"L	10½"W x 9½"L	11⅜"W x 13⅜"L	11⅜"W x 13⅜"L	9¾"W x 8½"L
OVERALL SIZE		26¾"H x 15½"W 23⅞"L	31"H x 17½"W 28⅜"L	26¾"H x 15½"W 24⅞"L	31"H x 17½"W 29⅜"L	31"H x 17½"W 28⅞"L	18"H x 13¾"W 21¾"L
WEIGHT	125 LBS.	22½ LBS.	34½ LBS.	19 LBS.	26 LBS.	25 LBS.	7½ LBS.
FIREBOX SIZE		20¼"H x 15½"W 22"L	24½"H x 17½"W 26½"L	20¼"H x 15½"W 22"L	24½"H x 17½"W 26½"L	24½"H x 17½"W 26½"L	12½"H x 13¾"W 18½"L
FLUE SIZE	6"	6"	6"	6"	6"	6"	6"
FLUE LOCATION	TOP	TOP	TOP	TOP	TOP	TOP	TOP
BODY FINISH							
BODY MATERIALS	CAST IRON						
THERMOSTATIC DRAFT CONTROL							
BTU RATING							
FEATURES							

MANUFACTURER	AUTOCRAT CORPORATION			BOW AND ARROW STOVE CO.			
PRODUCT	PINEWOOD 2218	PINEWOOD 2222	PINEWOOD 2226	FYRTØNDEN MODEL A	FYRTØNDEN MODEL B	FYRTØNDEN MODEL C	FYRTØNDEN MODEL D
WOOD LENGTH				18"	16"	14"-16"	14"
FIREDOOR SIZE	9¾"W x 8½"L	10½"W x 9½"L	11⅜"W x 13⅜"L				
OVERALL SIZE	18"H x 13¾"W 21¾"L	26¾"H x 15½"W 25¾"L	31"H x 17½"W 29¾"L	34¼"H x 23½"D	31½"H x 21½"D	37¾"H x 19"D	27½"H x 19"D
WEIGHT	9½ LBS.	15½ LBS.	22 LBS.	287 LBS.	265 LBS.	221 LBS.	118 LBS.
FIREBOX SIZE	12½"H x 13¾"W 18½"L	20¼"H x 15½"W 22"L	24½"H x 17½"W 26½"L	4600 CU. IN.	4100 CU. IN.	4000 CU. IN.	2700 CU. IN.
FLUE SIZE	6"	6"	6"	7"	6"	6"	6"
FLUE LOCATION	TOP	TOP	TOP	TOP; REAR	TOP; REAR	TOP; REAR	TOP; REAR
BODY FINISH				STOVE BLACK	STOVE BLACK	STOVE BLACK	STOVE BLACK
BODY MATERIALS				4 MM STEEL	4 MM STEEL	4 MM STEEL	4 MM STEEL
THERMOSTATIC DRAFT CONTROL							
BTU RATING							
FEATURES				FOOD WARMING GRATES	FOOD WARMING GRATES	FOOD WARMING GRATES	FOOD WARMING GRATES
						RING LID	RING LID
				SCREEN, GRATE TONGS, SHOVEL			
				FIREBRICK FLOOR	FIREBRICK FLOOR	FIREBRICK FLOOR	FIREBRICK FLOOR
				SECONDARY COMBUSTION CHAMBER			
				FIREBRICK LINED	FIREBRICK LINED	FIREBRICK LINED	FIREBRICK LINED

II DIVISION

HEATING STOVES SPECIFICATIONS

MANUFACTURER	BOW AND ARROW STOVE CO.			DAMSITE DYNAMITE STOVE CO.			
PRODUCT	PETIT GODIN #3720	PETIT GODIN #3721	PARRA SAUNA STOVE	DYNAMITE BOX STOVE	GREENWOOD DYNAMITE STOVE	DOUBLE DYNAMITE STOVE	DYNAMITE KITCHEN RANGE
WOOD LENGTH	16"	20"	18"	24"	24"	24"	24"
FIREDOOR SIZE							
OVERALL SIZE	32 1/4"H x 16"W x 21"D	39"H x 21"W x 27"D	29 1/2"H x 17"W x 18"D	27"H x 18"W x 34"L	38"H x 18"W x 34"L	38"H x 34 1/2"W x 26"	38"H x 34 1/2"W x 26"L
WEIGHT	128 LBS.	194 LBS.	133 LBS. (WITH TANK)	100 LBS.	200 LBS.	200 LBS.	265 LBS.
FIREBOX SIZE	470 CU. IN.	1700 CU. IN.	2700 CU. IN.	17" x 25" x 19"	17" x 25" x 21"	17" x 25" x 19"	17" x 26" x 19"
FLUE SIZE	4"	4"	5 1/4"	6"	6"	6"	8"
FLUE LOCATION	REAR	REAR	TOP	TOP	REAR	REAR	TOP
BODY FINISH	STOVE BLACK	STOVE BLACK					
BODY MATERIALS	STEEL/CAST IRON	STEEL/CAST IRON	ROSTALT STEEL STAINLESS STEEL	10 GA. STEEL	10 GA. STEEL	10 GA. STEEL	10 GA. STEEL
THERMOSTATIC DRAFT CONTROL							
BTU RATING	5 K.W.	5 K.W.					
FEATURES	SECONDARY AIR CHANNEL		STAINLESS STEEL WATER TANK				TWO OVENS FOR BAKING
	ALL PURPOSE TOOL	ALL PURPOSE TOOL	FIREBOX- 4MM STEEL PLATE				
	FIREBRICK LINED	FIREBRICK LINED					

MANUFACTURER	DOUBLE STAR		EDISON STOVE WORKS		FISHER STOVES		
PRODUCT	DOUBLE STAR COMFORT HEATER	DOUBLE STAR PARLOUR STOVE	EDISON POT BELLY STOVE PB-26	EDISON POT BELLY STOVE PB-34	PAPA BEAR	MAMA BEAR	BABY BEAR
WOOD LENGTH			11 1/2"	16"	30"	24"	18"
FIREDOOR SIZE					10"	10"	8"
OVERALL SIZE	52"H	31 1/4"H x 25"W x 22"D	25 1/4"H x 14"W x 14 1/2"D	31 1/4"H x 17 1/2"W x 19"D	30 1/4"H x 18"W x 32"L	30 1/4"H x 16"W x 27"L	26 1/2"H x 14"W x 21"L
WEIGHT	210 LBS.	210 LBS.	39 LBS.	59 LBS.	410 LBS.	345 LBS.	245 LBS.
FIREBOX SIZE	29"H x 16"D		8 1/2"D	9 1/2"D			
FLUE SIZE	6"	6"	5"	6"			
FLUE LOCATION			TOP	TOP	BACK	BACK	BACK
BODY FINISH	POLISHED CHROME TRIM		PAINT	PAINT	PAINTED WITH HIGH-TEMPERATURE FINISH		
BODY MATERIALS	CAST IRON	CAST IRON	CAST IRON	CAST IRON	1/4" / 5/16" MSPL STEEL		
THERMOSTATIC DRAFT CONTROL							
BTU RATING							
FEATURES					* HOT WATER COIL	HOT WATER COIL	HOT WATER COIL
					* PORTABLE OVEN	PORTABLE OVEN	PORTABLE OVEN
					* STEEL COVERED ASBESTOS FLOOR / WALL PADS		
					FIREBRICK FLOOR	FIREBRICK FLOOR	FIREBRICK FLOOR

MANUFACTURER	FISHER STOVES	GARDEN WAY	L.W. GAY STOVEWORKS, INC.		HOME FIREPLACES		
PRODUCT	GRANDPA BEAR	GARDEN WAY BOX STOVE	INDEPENDENCE STOVE	INDEPENDENCE II	PARLOUR STOVE #PS-32	COMFORT HEATER #CH-52	BOX STOVE #BS-32
WOOD LENGTH	24"	24"	30"				23"
FIREDOOR SIZE	24" x 13"	(TOP) 9"H x 10"W (BOTTOM) 13"W	12" x 12"	12" x 12"	(SIDE) 14 1/2" x 9" (FRONT) 10" x 13"	9" x 10"	8" x 10"
OVERALL SIZE	33 1/2"H x 29 1/2"W 28 1/2"D	22"H x 13 1/2"W (TOP) 18 1/2"W (B) x 34"D	30"H x 16"W x 36"L	30"H x 20"W x 36"L	31"H x 23"W x 26"L	52"H x 15 1/2"W	24"H x 32"L
WEIGHT	420 LBS.	102 LBS.	270 LBS.	390 LBS.	197 S.W.	210 S.W.	105 S.W.
FIREBOX SIZE	26"W x 21"D		30"L				
FLUE SIZE	8"	6"	6"	6"		6"	6"
FLUE LOCATION	TOP	TOP	TOP; REAR	TOP; REAR			TOP
BODY FINISH	PAINT		HIGH-TEMPERATURE PAINT		BLACK NICKEL WINDOW FRAME	CHROME HIGHLIGHTS	
BODY MATERIALS		11 GA. STEEL	3/16" STEEL	3/16" STEEL	CAST IRON		CAST IRON
THERMOSTATIC DRAFT CONTROL							
BTU RATING							
FEATURES	* PORTABLE OVEN		* HOT WATER COIL	HOT WATER COIL	* NICKEL FOOT-RAIL; TOPRAIL; SWINGTOP; URN		FLAT-TOP FOR COOKING
	* STEEL COVERED ASBESTOS FLOOR WALL PADS			15" x 30" SOAPSTONE SLAB			
					FRONT / SIDE LOADING		
	FIREBRICK FLOOR						

II DIVISION

HEATING STOVES SPECIFICATIONS

MANUFACTURER	HOME FIREPLACES	HYDRAFORM PRODUCTS		INGLEWOOD STOVE CO.			KICKAPOO STOVE WORKS LTD.
PRODUCT	POT BELLY # PB-31	LARGER EAGLE	SMALLER EAGLE	TORTOISE #EN WHITE	TORTOISE #EN BLACK	TORTOISE #PT. BLACK	BBR-1
WOOD LENGTH		30"	20"	15"	15"	15"	24"
FIREDOOR SIZE							11½"H x 8¾"(TOP) 12"(BOTTOM)
OVERALL SIZE	31"H x 13½"W	36"H x 26"D x 36"L		25½"H x 13⅛"W 22½"L	25½"H x 13⅛"W 22½"L	25½"H x 13⅛"W 22½"L	34"H x 20"W x 27"L
WEIGHT	80 S.W.	435 LBS.		170 LBS.	170 LBS.	168 LBS.	260 LBS.
FIREBOX SIZE							5.8 CU. FT.
FLUE SIZE	6"	6"					6"
FLUE LOCATION	TOP	BACK		BACK	BACK	BACK	TOP
BODY FINISH				WHITE ENAMEL	BLACK ENAMEL	BLACK PAINT	
BODY MATERIALS	CAST IRON			OUTER STEEL LINING/DECORATIVE CAST IRON PANEL			12 GA. STEEL/ CAST IRON
THERMOSTATIC DRAFT CONTROL							
BTU RATING		135,000 BTU		28,000 BTU	28,000 BTU	28,000 BTU	
FEATURES	ADJUSTABLE DRAFT CONTROL	*HUMIDIFIERS	*HUMIDIFIERS	VICTORIAN DESIGN	VICTORIAN DESIGN	VICTORIAN DESIGN	COOKING SURFACE
	6" COOKING PLATE (TOP)	*SMOKING/ COOKING OVENS	*SMOKING/ COOKING OVENS				FIREBRICK FIREBOX
		*GRILLE	*GRILLE				* 10 GA. STEEL HULL W/ INTERIOR FINS
		*WATER HEATER	*WATER HEATER				
		*GLASS DOORS	*GLASS DOORS				
		*DECORATION FOR ANY DECOR	*DECORATION FOR ANY DECOR				

MANUFACTURER	KICKAPOO STOVE WORKS LTD.	KRISTIA ASSOCIATES					LOCKE STOVE CO.
PRODUCT	BB-2 (BABY BBR)	JØTUL COMBI-FIRE #4	JØTUL WOODSTOVE #602	JØTUL WOODSTOVE #118	JØTUL COOKSTOVE #380	JØTUL WOODSTOVE #606	WARM-EVER W-15
WOOD LENGTH		20"	12"	24"	20"	12"	15"
FIREDOOR SIZE							
OVERALL SIZE		41¼"H x 23½"W 22¾"L	25¼"H x 12¾"W 19¼"L	30¼"H x 14¼"W 29½"L	27½"H x 16"W 22½"L	40½"H x 11¾"W 18¾"L	30½"H x 23"W 21¼"L
WEIGHT	235 LBS.	286 S.W.	117 S.W.	231 S.W.	131 S.W.	175 S.W.	106 S.W.
FIREBOX SIZE	4.2 CU. FT.						
FLUE SIZE	6"	7"	5"	5"	5"	5"	
FLUE LOCATION	TOP	BACK; TOP	BACK; TOP	BEHIND SIDE OF TOP PLATE	TOP	BACK; SIDE	
BODY FINISH		ENAMEL AND BLACK	ENAMEL	ENAMEL		BLACK	
BODY MATERIALS	12 GA. STEEL/ CAST IRON		CAST IRON	CAST IRON	CAST IRON		CAST IRON
THERMOSTATIC DRAFT CONTROL							
BTU RATING			27,000 BTU	44,500 BTU			
FEATURES	COOKING SURFACE	FIREPLACE OPTION	SEC. COMBUSTION CHAMBER	SEC. COMBUSTION CHAMBER	DOUBLE SIDE WALL CONSTRUCT.		FIREBRICK LINED
	FIREBRICK FIREBOX		BAFFLES IN FIREBOX	BAFFLES IN FIREBOX			
	* 10 GA. STEEL HULL W/ INTERIOR FIN						

MANUFACTURER	LOCKE STOVE CO.	LOUISVILLE TIN AND STOVE CO.					
PRODUCT	WARM-EVER W-24	FERNO #18F	FERNO #20F	FERNO #22F	FERNO #24F	FERNO #26F	DERBY DRUM STOVE #22D
WOOD LENGTH	24"						
FIREDOOR SIZE		8⅛"	8⅛"	10¼"	10¼"	10¼"	8⅜" x 11"
OVERALL SIZE	30½"H x 23"W x 30¾	18"H x 12"W x 18"L	20"H x 13¾"W x 20"L	22½"H x 16"W x 22"L	24¾"H x 17¼"W x 24	26¾"H x 18½"W x 26"	28½"H x 15½" x 22"L
WEIGHT	142 S.W.	9 S.W.	10 S.W.	13 S.W.	15 S.W.	18 S.W.	18 S.W.
FIREBOX SIZE							
FLUE SIZE		6"	6"	6"	6"	6"	6"
FLUE LOCATION		TOP	TOP	TOP	TOP	TOP	TOP
BODY FINISH							
BODY MATERIALS	CAST IRON	POLISHED BLUE STEEL	POLISHED BLUE STEEL	POLISHED BLUE STEEL	POLISHED BLUE STEEL	POLISHED BLUE STEEL	POLISHED BLUE STEEL
THERMOSTATIC DRAFT CONTROL							
BTU RATING							
FEATURES	FIREBRICK LINED	ALL MODELS LINED	ALL MODELS LINED	ALL MODELS LINED	ALL MODELS LINED	ALL MODELS LINED	CORRUGATED STEEL FIREBOX
		WHEEL DRAFT REGULATOR	WHEEL DRAFT REGULATOR	WHEEL DRAFT REGULATOR	WHEEL DRAFT REGULATOR	WHEEL DRAFT REGULATOR	SLIDING DRAFT

II DIVISION — HEATING STOVES SPECIFICATIONS

MANUFACTURER	LOUISVILLE TIN AND STOVE CO.		MARKADE - WINNWOOD				
PRODUCT	DERBY DRUM STOVE #24D	DERBY DRUM STOVE #26D	WINNWOOD #24	WINNWOOD #36	WINNWOOD #30	WINNWOOD #48	WINNWOOD #30-P
WOOD LENGTH			24"	36"			
FIREDOOR SIZE	8⅜" x 11"	8⅜" x 11	12" x 12"	12" x 12"			
OVERALL SIZE	28½"H x 17¼"W 24"L	28½"H x 18½"W 26"L	33¾"H x 16"W 29¼"L	33¾"H x 16"W 41½"L	28"H x 16"W x 36"L	28"H x 16"W x 51"L	28"H x 16"W x 36"L
WEIGHT	20 S.W.	22 S.W.			65 S.W.	85 S.W.	65 S.W.
FIREBOX SIZE					19½"H x 16"W x 30"L	19½"H x 16"W x 48"L	19½"H x 16"W x 30"L
FLUE SIZE	6"	6"	6"	6"	6"	6"	6"
FLUE LOCATION	TOP	TOP			REAR	REAR	REAR
BODY FINISH			16 GA. STEEL	16 GA. STEEL	16 GA. STEEL	16 GA. STEEL	16 GA. STEEL
BODY MATERIALS	POLISHED BLUE STEEL	POLISHED BLUE STEEL					
THERMOSTATIC DRAFT CONTROL							
BTU RATING							
FEATURES	CORRUGATED STEEL FIREBOX	CORRUGATED STEEL FIREBOX	BUILT-IN FLUE DAMPER	BUILT-IN FLUE DAMPER	DESIGNED FOR FIREBRICK LINER	DESIGNED FOR FIREBRICK LINER	DESIGNED FOR FIREBRICK LINER
	SLIDING DRAFT	SLIDING DRAFT					

MANUFACTURER	MARTIN INDUSTRIES						
PRODUCT	CORONA #180	MERIT #520	MARCO SUN 3-28 W	MARCO SUN 3-28 L	BOX HEATER #624	BOX HEATER #628	BOX HEATER #632
WOOD LENGTH					23"	26"	30"
FIREDOOR SIZE		17¼" x 8½"					
OVERALL SIZE	25"H	40"H x 16"W x 21½"L	34½"H x 12½"W x 27"L	27⅝"H x 12½"W x 27"L	23¼"H	24¼"H	25¼"H
WEIGHT	100 LBS.	130 S.W.	99 LBS.	93 LBS.	102 LBS.	123 LBS.	132 LBS.
FIREBOX SIZE	18½"	19" x 11¾"					
FLUE SIZE	6"	6"	6"	6"	6"	6"	6"
FLUE LOCATION	TOP						
BODY FINISH			CAST TOP/BOTTOM	CAST TOP/BOTTOM			
BODY MATERIALS		CAST IRON MAIN FRONT	HEAVY GA. BLUE STEEL	HEAVY GA. BLUE STEEL			
THERMOSTATIC DRAFT CONTROL							
BTU RATING							
FEATURES		CAST LINERS IN FIREBOX	URN/FOOTRAIL	URN/FOOTRAIL			
			CAST LINING	CAST LINING			

MANUFACTURER	MARTIN INDUSTRIES				THE MERRY MUSIC BOX		
PRODUCT	6600 AUTOMATIC WOOD HEATER	WOOD KING #2600-1	WOOD KING #2600-2	WOOD KING #2600-3	MODEL #2 HEATER	MODEL #3 HEATER	MODEL #4 HEATER
WOOD LENGTH							
FIREDOOR SIZE					7¼" SQ.	7¼" SQ.	7¼" SQ.
OVERALL SIZE	44"H	44"H	44"H	44"H	39"H x 15"D x 18"L	42½"H x 15"D x 18"L	48½"H x 18½"D 21¾"L
WEIGHT	144 S.W.	157 LBS.	152 LBS.	147 LBS.	289 LBS.	349 LBS.	533 LBS.
FIREBOX SIZE	19"H x 25"L	19"W x 25"L	19"W x 25"L	19"W x 25"L	19"H x 10½"D x 13"W	22"H x 10½"D x 13"W	26"H x 13¾"D x 15"W
FLUE SIZE							
FLUE LOCATION					BACK	BACK	BACK
BODY FINISH					ENAMEL/ NICKEL-PLATING	BLACK STEEL/ BLACK IRON	BLACK STEEL/ BLACK IRON
BODY MATERIAL		CAST IRON TOP; BOTTOM; DOOR FRAMES; DOORS					
THERMOSTATIC DRAFT CONTROL	YES	YES	YES	YES			
BTU RATING							
FEATURES	HEAVY-DUTY STEEL LINER	URN/FOOTRAIL	URN		FULL BRICK LINED	FULL BRICK LINED	FULL BRICK LINED
					FIREBRICK FLOOR	FIREBRICK FLOOR	FIREBRICK FLOOR
					FUEL DOOR W/LOCK-KEY	FUEL DOOR W/LOCK-KEY	FUEL DOOR W/LOCK-KEY
					HUMIDIFICATION CHAMBER	HUMIDIFICATION CHAMBER	HUMIDIFICATION CHAMBER

II DIVISION — HEATING STOVES SPECIFICATIONS

MANUFACTURER	MOHAWK INDUSTRIES	PORTLAND STOVE FOUNDRY					
PRODUCT	TEMPWOOD	HOME ATLANTIC #122	ATLANTIC BOX STOVE #124	ATLANTIC BOX STOVE #130	ATLANTIC BOX STOVE #136	MONITOR BOX STOVE #31	MONITOR BOX STOVE #38
WOOD LENGTH		18"					
FIREDOOR SIZE							
OVERALL SIZE	28"H x 18"D x 28"L	37"H x 19"D x 22½"L	24½"H x 36"L **	26½"H x 41½"L **	30"H x 47"L **	32"H x 43"L **	32"H x 51"L **
WEIGHT	130 LBS.						
FIREBOX SIZE		12½" x 18"	15¾"H x 11¾"W 23½"L	18"H x 14"W x 29"L	21"H x 13½"W x 35"L	21"H x 18"W x 31"L	22¼"H x 20"W x 38"L
FLUE SIZE	6"	6"	6"	6"	7"	7"	7"
FLUE LOCATION	TOP		TOP	TOP	TOP	TOP	TOP
BODY FINISH							
BODY MATERIALS	13GA./¼" STEEL						
THERMOSTATIC DRAFT CONTROL							
BTU RATING	60,000 BTU						
FEATURES	DOWN-DRAFT PRINCIPLE						

MANUFACTURER	PORTLAND STOVE FOUNDRY CO.						
PRODUCT	B+M STOVE #1	B+M STOVE #2	B+M STOVE #3	B+M STOVE #4	B+M STOVE #5	TROLLA #102	TROLLA #105
WOOD LENGTH						12"	18"
FIREDOOR SIZE						6½" x 9"	6½" x 9"
OVERALL SIZE	30½"H	33"H	38"H	39"H	51½"H	24½"H x 17½"L *	25½"H x 24¼"L *
WEIGHT	104 LBS.	152 LBS.	234 LBS.	296 LBS.	426 LBS.	79 LBS.	178 LBS.
FIREBOX SIZE	12"D	14"D	15"D	17"D	20"D	11"W x 16"L	13"W x 21"L
FLUE SIZE	5"	6"	6"	6"	6½"	4"	5"
FLUE LOCATION						TOP; BACK	TOP; BACK
BODY FINISH						STOVE-BLACK SILICONE	STOVE-BLACK SILICONE
BODY MATERIALS						CAST IRON	CAST IRON
THERMOSTATIC DRAFT CONTROL							
BTU RATING							
FEATURES							FIREBRICK SIDES/FLOOR

MANUFACTURER	PORTLAND STOVE FOUNDRY		RAM + FORGE	SCANDINAVIAN STORES INC.			
PRODUCT	TROLLA #107	TROLLA #800 FIREPLACE STOVE	WOOD STOVE	LANGE MODEL #6303A	LANGE MODEL #6303	LANGE MODEL #6302A	LANGE MODEL #6302K
WOOD LENGTH	24"		28"				
FIREDOOR SIZE	6½" x 9"	6" x 10"					
OVERALL SIZE	28¼"H x 31"L **	41"H x 25"W x 20"D	30"H x 14"W x 36"L	23½"H x 16"W x 25"L	37½"H x 16"W x 25"L	34"H x 16"W x 34"L	50½"H x 16"W x 34"L
WEIGHT	253 LBS.	250 LBS.		145 S.W.	220 S.W.	272 S.W.	370 S.W.
FIREBOX SIZE	13"W x 27¼"L			20½"L	20½"L	26"L	26"L
FLUE SIZE	5"	8"		5"	5"	5"	5"
FLUE LOCATION	TOP; BACK	TOP		TOP	SIDE	TOP	SIDE
BODY FINISH	STOVE-BLACK SILICONE			BLACK/ENAMEL FINISH	BLACK/ENAMEL FINISH	BLACK/ENAMEL FINISH	BLACK/ENAMEL FINISH
BODY MATERIALS	CAST IRON	CAST IRON	¼" STEEL PLATE	CAST IRON	CAST IRON	CAST IRON	CAST IRON
THERMOSTATIC DRAFT CONTROL							
BTU RATING							
FEATURES		FIREPLACE OPTION; RAISED RELIEF INSIDE WALLS; FIREBRICK FLOOR		COOKING OPTION	EXTRA HEAT CHAMBER	COOKING OPTION; BAFFLES IN FIREBOX	EXTRA HEAT CHAMBER; BAFFLES IN FIREBOX; OVEN/COOK PLATE

II DIVISION

HEATING STOVES SPECIFICATIONS

MANUFACTURER	SCANDINAVIAN STORES INC.			SCOT'S STOVE COMPANY			SHENANDOAH MFG. CO.
PRODUCT	LANGE MODEL #6203	LANGE MODEL #6204	LANGE MODEL #61MF	SIZE B	SIZE C	SIZE D	MODEL #R-55
WOOD LENGTH			16"	18"	22"	25"	
FIREDOOR SIZE							
OVERALL SIZE	41"H x 13 1/4"W x 20"L	41"H x 13 1/4"W x 25"L	38"H x 20 1/2"W x 19"D				37"H x 18"D
WEIGHT	213 S.W.	250 S.W.	286 S.W.				158 LBS.
FIREBOX SIZE	16"L	20"L	11" x 14"				23"H
FLUE SIZE	5"	5"	6 3/4"				5"
FLUE LOCATION	TOP	TOP	BACK	TOP	TOP	TOP	
BODY FINISH	BLACK/ENAMEL FINISH	BLACK/ENAMEL FINISH	BLACK/ENAMEL FINISH				
BODY MATERIALS	CAST IRON	CAST IRON	CAST IRON	GA. SHEET STEEL	GA. SHEET STEEL	GA. SHEET STEEL	11/18 GA. METAL
THERMOSTATIC DRAFT CONTROL				YES	YES	YES	
BTU RATING							
FEATURES	BAFFLES IN FIREBOX	BAFFLES IN FIREBOX	BAFFLES IN FIREBOX	COOKING OPTION	COOKING OPTION	COOKING OPTION	FIREBRICK FIREBOX
	EUROPEAN TILE PATTERN	EUROPEAN TILE PATTERN	FIREBRICK FIREBOX				
			FIREPLACE OPTION				

MANUFACTURER	SHENANDOAH MFG. CO.	SOUTHEASTERN VERMONT COMMUNITY ACTION	SOUTHPORT STOVES				
PRODUCT	MODEL #R-75	S.E.V.C.A. STOVE	MORSØ MODEL #280	MORSØ MODEL #28	MORSØ MODEL #1B	MORSØ MODEL #180	MORSØ MODEL #6B
WOOD LENGTH		30"					
FIREDOOR SIZE	12" x 13"						
OVERALL SIZE	35"H x 24"D	31"H x 16"W x 36"L	40"H x 13"W 27 1/2"D	28"H x 13"W 27 1/2"D	34"H x 14 1/4"W 30 1/4"D	51 1/4"H x 14 1/4"W 30 1/4"D	24 1/2"H x 14"W 23 1/2"D
WEIGHT	164 LBS.		164 LBS.	124 LBS.	254 LBS.	353 LBS.	146 LBS.
FIREBOX SIZE					5,186 CU. IN.	5,186 CU. IN.	
FLUE SIZE	6"	6"	4 3/4"	4 3/4"	4 3/4"	4 3/4"	4 3/4"
FLUE LOCATION		TOP	TOP	TOP	TOP	TOP	TOP
BODY FINISH		HEAT-RESISTANT PAINT	ENAMEL/OPAQUE PORCELAIN	ENAMEL/OPAQUE PORCELAIN	ENAMEL/OPAQUE PORCELAIN	ENAMEL/OPAQUE PORCELAIN	ENAMEL/OPAQUE PORCELAIN
BODY MATERIALS	11/18 GA. METAL	3/16" ROLLED STEEL PLATE	CAST IRON	CAST IRON	CAST IRON	CAST IRON	CAST IRON
THERMOSTATIC DRAFT CONTROL							
BTU RATING							
FEATURES	FIREBRICK FIREBOX	COOKING SURFACE					
		SEC. COMBUSTION CHAMBER					

MANUFACTURER	SOUTHPORT STOVES	SUNSHINE STOVE WORKS	U.S. STOVE CO.		VERMONT CASTINGS, INC.	VERMONT IRON STOVE WORKS	VERMONT WOODSTOVE CO.
PRODUCT	MORSØ #1125	SUNSHINE STOVE	BOXWOOD MODEL #126-S	BOXWOOD MODEL #132-S	DEFIANT PARLOUR STOVE	"THE ELM"	DOWNDRAFTER MODEL #DD1
WOOD LENGTH		24"			26"		
FIREDOOR SIZE		8" x 10"			10 1/2" x 14 3/4"		
OVERALL SIZE	41 3/4"H x 29 1/2"W 22 3/4"D	28"H x 14"W x 34"L	24 1/2"H x 15"W 34 1/2"L	24 1/2"H x 15"W 40 1/2"L	32"H x 23"D x 34"L	25"H x 23"W x 33"L	32"H x 26"W x 34"L
WEIGHT	310 LBS.	225 LBS.	95 S.W.	110 S.W.	285 S.W.	200 LBS.	300 LBS.
FIREBOX SIZE			15"H x 12"W x 26"L	15"H x 12"W x 32"L		19"D x 24"L	22"H x 16"W x 26"L
FLUE SIZE	8"	6"	6"	6"	7"-8"	6"	8"
FLUE LOCATION	TOP; BACK		TOP	TOP	TOP	BACK	BACK
BODY FINISH	OPAQUE/TRANSLUCENT POR.		HIGH TEMP. FINISH	HIGH TEMP. FINISH		BLACK WITH CHROME TRIM	BLACK
BODY MATERIALS	CAST IRON	GA. STEEL - CAST IRON	HEAVY GA. STEEL	HEAVY GA. STEEL	CAST IRON	CAST IRON	STEEL PLATE STAINLESS STEEL
THERMOSTATIC DRAFT CONTROL					YES	YES	YES
BTU RATING							60,000 BTU
FEATURES	FIREPLACE OPTION	COOKING OPTION			FIREPLACE + COOKING OPTION	FIREBRICK FIREBOX	* WATER COILS
		FIREBRICK FIREBOX			BAFFLES IN FIREBOX		FIREBRICK FIREBOX
		BAFFLES IN FIREBOX			* HEAT SHIELD		COOKING OPTION
		SEC. COMBUSTION CHAMBER			* FIREPLACE SCREEN		
					* TOOLS		
					SEC. COMBUSTION CHAMBER		

MANUFACTURER	VERMONT WOODSTOVE CO.	WASHINGTON STOVE WORKS					
PRODUCT	DOWNDRAFTER MODEL #DDII	CANNONBALL OLYMPIC #13	CANNONBALL OLYMPIC #17	CANNONBALL MARTIN #S-300	CANNONBALL MARTIN #S-40	CANNONBALL MARTIN #S-50	CANNONBALL MARTIN #S-60
WOOD LENGTH							
FIREDOOR SIZE							
OVERALL SIZE	28"H x 22"W x 25"L	41"H x 13"D	47"H x 17"D	27"H x 9"D	29"H x 12"D	30 3/4"H x 14"D	36"H x 15 1/2"D
WEIGHT	180 LBS.	124 LBS.	220 LBS.	52 LBS.	66 LBS.	86 LBS.	118 LBS.
FIREBOX SIZE	18"H x 16"W x 20"L						
FLUE SIZE	6"	6"	7"	5"	6"	6"	6"
FLUE LOCATION	BACK	TOP	TOP	TOP	TOP	TOP	TOP
BODY FINISH	BLACK						
BODY MATERIALS	STEEL PLATE STAINLESS STEEL	CAST IRON	CAST IRON	CAST IRON	CAST IRON	CAST IRON	CAST IRON
THERMOSTATIC DRAFT CONTROL	YES						
BTU RATING	20,000 BTU						
FEATURES	FIREBRICK FIREBOX	COOKING SURFACE	COOKING SURFACE	COOKING SURFACE	COOKING SURFACE	COOKING SURFACE	COOKING SURFACE
	COOKING OPTION						

MANUFACTURER	WASHINGTON STOVE WORKS						
PRODUCT	PARLOR STOVE (BASIC)	JUMBO HEATER #24	FOC'SLE HEATER GYPSY #1	FOC'SLE HEATER SKIPPY #1	FOC'SLE HEATER MIDGET #1	ARCTIC #25	ARCTIC #30
WOOD LENGTH							
FIREDOOR SIZE							
OVERALL SIZE	31 1/4"H x 25 1/2"W 22 1/2"D	24"H x 24"L	12 5/8"H x 13"W 18"L	13 1/2"H x 13"W 18"L	11 1/4"H x 12 1/2"W 16 1/2"L	26 1/2"H x 15 1/4"W 34 1/2"D	26 1/2"H x 16 1/2"W 37 1/4"D
WEIGHT	180 LBS.	156 LBS.	55 S.W.	43 1/2 S.W.	35 S.W.	110 LBS.	140 LBS
FIREBOX SIZE							
FLUE SIZE	6"	7"	4"-5"	5"	4"	6"	6"
FLUE LOCATION		TOP	TOP	TOP	TOP	TOP	TOP
BODY FINISH	BLACK						
BODY MATERIALS	CAST IRON	16 GA. STEEL / IRON CASTINGS				CAST IRON	CAST IRON
THERMOSTATIC DRAFT CONTROL							
BTU RATING							
FEATURES	FRONT; TOP SIDE OPENINGS						

MANUFACTURER	WASHINGTON STOVE WORKS						
PRODUCT	ARCTIC #36	KING #624	KING #628	KING #632	MARCO SUN	LAUNDRY HEATER	CAMPER'S COMPANION #12
WOOD LENGTH							
FIREDOOR SIZE							
OVERALL SIZE	30"H x 19 1/2"W 44 1/2"D	23 1/4"H x 11"W 31"D	24 1/4"H x 13 1/2"W 32 1/2"D	25 1/4"H x 14 1/2"W 36"D	27 5/8"H x 12 1/2"W 27"D	25 1/2"H x 21"W 17"D	28"H x 12"W
WEIGHT	220 LBS.	93 LBS.	115 LBS.	140 LBS.	95 LBS.	112 LBS.	30 LBS
FIREBOX SIZE							
FLUE SIZE	6"	6"	6"	6"	6"	6"	
FLUE LOCATION	TOP	TOP	TOP	TOP	TOP	TOP	
BODY FINISH							
BODY MATERIALS	CAST IRON	CAST IRON	CAST IRON	CAST IRON	CAST IRON + BLUED STEEL	CAST IRON	
THERMOSTATIC DRAFT CONTROL							
BTU RATING							
FEATURES							

II DIVISION	MANUFACTURER	WASHINGTON STOVE WORKS	YANKEE WOODSTOVES					
	PRODUCT	KITCHEN HEATER #1600	MODEL #15	MODEL #18	MODEL #28	MODEL #28CR		
HEATING STOVES SPECIFICATIONS	WOOD LENGTH		12"	18"	24"	24"		
	FIREDOOR SIZE		7"x 9½"	8" x 11"	7½"x 11"	7½"x 11"		
	OVERALL SIZE	36"H x 12"W x 25"L	31½"H x 18"W 18"D	35½"H x 18½"W 20¾"D	27"H x 19"W 29"D	27"H x 19"W 29"D		
	WEIGHT	164 S.W.						
	FIREBOX SIZE	17½"x 9" 9½"						
	FLUE SIZE	6"	5"	5"	5"	5"		
	FLUE LOCATION	TOP; BACK	TOP; BACK	TOP; BACK	TOP	TOP		
	BODY FINISH	PORCELAIN ENAMEL						
	BODY MATERIALS	CAST IRON						
	THERMOSTATIC DRAFT CONTROL							
	BTU RATING							
	FEATURES	COOKING OPTION	DESIGNED FOR MOBILE HOMES	COOKING OPTION	BAFFLES IN FIREBOX	BAFFLES IN FIREBOX		
	*	WATER COILS	BAFFLES IN FIREBOX	BAFFLES IN FIREBOX	FIREBRICK FIREBOX	FIREBRICK FIREBOX		
			FIREBRICK FIREBOX	FIREBRICK FLOOR				

Division III

Circulating Heaters

EDITOR'S NOTE

Circulating Heaters are defined as a heating stove with an exterior cabinet allowing air circulation around the heater by convection or by a blower unit.

*indicates optional at extra cost

III DIVISION	MANUFACTURER	ASHLEY PRODUCTS DIVISION		ATLANTA STOVE WORKS, INC.		AUTOCRAT CORP.		BOW AND ARROW STOVE COMPANY
	PRODUCT	C-60 IMPERIAL	C-62 COMPACT CONSOLE	HOMESTEADER 240	HOMESTEADER 24WGE	MODEL #6724	MODEL #FF65	SUPRA #401
CIRCULATING HEATERS SPECIFICATIONS	WOOD LENGTH	24"	24"		24"	24"	25"	12"-14"
	FIREDOOR SIZE	13⅜"x 9¾"	13⅜"x 9¾"		10"x 13½"	11⅜"x 12"	23½"x 9¾"	
	OVERALL SIZE	36"H x 35¼"W 21¼"D	35"H x 28¼"W 20¾"D	33⅜"H x 32¼"W 19¼"D	36½"H x 35"L	33"H x 31"W 20½"D	34¼"H x 32⅜"W 21¾"D	26"H x 21"W x 12"D
	WEIGHT	267 S.W.	223 S.W.	240 S.W.	317 S.W.	215 S.W.	245 S.W.	117 S.W.
	FIREBOX SIZE	23"H x 14"W x 27"L	20½"H x 13½"W 19½"L		21"H x 12"W x 24½"	22⅞"H x 25¼"W 14½"D	22"H x 28¼"W 15½"D	1,500 CU. IN.
	FLUE SIZE			6"	6"	6"	6"	5"
	BODY MATERIAL			CAST IRON	CAST IRON	CAST IRON + 18 GA. STEEL	CAST IRON + 18 GA. STEEL	STEEL + CAST IRON
	FIREBOX MATERIAL						ASBESTOS-LINED DOORS	FIREBRICK
	CABINET FINISH			BONDERIZED ENAMEL		ENAMEL SILICONE	ENAMEL SILICONE	ENAMELED STEEL
	THERMOSTATIC DRAFT CONTROL	YES	YES	YES	YES	YES	YES	
	GRATE			YES	YES	YES	YES	YES
	ASHPAN	YES	YES	YES	YES			YES
	BLOWER (CFM)	* 150 CFM	* 150 CFM	* YES	* 150 CFM	* YES	* YES	
	ACCESSORIES	DRAFT EQUALIZER	DRAFT EQUALIZER					
	FEATURES				COOKING OPTION			BAFFLES IN FIREBOX

III DIVISION

CIRCULATING HEATERS

SPECIFICATIONS

MANUFACTURER	HOME FIREPLACES	HUNTER		LOCKE STOVE COMPANY	MARTIN INDUSTRIES		PORTLAND STOVE FOUNDRY
PRODUCT	HOMESTEADER HS-36	HEATER #C-26	HEATER #C-31	WARM MORNING #7018	KING AUTOMATIC #7801-C	WOOD CIRCULATOR #7801-B	TROLLA #530
WOOD LENGTH		18"	24"	26"	25 1/2"	25 1/2"	
FIREDOOR SIZE		10"x 11"	10"x 11"	10"x 14"			
OVERALL SIZE	34"H x 32 1/4"W	35"H x 28"W 22 1/4"D	35"H x 34 1/2"W 22 1/4"D	33 1/4"H x 36"W 18"D	32"H x 33"L 21"D	32"H x 33"L 31"D	37 1/2"H x 26"W 16 1/2"D
WEIGHT	240 S.W.	170 S.W.	195 S.W.	290 S.W.	230 S.W.	215 S.W.	
FIREBOX SIZE		4 1/2 CU.FT.	6 CU.FT.		18 3/4"H x 25 1/4"L 14 5/8"D	18 3/4"H x 25 1/4"L 14 5/8"D	
FLUE SIZE		6"	6"	6"	6"	6"	5"
BODY MATERIAL					CAST IRON + 18 GA. STEEL	CAST IRON + 18 GA. STEEL	CAST IRON
FIREBOX MATERIAL	CAST IRON LINING	STAINLESS STEEL LINERS	STAINLESS STEEL LINERS	FIREBRICK	CAST IRON	REFRACTORY BRICK	FIREBRICK
CABINET FINISH			TWO-TONED ENAMEL	PORCELAIN ENAMEL	PORCELAIN ENAMEL	PORCELAIN ENAMEL	TWO-TONED ENAMEL
THERMOSTATIC DRAFT CONTROL		YES	YES	YES	YES	YES	YES
GRATE				YES	YES	YES	YES
ASHPAN				YES	YES	YES	
BLOWER (CFM)				* YES	* YES	* YES	
ACCESSORIES					BAROMETRIC DAMPER	BAROMETRIC DAMPER	
FEATURES		BAFFLES IN FIREBOX	BAFFLES IN FIREBOX		COOKING OPTION	COOKING OPTION	FIREPLACE-COOKING OPTION
		SEC. COMBUSTION CHAMBER	SEC. COMBUSTION CHAMBER				20,000 BTU

MANUFACTURER	PRESTON DISTRIBUTING CO.		RITEWAY MANUFACTURING CO.		SHENANDOAH MFG. CO.		SUBURBAN MFG., CO.
PRODUCT	POELE CHARBON BOIS #8033	POELE A BOIS #8008	WOOD HEATER #2000	MODEL #37	MODEL #R-76	MODEL #R-76L	WOODMASTER #W4-75
WOOD LENGTH	16"		24"	24"			
FIREDOOR SIZE			12"x 12"	12"x 12"	13"x 12"	13"x 12"	12"H x 10"W
OVERALL SIZE			33"H x 21"W 33"L	40"H x 24"W 33"L	36"H x 24"W 35 1/2"L	33"H x 21"W 29"L	32 1/2"H x 32"W 19"D
WEIGHT			200 S.W.	400 S.W.			225 LBS.
FIREBOX SIZE			4 CU. FT.	7 1/2 CU. FT.	23"H x 15 1/2"W 26"L	23"H x 15 1/2"W 26"L	20 1/2"H x 11"W 25 1/4"L
FLUE SIZE	5"	5"	6"	6"	6"	6"	6"
BODY MATERIAL			14 GA. STEEL	14 GA. STEEL	IRON	IRON	14 GA. STEEL
FIREBOX MATERIAL				FIREBRICK	FIREBRICK LINING	FIREBRICK LINING	FIREBRICK-STEEL
CABINET FINISH	ENAMEL	ENAMEL			PORCELAIN AND TRIM		HIGH SILICONE FINISH
THERMOSTATIC DRAFT CONTROL			YES	YES	YES	YES	YES
GRATE	YES		YES	YES	YES	YES	YES
ASHPAN	YES		YES	YES	YES	YES	YES
BLOWER (CFM)							* 230 CFM
ACCESSORIES			CABINETS	CABINETS			
FEATURES			SEC. COMBUSTION CHAMBER	SEC. COMBUSTION CHAMBER			COOKING OPTION
			50,000 BTU	73,000 BTU			

MANUFACTURER	SUBURBAN MFG., CO.	U.S. STOVE CO.	WASHINGTON STOVE WORKS				
PRODUCT	WOODMASTER #MM 76-H	WONDERWOOD #726	NORWESTER #38	NORWESTER #40			
WOOD LENGTH		26"					
FIREDOOR SIZE	12"H x 10"W		10"x 13 1/2"	10"x 13 1/2"			
OVERALL SIZE	32 1/4"H x 32"W 19"D	33"H x 19"W 32 1/2"L	32"H x 29"W 21"D	31"H x 29"W 19"D			
WEIGHT	315 LBS.	220 S.W.	270 S.W.	206 S.W.			
FIREBOX SIZE	20 1/2"H x 11"W 25 1/4"L		12"x 24"	12"x 24"			
FLUE SIZE		6"	6"	6"			
BODY MATERIAL	14 GA. STEEL						
FIREBOX MATERIAL	FIREBRICK AND STEEL	FIREBRICK		CAST IRON LININGS			
CABINET FINISH	HIGH SILICONE FINISH		PORCELAIN ENAMEL	BAKED ENAMEL			
THERMOSTATIC DRAFT CONTROL	YES	YES		YES			
GRATE	YES	YES					
ASHPAN		YES					
BLOWER (CFM)	*230 CFM	* YES	* YES				
FEATURES	COOKING OPTION						

Division IV — Furnaces

EDITOR'S NOTE

Furnaces are defined as any appliance designed to operate as a central heating system.

*indicates optional at extra cost

IV DIVISION — FURNACES SPECIFICATIONS

MANUFACTURER	BELLWAY MANUFACTURER			CARLSON MECH. CONTR'S., INC.	WALDO G. CUMINGS	DAMSITE STOVE	SAM DANIELS COMPANY
PRODUCT	MODEL #F50	MODEL #F75	MODEL #F125	CARLSON SOLID FUEL BOILER #SF-1	NORTHEASTER #101-B	DYNAMITE FURNACE STOVE	CHUNK FURNACE #R30W
WOOD LENGTH	24"	36"	48"	20"	16"	24"	
FIREDOOR SIZE	14"x17"	15"x20"	18"x23"	14"x15"			15"x17"
OVERALL SIZE	48"H x 30"W 44"L	55"H x 36"W 54"L	60"H x 46"W 72"L	54"H x 23"W 28"L	48"H x 36"W 29"D	38"H x 18"W 34"L	56"H x 41"W 41"L
WEIGHT				700 LBS.		200 LBS.	
FIREBOX SIZE				16½"H x 15"W 26"D	24"H x 24"W 21"D	30"H x 17"W 23"L	21½"H x 21⅝"W 30"L
FLUE SIZE	6"	7"	8"	8"	7"	6"-8"	8"
OUTER CABINET MATERIALS	STEEL	STEEL	STEEL	STEEL PLATE AND TUBING	⅛" ROLL. STEEL	GA. STEEL	
FIREBOX MATERIAL	BRICKS	BRICKS	BRICKS		⅛" STEEL LINERS		STAINLESS STEEL
THERMOSTATIC DRAFT CONTROL	YES			YES	YES		YES
FANS				* YES	YES		* YES
BTU RATING	25,000 TO 35,000 BTU	25,000 TO 35,000 BTU	25,000 TO 35,000 BTU	NET 170,000 BTU	100,000 BTU		100,000 BTU
GRATE	YES	YES	YES				YES
WATER HEATER			YES	YES	YES		
COMBINATION	COAL/OIL ELECTRIC	COAL/OIL ELECTRIC	COAL/OIL ELECTRIC				
ACCESSORIES							THERMOSTAT CONTROLS
FEATURES	POKER/SHAKER ASH SHOVEL SEC. COMBUSTION CHAMBER	POKER/SHAKER ASH SHOVEL SEC. COMBUSTION CHAMBER	POKER/SHAKER ASH SHOVEL SEC. COMBUSTION CHAMBER				COOL AIR OPTION

MANUFACTURER	SAM DANIELS COMPANY					HUNTER	
PRODUCT	FURNACE #R36W	FURNACE #R42W	FURNACE #R48W	FURNACE #R-1-60W	FURNACE #R-2-60W	VALLEY COMFORT #F-51	VALLEY COMFORT #F-71
WOOD LENGTH						36"	48"
FIREDOOR SIZE						12¾"x15¾"	12¾"x15¾"
OVERALL SIZE	56"H x 41"W 47"L	56"H x 41"W 53"L	56"H x 41"W 60"L	56"H x 41"W 70½"L	65"H x 53½"W 73"L	50"H x 32½"W 65½"L	50"H x 32½"W 77½"L
WEIGHT						860 S.W.	950 S.W.
FIREBOX SIZE	25½"H x 21⅝"W 36"L	25½"H x 21⅝"W 42"L	25½"H x 21⅝"W 48"L	25½"H x 21⅝"W 60"L	29"H x 25"W 60"L	36"L	48"L
FLUE SIZE	8"	8"	9"	9"	9"	8"	8"
OUTER CABINET MATERIALS						GA. STEEL PLATE	GA. STEEL PLATE
FIREBOX MATERIAL	STAINLESS STEEL					FIREBRICK LINED	FIREBRICK LINED
THERMOSTATIC DRAFT CONTROL	YES	YES	YES	YES	YES	YES	YES
FANS	* YES	* YES	* YES	* YES	* YES	YES	YES
BTU RATING	128,000 BTU	157,000 BTU	225,000 BTU	300,000 BTU	400,000 BTU	90,000 BTU	130,000 BTU
GRATE							
WATER HEATER							
COMBINATION							
ACCESSORIES	THERMOSTAT CONTROLS	THERMOSTAT CONTROLS	THERMOSTAT CONTROLS	THERMOSTAT CONTROLS	THERMOSTAT CONTROLS		
FEATURES							

IV DIVISION — FURNACES — SPECIFICATIONS

MANUFACTURER	HUNTER		LYNNDALE MANUFACTURING CO.		MARKADE-WINNWOOD	NEWMAC	
PRODUCT	VALLEY COMFORT #RB-3D	VALLEY COMFORT #RB-4D	FURNACE #810	FURNACE #910	WOOD-BURNING FURNACE CORE	OIL-WOOD FURNACE #CL-115	OIL-WOOD FURNACE #CL-140
WOOD LENGTH	36"	48"	30"	30"			
FIREDOOR SIZE	12¾"x15¾"	12¾"x15¾"				17½"x17½"	17½"x17½"
OVERALL SIZE	46"H X 30"W 45"L	46"H X 30"W 57"L	51"H X 32"W 68"L	51"H X 36"W 73"L		51¾"H X 46"W 54"L	51¾"H X 46"W 54"L
WEIGHT	474 S.W.	550 S.W.	1,000 LBS.	1,000 LBS.	400 S.W.	775 LBS.	780 LBS.
FIREBOX SIZE	36"L x 23"D	48"L x 23"D					
FLUE SIZE	8"	8"	7"	8"			8"
OUTER CABINET MATERIALS	GA. STEEL PLATE	GA. STEEL PLATE	24 GA. SHEET METAL	24 GA. SHEET METAL	12 GA. STEEL		
FIREBOX MATERIAL			2½" FIREBRICK	2½" FIREBRICK	FIREBRICK FLOOR	STEEL LINER	STEEL LINER
THERMOSTATIC DRAFT CONTROL	YES	YES	YES	YES		YES	YES
FANS	* YES	* YES	YES	YES		YES	YES
BTU RATING	90,000 BTU		125,000 BTU	200,000 BTU		113,000 BTU	137,000 BTU
GRATE			YES	YES			
WATER HEATER			* YES	* YES			
COMBINATION			OIL / GAS ELECTRIC	OIL / GAS ELECTRIC		OIL; WOOD	OIL; WOOD
ACCESSORIES			AIR CONDITIONING	AIR CONDITIONING			
			AUXILARY HEAT SOURCE	AUXILARY HEAT SOURCE			
FEATURES	SEC. COMBUSTION CHAMBER	SEC. COMBUSTION CHAMBER					

MANUFACTURER	NEWMAC		RAMANDFORGE	RITEWAY MANUFACTURING COMPANY			
PRODUCT	OIL-WOOD FURNACE #CL-155	OIL-WOOD FURNACE #CL-170	WOOD FURNACE	FURNACE #LF-20	FURNACE #LF-30	FURNACE #LF-50	FURNACE #LF-70
WOOD LENGTH		24"	28"	24"	24"	36"	48"
FIREDOOR SIZE	17½"x17½"	17½"x17½"		13"x14"	14"x16"	14"x16"	16"x16"
OVERALL SIZE	51¾"H X 46"W 54"L	51¾"H X 46"W 54"L		53"H X 34"W 68"L	65"H X 36"W 73"L	65"H X 36"W 85"L	65"H X 40"W 103"L
WEIGHT	785 LBS.	790 LBS.		1,000 S.W.	1,800 S.W.	2,400 S.W.	3,050 S.W.
FIREBOX SIZE				13.5 CU.FT.	20 CU.FT.	30 CU.FT.	45 CU.FT.
FLUE SIZE	8"	8"	5"	7"	8"	8"	10"
OUTER CABINET MATERIALS				GA. CASING	GA. CASING	GA. CASING	GA. CASING
FIREBOX MATERIAL	STEEL LINER	STEEL LINER		FIREBRICK	FIREBRICK	FIREBRICK	FIREBRICK
THERMOSTATIC DRAFT CONTROL	YES	YES	* YES	YES	YES	YES	YES
FANS	YES	YES	YES	YES	YES	YES	YES
BTU RATING	149,000 BTU	168,000 BTU		125,000 BTU	160,000 BTU	200,000 BTU	350,000 BTU
GRATE				YES	YES	YES	YES
WATER HEATER				* YES	* YES	* YES	* YES
COMBINATION	OIL; WOOD	OIL; WOOD	OIL; WOOD	OIL; GAS BURNERS	OIL; GAS BURNERS	OIL; GAS BURNERS	OIL; GAS BURNERS
ACCESSORIES			FAN SWITCH	HUMIDIFIER	HUMIDIFIER	HUMIDIFIER	HUMIDIFIER
FEATURES			SEC. COMBUSTION CHAMBER	HEAT EXCHANGER	HEAT EXCHANGER	HEAT EXCHANGER	HEAT EXCHANGER
			BAFFLES IN FIREBOX				

MANUFACTURER	RITEWAY MANUFACTURING COMPANY			TEKTON DESIGN CORPORATION			
PRODUCT	BOILER #LB-30	BOILER #LB-50	BOILER #LB-70	HS #OT-35	HS #OT-50	HS #OT-70	TASSO #A3
WOOD LENGTH	24"	36"	48"				
FIREDOOR SIZE	14"x16"	14"x16"	16"x16"				
OVERALL SIZE	66"H X 36"W 56"L	66"H X 36"W 68"L	66"H X 40"W 84"L	51"H X 39½"W 30"L	51"H X 46¾"W 30"L	51"H X 46¾"W 39¾"L	39½"H X 18"W 27½"L
WEIGHT	2,000 S.W.	2,600 S.W.	3,200 S.W.	1089 LBS	1444 LBS.	1800 LBS.	860 LBS.
FIREBOX SIZE	17.5 CU.FT.	26.2 CU.FT.	43.4 CU.FT.	30"H X 10½"W 21½"L	30"H X 10½"W 21½"L	30"H X 13¼"W 31"L	24"H X 13"W 23"L
FLUE SIZE	8"	8"	10"				
OUTER CABINET MATERIALS	GA. CASING	GA. CASING	GA. CASING	ENAMEL	ENAMEL	ENAMEL	
FIREBOX MATERIAL							
THERMOSTATIC DRAFT CONTROL	YES	YES	YES	YES	YES	YES	YES
FANS	YES	YES	YES				
BTU RATING	160,000 BTU	200,000 BTU	350,000 BTU	112,000 BTU	140,000 BTU	200,000 BTU	
GRATE	YES	YES	YES	YES	YES	YES	YES
WATER HEATER	* YES	* YES	* YES	YES	YES	YES	
COMBINATION	OIL; GAS BURNERS	OIL; GAS BURNERS	OIL; GAS BURNERS	OIL; GAS; COAL ELECTRIC	OIL; GAS; COAL ELECTRIC	OIL; GAS; COAL ELECTRIC	COAL
ACCESSORIES	HUMIDIFIER	HUMIDIFIER	HUMIDIFIER	SWITCHBOARD	SWITCHBOARD	SWITCHBOARD	
FEATURES	FUEL SELECTOR DAMPER	FUEL SELECTOR DAMPER	FUEL SELECTOR DAMPER	AUTOMATIC HEAT REGULATOR	AUTOMATIC HEAT REGULATOR	AUTOMATIC HEAT REGULATOR	

IV DIVISION	MANUFACTURER	TEKTON DESIGN CORP.	
	PRODUCT	HS #A-16	HS #A-35
FURNACES — SPECIFICATIONS	WOOD LENGTH		
	FIREDOOR SIZE		
	OVERALL SIZE	35½"H x 28¾"W 51½"L	48"H x 41½"W 56¾"L
	WEIGHT	640 LBS.	1213 LBS.
	FIREBOX SIZE	24"H x 24"W 43½"L	37"H x 37"W 48"L
	FLUE SIZE		
	OUTER CABINET MATERIALS		
	FIREBOX MATERIAL		
	THERMOSTATIC DRAFT CONTROL	YES	YES
	FANS		
	BTU RATING		
	GRATE	YES	YES
	WATER HEATER		
	COMBINATION		
	ACCESSORIES		
	FEATURES		

Division V Free-Standing Fireplaces

EDITOR'S NOTE

Free-Standing Fireplaces are defined as any appliance designed for heating and having the ability to be used as an open fireplace. These products do not require additional construction for their installation (with the exception of normal flue connection).

V DIVISION	MANUFACTURER	ATLANTA STOVE WORKS, INC.						DOUBLE STAR
	PRODUCT	FRANKLIN #32	FRANKLIN #26	FRANKLIN #22	MODEL #39	MODEL #34	MODEL #39 P	FRANKLIN FIREPLACE #26
FREE-STANDING FIREPLACES — SPECIFICATIONS	TYPE	FRANKLIN	FRANKLIN	FRANKLIN	TEPEE	TEPEE	TEPEE	FRANKLIN
	OVERALL SIZE	41½"H x 44"W 30"D	38½"H x 38"W 25"D	36½"H x 34"W 23"D	41"H x 38½"W 28¼"D	38"H x 33½"W 24"D		39"H x 38"W 25"D
	WEIGHT	400 S.W.	310 S.W.	248 S.W.	165 S.W.	193 S.W.		308 LBS.
	FIREBOX SIZE	19¾"H x 26"W (REAR) 32"W (FRONT)	18¼"H x 20"W (R) 25½"W (F)	17¼"H x 17"W (R) 21"W (F)				25½"W
	FLUE SIZE	10"	8"	8"	8"	8"		8"
	FLUE LOCATION	BACK; TOP	BACK; TOP	BACK; TOP	TOP	TOP		
	BODY MATERIALS	CAST IRON	CAST IRON	CAST IRON				CAST IRON
	BODY FINISH				BONDERIZED METAL COATING	ENAMEL	PORCELAIN ENAMEL	
	ACCESSORIES	HEATSHIELD KIT	HEATSHIELD KIT	HEATSHIELD KIT	GRATE	GRATE		SWING-OUT BEANPOT
		SWING-OUT BEANPOT	SWING-OUT BEANPOT	SWING-OUT BEANPOT				SWING-OUT GRILLE
		SWING-OUT GRILLE	SWING-OUT GRILLE	SWING-OUT GRILLE				FIRESCREEN
		FLUE ADAPTER	FLUE ADAPTER	FLUE ADAPTER				
		STOVE PIPE KIT	STOVE PIPE KIT	STOVE PIPE KIT				
		FIRESCREEN W/BRASS HANDLE	FIRESCREEN W/BRASS HANDLE	FIRESCREEN W/BRASS HANDLE				
	FEATURES				MESH SCREEN	MESH SCREEN		
					COAL; GAS ELECTRIC LOGS	COAL; GAS ELECTRIC LOGS		
					REFRACTORY LINER	REFRACTORY LINER		
					BAFFLES IN FIREBOX	BAFFLES IN FIREBOX		

I DIVISION

FREE-STANDING FIREPLACES

SPECIFICATIONS

MANUFACTURER	HOME FIREPLACES			K.N.T., INC.			KRISTIA
PRODUCT	HOMEGUARD FRANKLIN HG-268	HOMEGUARD FRANKLIN HGF-308	SUNFIRE	IMPRESSION 100	IMPRESSION 101	IMPRESSION 102	JØTUL # 3
TYPE	FRANKLIN	FRANKLIN	TEPEE	FRANKLIN	FRANKLIN	FRANKLIN	RECTANGULAR
OVERALL SIZE	30"H X 36"W 25"D	31"H X 44"W 27"D	40"H X 30"W	31"H X 31"W 30"L	31"H X 31"W 30"L	31"H X 31"W 30"L	34 1/2"H X 20"W 19"L
WEIGHT	290 LBS.	330 LBS.	88 S.W.				183 S.W.
FIREBOX SIZE	18 1/4"H X 26"W (FRONT) 20"W (BACK)	19"H X 30"W (F) 26"W (B)					
FLUE SIZE	8"	8"	7"	8"	8"	8"	7"
FLUE LOCATION	TOP; BACK	TOP; BACK	TOP	TOP	TOP	TOP	BACK
BODY MATERIALS	CAST IRON	CAST IRON	GA. STEEL	GA. STEEL	GA. STEEL	GA. STEEL	CAST IRON
BODY FINISH				CERAMIC TILE (FRONT)	CERAMIC TILE (TOP / FRONT)	FLAT BLACK	ENAMEL
ACCESSORIES	BARBEQUE GRILL	BARBEQUE GRILL		BLOWER	BLOWER	BLOWER	
	SAFETY SPARK GUARDS	SAFETY SPARK GUARDS					
	FIRE TENDING TOOLS	FIRE TENDING TOOLS					
	STOVE POLISH AND CEMENT	STOVE POLISH AND CEMENT					
	PIPE COLLARS AND WIRE	PIPE COLLARS AND WIRE					
	CAST IRON DAMPERS	CAST IRON DAMPERS					
	CAST IRON SKILLETS	CAST IRON SKILLETS					
	STOVE SCRAPER	STOVE SCRAPER					
	COAL SHOVEL	COAL SHOVEL					
FEATURES			BUILT-IN HUMIDIFIER	REPLACEABLE HEAT SHIELD	REPLACEABLE HEAT SHIELD	REPLACEABLE HEAT SHIELD	ASHPAN
			SPARK SAFETY SCREEN	BARBEQUE GRILL	BARBEQUE GRILL	BARBEQUE GRILL	FIREBRICK FIREBOX
			FRESH-AIR REGULATOR	SURFACE PLATE (COOKING)	SURFACE PLATE (COOKING)	SURFACE PLATE (COOKING)	
				FIREBRICK FLOOR	FIREBRICK FLOOR	FIREBRICK FLOOR	

MANUFACTURER	KRISTIA	MAJESTIC					
PRODUCT	JØTUL # 6	FIREHOOD FFW 38	FIREHOOD FFW 45	AZTEC AZT 32	AZTEC AZT 36	MANCHESTER-PIERCE MPW 34T	MANCHESTER-PIERCE MPW 34R
TYPE	CIRCULAR	TEPEE	TEPEE	TEPEE	TEPEE	RECTANGULAR	RECTANGULAR
OVERALL SIZE	43 1/4"H X 29"W 22"L	38 1/2"H X 23 1/2"W 38"L	50"H X 23 3/4"W 45"L	28"H X 24"W 32"L	33 1/2"H X 24"W 36"L	26 3/4"H X 21 3/4"W 34"L	26 3/4"H X 21 3/4"W 34"L
WEIGHT	319 S.W.						
FIREBOX SIZE				32"D	36"D		
FLUE SIZE	7"	9"	11"-12"	7"	9"	7"	7"
FLUE LOCATION	TOP; BACK	TOP	TOP	TOP	TOP	TOP; BACK	TOP; BACK
BODY MATERIALS				DOUBLE-WALL CONSTRUCTION	DOUBLE-WALL CONSTRUCTION	DOUBLE-WALL CONSTRUCTION	DOUBLE-WALL CONSTRUCTION
BODY FINISH	BLACK	PORCELAIN	MATTE BLACK	PORCELAIN	PORCELAIN	MATTE BLACK	MATTE BLACK
ACCESSORIES	GRATE	GRATE	GRATE	GRATE	GRATE	GRATE	GRATE
FEATURES		SCREEN	SCREEN	SCREEN	SCREEN	SCREEN	SCREEN
		UNIVERSAL ADAPTER	UNIVERSAL ADAPTER	TELESCOPING PIPE	TELESCOPING PIPE	UNIVERSAL ADAPTER	UNIVERSAL ADAPTER
		HEARTH MATERIAL	HEARTH MATERIAL	PIPE FOR 8 FT. CEILING	PIPE FOR 8 FT. CEILING	PIPE FOR 8 FT. CEILING	PIPE FOR 8 FT. CEILING
		PIPE FOR 8 FT. CEILING	PIPE FOR 8 FT. CEILING			STEEL FIREBOX	STEEL FIREBOX
		REFRACTORY FIREBOX	REFRACTORY FIREBOX				

MANUFACTURER	MAJESTIC						
PRODUCT	JUPITER JFW 24	JUPITER JFW 30	JUPITER JFW 38	VENUS VFW 36	ATHENA AFW 39	TUDOR WHW 32	MERCURY MFW 36
TYPE	TEPEE	TEPEE	TEPEE	RECTANGULAR	RECTANGULAR	RECTANGULAR	CIRCULAR
OVERALL SIZE	43 1/4"H X 18 1/4"W 24"L	56"H X 24"W 30"L	77"H X 24"W 38"L	45 3/4"H X 25"W 36"L	45 3/4"H X 28"W 37 1/2"L	81 3/8"H X 29"W 49"L	65 1/2"H X 36"D
WEIGHT							
FIREBOX SIZE							
FLUE SIZE	7"	7"	10"	10"	9"	16"	10"
FLUE LOCATION	TOP	TOP	TOP	TOP	TOP	TOP	TOP
BODY MATERIAL						STAINLESS STEEL	
BODY FINISH	PORCELAIN	MATTE BLACK	MATTE BLACK	PORCELAIN OR MATTE BLACK		PEWTER OR MATTE BLACK	PORCELAIN OR MATTE BLACK
ACCESSORIES	GRATE	GRATE	GRATE	GRATE	GRATE		HANGING INSTALLATION
							PEDESTAL BASE
FEATURES	SCREEN	SCREEN	SCREEN	SCREEN	SCREEN	ADAPTS TO CEILING HEIGHT	HOOD
	UNIVERSAL ADAPTER	UNIVERSAL ADAPTER	UNIVERSAL ADAPTER	UNIVERSAL ADAPTER	UNIVERSAL ADAPTER		ADAPTER
	PIPE FOR 8 FT. CEILING	PIPE FOR 8 FT. CEILING	PIPE FOR 8 FT. CEILING	PIPE FOR 8 FT. CEILING	PIPE FOR 8 FT. CEILING		CIRCLE SCREEN
				READY MIX HEARTH MATERIAL	READY MIX HEARTH MATERIAL		HEARTH PAN
							PIPE FOR 8 FT. CEILING
							REFRACTORY MATERIAL

VI DIVISION

FREE-STANDING FIREPLACES

SPECIFICATIONS

MANUFACTURER	MAJESTIC		MALM FIREPLACES, INC.				
PRODUCT	MERCURY MFW 42	REGENCY RFW 36	CUBE # 75	DIAMOND FIRE WALL MODEL	DIAMOND FIRE CORNER MODEL	IMPERIAL CAROUSEL	ROYAL CAROUSEL
TYPE	CIRCULAR	TEPEE	RECTANGULAR	TEPEE	TEPEE	CIRCULAR	CIRCULAR
OVERALL SIZE	65½"H x 42"D	45½"H x 24⅛"W 36"L	38"H x 25½"W 25½"D	29"H x 30"W 30"D	29"H x 30"W 30"D	54"H x 40"W 40"D	70"H x 32"W 32"D
WEIGHT							
FIREBOX SIZE							
FLUE SIZE	12"	8"	8"	8"	8"	8"	7"
FLUE LOCATION	TOP	TOP	TOP	TOP	TOP	TOP	TOP
BODY MATERIAL			GA. STEEL	GA. STEEL	GA. STEEL	GA. STEEL	GA. STEEL
BODY FINISH	MATTE BLACK	PORCELAIN OR MATTE BLACK	PORCELAIN OR MATTE BLACK	PORCELAIN OR MATTE BLACK	PORCELAIN OR MATTE BLACK		
ACCESSORIES	HANGING INSTALLATION	BASE OR TRIPOD LEGS	PORCELAIN PIPE				
	PEDESTAL BASE						
FEATURES	HOOD	ASH DRAWER		BLACK PIPE	BLACK PIPE	HEAT RESISTANT GLASS FIREBOX	HEAT RESISTANT GLASS FIREBOX
	ADAPTER	SCREENS		FIRE CLAY MATERIAL	FIRE CLAY MATERIAL		
	CIRCLE SCREEN	UNIVERSAL ADAPTER					
	HEARTH PAN	PIPE FOR 8 FT. CEILING					
	PIPE FOR 8 FT. CEILING	FIREBRICK FIREBOX					
	REFRACTORY MATERIAL						

MANUFACTURER	MALM FIREPLACES, INC.						
PRODUCT	LANCER	FIRE DUKE 24" MODEL	FIRE DUKE 30" MODEL	GRAND DUKE 38" MODEL	GRAND DUKE 44" MODEL	ROYAL FRANKLIN	FIRE DUCHESS 36" FREE STAND
TYPE	TEPEE	TEPEE	TEPEE	TEPEE	TEPEE	FRANKLIN	CIRCULAR
OVERALL SIZE	34"H x 30"W 30"D	40"H x 24"W 24"D	50"H x 30"W 30"D	90"H x 38"W 38"D	90"H x 44"W 44"D	34¾"H x 38"W 23"D	51½"H x 36"W 36"D
WEIGHT							
FIREBOX SIZE							
FLUE SIZE	7"	7"	7"	10"	12"	8"	10"
FLUE LOCATION	TOP	TOP	TOP	TOP	TOP	TOP	TOP
BODY MATERIAL	GA. STEEL	GA. STEEL	GA. STEEL	GA. STEEL	GA. STEEL	GA. STEEL	GA. STEEL
BODY FINISH	PORCELAIN EN. OR MATTE BLACK					HEAT RESISTANT PAINT	PORCELAIN ENAMEL
ACCESSORIES	COLORED PIPE					GLASS DOOR	
FEATURES	STRAIGHT WALL-CORNER INSTALL.					STRAIGHT WALL-CORNER INSTALL.	
						DOUBLE-WALL FIREBOX	

MANUFACTURER	MALM FIREPLACES, INC.						
PRODUCT	FIRE DUCHESS 36" SUSPENDED	FIRE DUCHESS 42" FREE STAND.	FIRE DUCHESS 42" SUSPENDED	FIRE DUCHESS 48" FREE STAND.	FIRE DUCHESS 48" SUSPENDED	FIRE BARON	FIRE QUEEN
TYPE	CIRCULAR	CIRCULAR	CIRCULAR	CIRCULAR	CIRCULAR	TEPEE	RECTANGULAR
OVERALL SIZE	54"H x 36"W 36"D	53"H x 42"W 42"D	56"H x 42"W 42"D	56½"H x 48"W 48"D	59½"H x 48"W 48"D	44½"H x 37"W 30¼"D	48"H x 36"W 25"D
WEIGHT							
FIREBOX SIZE							
FLUE SIZE	10"	10"	10"	12"	12"	10"	10"
FLUE LOCATION	TOP	TOP	TOP	TOP	TOP	TOP	TOP
BODY MATERIAL	GA. STEEL	GA. STEEL	GA. STEEL	GA. STEEL	GA. STEEL	GA. STEEL	GA. STEEL
BODY FINISH	PORCELAIN ENAMEL	PORCELAIN ENAMEL	PORCELAIN ENAMEL	PORCELAIN ENAMEL	PORCELAIN ENAMEL	PORCELAIN ENAMEL	PORCELAIN ENAMEL
ACCESSORIES							
FEATURES							

V DIVISION — FREE-STANDING FIREPLACES — SPECIFICATIONS

MANUFACTURER	MALM FIREPLACES, INC.					MARKADE-WINNWOOD	MARTIN IND.
PRODUCT	FIRE JESTER 39" MODEL	FIRE JESTER 47" MODEL	FIRE COUNT	FIRE KNIGHT	BARREL FIREPLACE	WINNWOOD FIREPLACE STOVE #53	FRANKLIN #98-1800
TYPE	RECTANGULAR	RECTANGULAR	CIRCULAR	CIRCULAR	DRUM	RECTANGULAR	FRANKLIN
OVERALL SIZE	48"H x 39"W 28"D	48"H x 47"W 33½"D	44"H x 32"D	27"H x 28"L	24"H x 28"L	45½"H x 24"W 36"L	29½"H x 26"W 37"L
WEIGHT							295 LBS.
FIREBOX SIZE			32"D				24½"W (FRONT) 20"W (B) x 12½"D
FLUE SIZE	10"	10"	8¼"	8"	7"	8"	8"-10"
FLUE LOCATION	TOP	TOP	TOP	TOP	TOP	TOP	TOP
BODY MATERIAL	GA. STEEL	GA. STEEL	GA. STEEL	GA. STEEL	GA. STEEL	16 GA. STEEL	
BODY FINISH	PORCELAIN ENAMEL	PORCELAIN ENAMEL	MATTE BLACK OR PORCELAIN ENAM.				
ACCESSORIES							REDUCERS
							FIRESCREEN
							BARBEQUE GRILL
							ELBOW AND STOVEPIPE
							GAS FIRELOG
							BEANPOT
FEATURES	CORNER INSTALLATION	CORNER INSTALLATION	MOBILE HOME DESIGN	COOKING OPTION	HEAT STOVE OPTION	HEAT STOVE OPTION	
			STORM COLLAR	SLIDING METAL DOORS	COOKING PLATES	FIRESCREEN	
			REFRACTORY FIREBOX	BRASS TRIM	ADJUSTABLE SIDE VENTS		
				PIPE FOR 8 FT. CEILING	GLASS PANEL		
				REFRACTORY FIREBOX	STEEL DOOR		
				HEAT STOVE OPTION	FIREBRICK FLOOR		

MANUFACTURER	MARTIN IND.	PORTLAND STOVE FOUNDRY CO.					
PRODUCT	FRANKLIN #98-1880	ANTIQUE '76	EAGLE '42	CHALET #7 STOVE	1812 FRANKLIN	CONSTITUTION #1	CONSTITUTION #2
TYPE	FRANKLIN	FRANKLIN	FRANKLIN	CIRCULAR	FRANKLIN	FRANKLIN	FRANKLIN
OVERALL SIZE	30½"H x 26½"W 42½"L	31½"H x 35"W	32½"H x 39⅝"W	39"H	27 3/16"H x 30¾"W	26½"H x 31⅜"W	30½"H x 35⅛"W
WEIGHT	350 LBS.	300 S.W.	290 S.W.	152 LBS.	200 S.W.	185 S.W.	230 S.W.
FIREBOX SIZE	31"W(F) x 27½"W 14"D	10¼"D x 18⅝"W (BACK)	14¼"D x 21⅜"W (BACK)	15¼" DIAM. 3" DEPTH	9¾"D 17¾"W (BACK)	10½"D 15"W (BACK)	12½"D 18⅛"W (BACK)
FLUE SIZE	8"-10"	8"	8"	8"	7"	7"	8"
FLUE LOCATION	TOP	TOP; BACK	TOP; BACK	TOP	TOP; BACK	TOP; BACK	TOP; BACK
BODY MATERIAL		CAST IRON					
BODY FINISH							
ACCESSORIES	REDUCERS	ASHPAN	DOORS		GRATE AND STEAK GRILL	ASHPAN	ASHPAN
	FIRESCREEN		ASHPAN		ASHPAN		
	BARBEQUE GRILL						
	ELBOW AND STOVEPIPE						
	GAS FIRELOG						
	BEANPOT						
FEATURES				DO-IT-YOURSELF ASSEMBLAGE			
				CHARCOAL BROILER			
				COOKING OPTION			

VI DIVISION

FREE-STANDING FIREPLACES

SPECIFICATIONS

MANUFACTURER	PORTLAND FOUNDRY CO.	PREWAY INC.			TEMPCO		
PRODUCT	CONSTITUTION #3	CAPRI C30	MODERNE C38	CONTINENTAL FB24	THE HUNTER	THE FORESTER	THE WOODSMAN
TYPE	FRANKLIN	TEPEE	TEPEE	RECTANGULAR	TEPEE	TEPEE	CIRCULAR
OVERALL SIZE	32½"H×39⅝"W	41"H×23¼"W 30⅝"L	41½"H×28¾"W 38"L	41¼"H×28⁹⁄₁₆" 42¼"L	38"H×38"L	33⅜"H×29"L	29¼"H×29"L
WEIGHT	290 S.W.	95 S.W.	113 S.W.		211 S.W.	105 S.W.	155 S.W.
FIREBOX SIZE	14¼"D 21⅜"W(BACK)						
FLUE SIZE	8"	8"	8"	8"	9"	7"	8"
FLUE LOCATION	TOP; BACK	TOP	TOP	TOP	TOP	TOP	TOP
BODY MATERIAL			GA. STEEL				
BODY FINISH		HI-TEMP PORCELAIN	HI-TEMP PORCELAIN	HI-TEMP PORCELAIN	ENAMEL/PAINT	ENAMEL/PAINT	ENAMEL/PAINT
ACCESSORIES	ASHPAN	GRATE	GRATE	GRATE			
			HEATSHIELD KIT				
		EXTENSION PIPE COLLAR, ELBOW	EXTENSION PIPE, COLLAR, ELBOW	EXTENSION PIPE COLLAR, ELBOW			
		SIMPLIFIED CHIMNEY SYSTEM	SIMPLIFIED CHIMNEY SYSTEM	SIMPLIFIED CHIMNEY SYSTEM			
		ELECTRIC LOG	ELECTRIC LOG	ELECTRIC LOG			
FEATURES		FIRESCREEN	FIRESCREEN	FIRESCREEN	WOOD; GAS ELECTRIC LOGS	WOOD; GAS ELECTRIC LOGS	DRAWSCREEN
		PIPE FOR 8 FT. CEILING	PIPE FOR 8 FT. CEILING	PIPE FOR 9 FT. CEILING		LIFT-OFF SCREEN	
		CONVERTA-SKIRT	CONVERTA-SKIRT	ASH GUARD			
		DAMPER	DAMPER	DAMPER			
		WOOD; GAS ELECTRIC LOGS	WOOD; GAS ELECTRIC LOGS	WOOD; GAS ELECTRIC LOGS			

MANUFACTURER	TEMPCO	TORRID AIR	U.S. STOVE CO.		WASHINGTON STOVE WORKS		
PRODUCT	THE COLONIAL	CARIBE	FRANKLIN #261-ST	FRANKLIN #301-ST	OLYMPIC FRANKLIN #18	OLYMPIC FRANKLIN #22	OLYMPIC FRANKLIN #26
TYPE	RECTANGULAR	RECTANGULAR	FRANKLIN	FRANKLIN	FRANKLIN	FRANKLIN	FRANKLIN
OVERALL SIZE	78"H×19⁹⁄₁₆"W 32"L		31½"H×38"L	31½"H×43"L	25¼"H×20½" 30"L	29⅛"H×23"W 33¾"L	31¼"H×25"W 38⅛"L
WEIGHT	264 S.W.		214 LBS.	250 LBS.	175 LBS.	225 LBS.	290 LBS.
FIREBOX SIZE			20"W(BACK) 13¼"D	24⅝"W(BACK) 13¼"D			
FLUE SIZE	5"		8"	8"	7"	8"	8"
FLUE LOCATION	TOP				TOP; BACK	TOP; BACK	TOP; BACK
BODY MATERIAL	ENAMEL/PAINT		CAST IRON	CAST IRON	CAST IRON	CAST IRON	CAST IRON
BODY FINISH							
ACCESSORIES		HEATSAVER	REDUCER; ELBOW STOVEPIPE; COLLAR	REDUCER; ELBOW STOVEPIPE; COLLAR	SPARK GUARD	SPARK GUARD	SPARK GUARD
			GRATE BASKET	GRATE BASKET	LOG HOLDERS	LOG HOLDERS	LOG HOLDERS
			CONNECTOR ADAPTER	CONNECTOR ADAPTER	BRASS MANTLE RAIL; FINIAL; KNOB	BRASS MANTLE RAIL; FINIAL; KNOB	BRASS MANTLE RAIL; FINIAL; KNOB
			BASKET FIRESCREEN	BASKET FIRESCREEN	"CODE KIT"	"CODE KIT"	"CODE KIT"
			BARBEQUE GRILL	BARBEQUE GRILL	ASH DRAWER	ASH DRAWER	ASH DRAWER
			BEAN POT	BEAN POT		BARBEQUE	BARBEQUE
			HEATSHIELD	HEATSHIELD		3 QUART COOKING POT	3 QUART COOKING POT
			BRASS KNOBS AND PLAQUE	BRASS KNOBS AND PLAQUE			BRASS HEARTH RAIL
FEATURES	WOOD; GAS LOGS	FORCED AIR HEAT EXCHANGER	WOOD; COAL; GAS CHARCOAL LOGS	WOOD; COAL; GAS CHARCOAL LOGS			
		ADAPTABLE TO GAS	BUILT-IN DAMPER CONTROL	BUILT-IN DAMPER CONTROL			
		SMALLER MODEL					

V DIVISION	MANUFACTURER	WASHINGTON STOVE WORKS						
	PRODUCT	OLYMPIC FRANKLIN #30	OLYMPIC FRANKLIN #34	ZODIAC	DRUMMER	FRANKLIN 44-ER		
FREE-STANDING FIREPLACES SPECIFICATIONS	TYPE	FRANKLIN	FRANKLIN	CIRCULAR	DRUM	FRANKLIN		
	OVERALL SIZE	32½"H x 26½"W 42¼"L	33½"H x 26"W 45½"L	38"H x 28"W	33"H x 22"L 23"(DIAMETER)	31¼"H x 25"D 38"L		
	WEIGHT	355 LBS.	430 LBS.		120 LBS.	295 S.W.		
	FIREBOX SIZE					15½"D		
	FLUE SIZE	10"	10"	8" OR 10"	7"	8"		
	FLUE LOCATION	TOP; BACK	TOP; BACK	TOP	TOP	TOP; BACK		
	BODY MATERIAL	CAST IRON	CAST IRON	CAST IRON SHEET METAL		CAST IRON		
	BODY FINISH							
	ACCESSORIES	SPARKGUARD	SPARKGUARD	"PIPE KIT"		SPARK GUARD		
		LOG HOLDERS	LOG HOLDERS			LOG HOLDERS		
		BRASS MANTLE RAIL; FINIAL; KNOB	BRASS MANTLE RAIL; FINIAL; KNOB			BRASS MANTLE RAIL; FINIAL; KNOB		
		"CODE KIT"	"CODE KIT"			"CODE KIT"		
		ASH DRAWER	ASH DRAWER			BARBEQUE		
		BARBEQUE	BARBEQUE			3 QUART COOKING POT		
		3 QUART COOKING POT	3 QUART COOKING POT			BRASS HEARTH RAIL		
		BRASS HEARTH RAIL	BRASS HEARTH RAIL					
	FEATURES				ASHPAN			
					INSULATING BACK SHIELD			

Division VI Pre-Fabricated Fireplaces

EDITOR'S NOTE

Pre-Fabricated Fireplaces are defined as any product whose operation requires it to be built into the structure. The product, rather than being free-standing and independent of the building, must be installed as a component of the building.

*indicates optional at extra cost

VI DIVISION	MANUFACTURER	HOME FIREPLACES, LTD.		K.N.T., INC.	MAJESTIC			
	PRODUCT	NORTH. HEATLINER MODEL #33	NORTH. HEATLINER MODEL #37	MARK X	MODEL L28B	MODEL L36B	MODEL L42B	MODEL L36BL; L36BR
PRE-FABRICATED FIREPLACES SPECIFICATIONS	FRONT OPENING	25"H x 33"L	25"H x 37"L		22"H x 28"L	24"H x 36"L	27"H x 42"L	24"H x 36"L
	OVERALL SIZE	38"H x 22"W 40"L	38"H x 22"W 44"L		48"H x 24½"W 38"L	52"H x 24½"W 46"L	56"H x 24½"W 55"L	52"H x 41"W 24½"D
	WEIGHT	265 LBS.	285 LBS.		190 S.W.	270 S.W.	336 S.W.	260 S.W.
	FLUE SIZE	6"	6"		18"	18"	21 3/16"	18"
	FIREBOX MATERIAL	DOUBLE-WALL	DOUBLE-WALL		FIREBRICK	FIREBRICK	FIREBRICK	FIREBRICK
	CLEARANCE				0	0	0	0
	AIR CIRCULATOR	YES	YES	YES				
	FAN	* YES	* YES	YES				
	GRATE	* YES	* YES		* YES	* YES	* YES	* YES
	FIREBOX FLOOR	FIREBRICK	FIREBRICK	FIREBRICK				
	ACCESSORIES	FIREPLACE BARBEQUE	FIREPLACE BARBEQUE		FIRE B-Q	FIRE B-Q	FIRE B-Q	FIRE B-Q
					CHIMNEY COMPONENTS	CHIMNEY COMPONENTS	CHIMNEY COMPONENTS	CHIMNEY COMPONENTS
	FEATURES	COAL OR WOOD	COAL OR WOOD	COAL OR WOOD	SCREENS	SCREENS	SCREENS	SCREENS
				GLASS DOORS	FIRESTOP	FIRESTOP	FIRESTOP	FIRESTOP
					CORNER INSTALLATION	STARTER SECTION	STARTER SECTION	STARTER SECTION
					STARTER SECTION			

VI DIVISION

PRE-FABRICATED FIREPLACES SPECIFICATIONS

MANUFACTURER	MAJESTIC						
PRODUCT	MODEL L42BL, L42BR	CIRCULATOR FIREPLACE R2800	CIRCULATOR FIREPLACE R3200	CIRCULATOR FIREPLACE R3600	CIRCULATOR FIREPLACE R4000	CIRCULATOR FIREPLACE R4600	CIRCULATOR FIREPLACE R5400
FRONT OPENING	27"H x 42"L	22"H x 28"W	24"H x 32"W	25"H x 36"W	27"H x 40"W	29"H x 46"W	31"H x 54"W
OVERALL SIZE	56"H x 47 1/2"W 24 1/2"D	44"H x 34 3/4"W 18"D	48"H x 38 1/2"W 19"D	51"H x 42 1/2"W 20"D	55"H x 47 3/4"W 21"D	60"H x 55"W 22"D	65"H x 63"W 23"D
WEIGHT	345 S.W.	156 S.W.	181 S.W.	218 S.W.	252 S.W.	303 S.W.	360 S.W.
FLUE SIZE	21"	8 1/2" x 13"	8 1/2" x 13"	13" x 13"	13" x 13"	13" x 13"	13" x 18"
FIREBOX MATERIALS	FIREBRICK						
CLEARANCE	0						
AIR CIRCULATOR		YES	YES	YES	YES	YES	YES
FAN		* YES	* YES	* YES	* YES	* YES	* YES
GRATE	* YES	* YES	* YES	* YES	* YES	* YES	* YES
FIREBOX FLOOR							
ACCESSORIES	FIRE B-Q	ASH DUMPS	ASH DUMPS	ASH DUMPS	ASH DUMPS	ASH DUMPS	ASH DUMPS
	CHIMNEY COMPONENTS	CIRCULATOR GRILLES	CIRCULATOR GRILLES	CIRCULATOR GRILLES	CIRCULATOR GRILLES	CIRCULATOR GRILLES	CIRCULATOR GRILLES
		FIRE-B-Q	FIRE-B-Q	FIRE-B-Q	FIRE B-Q	FIRE-B-Q	FIRE-B-Q
		CORNER SUPPORT POSTS	CORNER SUPPORT POSTS	CORNER SUPPORT POSTS	CORNER SUPPORT POSTS	CORNER SUPPORT POSTS	CORNER SUPPORT POSTS
FEATURES	FIRESTOP	FIREBOX DAMPER	FIREBOX DAMPER	FIREBOX DAMPER	FIREBOX DAMPER	FIREBOX DAMPER	FIREBOX DAMPER
	STARTER SECTION	SMOKE CHAMBER	SMOKE CHAMBER	SMOKE CHAMBER	SMOKE CHAMBER	SMOKE CHAMBER	SMOKE CHAMBER
	SCREENS						

MANUFACTURER	MALM FIREPLACES	METAL BLDG. PRODUCTS, INC.	PREWAY INC.				
PRODUCT	WOODSIDE ZERO CLEARANCE FIRE.	FIREPLACE CIRCULATOR	WOODBURNING BUILT-INS B128	WOODBURNING BUILT-INS B136	WOODBURNING BUILT-INS B142	WOODBURNING BUILT-INS B142R	WOODBURNING BUILT-INS B142L
FRONT OPENING			16"H x 28"L	18"H x 36"L	20"H x 42"L	16 3/4"H x 42"L	16 3/4"H x 42"L
OVERALL SIZE			40 3/8"H x 24"W 38"L	42 3/8"H x 24"W 46"L	44 3/8"H x 24"W 52"L	44 3/8"H x 24"W 47 1/4"L	44 3/8"H x 24"W 47 1/4"L
WEIGHT							
FLUE SIZE							
FIREBOX MATERIALS			FIREBRICK	FIREBRICK	FIREBRICK	FIREBRICK	FIREBRICK
CLEARANCE	0		0	0	0	0	0
AIR CIRCULATOR	YES		YES	YES	YES	YES	YES
FAN							
GRATE			* YES	* YES	* YES	* YES	* YES
FIREBOX FLOOR							
ACCESSORIES			GLASS DOORS	GLASS DOORS	GLASS DOORS	GLASS DOORS	GLASS DOORS
FEATURES	SHUT-OFF DAMPER		CORNER/WALL INSTALLATION	CORNER/WALL INSTALLATION	CORNER/WALL INSTALLATION	CORNER/WALL INSTALLATION	CORNER/WALL INSTALLATION
	TWIST LOCK CHIMNEY						

MANUFACTURER	PREWAY INC.	RIDGWAY STEEL FABRICATORS, INC.			TEMCO		
PRODUCT	VALLEY FORGE B128-MH	HYDROPLACE #34	HYDROPLACE #38	HYDROPLACE #42	BUILT-IN FIRE. TBF-28-1	BUILT-IN FIRE. TBF-36-1	BUILT-IN FIRE. TBF-42-1
FRONT OPENING	28"W	27"H x 34"L	30"H x 37 3/8"L	32 1/2"H x 42"L	24"H x 28"L	24"H x 36"L	24"H x 42"L
OVERALL SIZE	37 1/4"H x 24"W 38"L	35 3/4"H x 19 1/2"W 36 1/8"L	37 3/4"H x 20 3/8"W 39 1/2"L	40 3/4"H x 22 3/8"W 44 1/8"L	44 1/2"H x 25"W 38"L	55"H x 25"W 46"L	58 1/2"H x 26"W 52"L
WEIGHT	350 LBS.	464 LBS.	500 LBS.	520 LBS.	187 S.W.	210 S.W.	317 S.W.
FLUE SIZE					15"	15"	15"
FIREBOX MATERIALS	FIREBRICK	STEEL	STEEL	STEEL			REFRACTORY LINED
CLEARANCE	0				0	0	0
AIR CIRCULATOR	YES	YES	YES	YES			
FAN		YES	YES	YES			
GRATE	* YES	YES	YES	YES	* YES	* YES	* YES
FIREBOX FLOOR		STEEL	STEEL	STEEL			REFRACTORY LINED
ACCESSORIES	FIREPLACE SCREEN				MESH FIRESCREEN	MESH FIRESCREEN	MESH FIRESCREEN
	ELECTRIC LOG				HEARTH EXTENSION	HEARTH EXTENSION	HEARTH EXTENSION
FEATURES	DESIGNED FOR MOBILE HOMES	NON-CORROSIVE	NON-CORROSIVE	NON-CORROSIVE	INSTALLATION FLEXIBILITY	INSTALLATION FLEXIBILITY	INSTALLATION FLEXIBILITY
	AIR INTAKE DUCT AND GRILL	MASONRY INSTALLATION	MASONRY INSTALLATION	MASONRY INSTALLATION			
	GLASS DOORS	ASH-PIT OPENING	ASH-PIT OPENING	ASH-PIT OPENING			
	CHIMNEY COMPONENTS	DOMESTIC WATER HEATER	DOMESTIC WATER HEATER	DOMESTIC WATER HEATER			
	SPARK ARRESTER	WATER FILLED STEEL FLOOR	WATER FILLED STEEL FLOOR	WATER FILLED STEEL FLOOR			
		WATER IN STEEL JACKET/GRATE	WATER IN STEEL JACKET/GRATE	WATER IN STEEL JACKET/GRATE			

EDITOR'S NOTE

Fireplace Accessories are defined as any product tht can be used with and integrated into the operation of a fireplace.

VII DIVISION — FIREPLACE ACCESSORIES SPECIFICATIONS

MANUFACTURER	AGLOW HEAT-X-CHANGER, INC.		AMERICAN STOVALATOR	C + D DISTRIBUTOR	CHIM-A-LATOR		
PRODUCT	HEAT-X-CHANGER	HEAT SINK	STOVALATOR	BETTER N' BEN'S	CHIM-A-LATOR	AIR-A-LATOR TYPE "NH"	AIR-A-LATOR TYPE "AO"
OVERALL SIZE	10"H LENGTH VARIES		THREE BASIC SIZES		SIZE VARIES	2 5/8"H x 4 3/8"W 17 3/4"L	2 1/2"H x 18"D 16"L
BODY MATERIAL	3/16" PLATE STEEL	STEEL PLATE CHAMBER	STEEL; TEMPER-GLASS DOORS	BLACK IRON	RUSTPROOF HIGH TEMP. MATERIAL		
ACCESSORIES			SPECIAL SIZES CUSTOM MADE	SCREEN DOOR	SIDE MOUNT MASONRY		
				FREE STANDING ADAPTER KIT	CLOSURE HOOD KIT		
				UNIVERSAL LOCKING BAR	EXTENSION BRACKET		
					FLEXIBLE CONDUIT		
					ASBESTOS SHINER ROW		
FEATURES	16,000 BTU		REFRACTORY CEMENT	COOKING OPTION	ADAPTS TO PRE-FAB./FREE-ST.	FOR NEW CONSTRUCTION	FOR EXISTING FIREPLACES
	FAN		SIMPLE CONVECTION SYSTEM	DAMPER/DRAFT CONTROL			
	ADAPTS TO GLASS ENCLOSURE			BAFFLES			
	ADAPTS TO WIRE CURTAINS						

MANUFACTURER	DE FORGE IND.	GOLDEN ENTERPRISE			LANCE INTERNATIONAL	LASSY TOOLS, INC.	
PRODUCT	VT. FIREPLACE HEATER	THERMAL GRATE SIX TUBE	THERMAL GRATE SEVEN TUBE	THERMAL GRATE EIGHT TUBE	FIREPLACE HEATING GRATE	"HEAT-CATCHER" STYLE E # E6	"HEAT-CATCHER" STYLE E # ER6
OVERALL SIZE	21"H x 21"W 18"D	25 1/2"H x 19 1/4"W 24"L	SIZE VARIES	SIZE VARIES	22"H x 21 1/2"W 18"D	25"H x 15 1/16"W 17 1/8"W(F) x 19 1/2"D	25"H x 15 1/16"W(B) 17 1/8"W(F) x 19 1/2"D
BODY MATERIAL	DOUBLE WALL CONSTRUCTION	GA. STEEL	GA. STEEL	GA. STEEL	STEEL	16 GA. STEEL	16 GA. STEEL
ACCESSORIES	ELBOWS	FAN	FAN	FAN	EXTENSION TUBE SET	ADJUSTABLE EXTENSION SLEEVE	EMERGENCY BLOWER
	EXTENSIONS				EMERGENCY DC ACCESSORY UNIT	SPEED CONTROL	
	CUSTOM BUILT MODELS					LOG RETAINER	
						1/2" RAILS (HEARTH)	
						EMERGENCY BLOWER	
FEATURES	ADJUSTABLE IN DEPTH/HEIGHT	2" DIAMETER TUBING	2" DIAMETER TUBING	2" DIAMETER TUBING	FAN	6 HEAT TUBES	SPEED CONTROL
	25,000 BTU	CUSTOM-BUILT	CUSTOM-BUILT	CUSTOM-BUILT		FAN	AIR FILTER
	FAN	FIRE-RETARDANT PAINT FINISH	FIRE-RETARDANT PAINT FINISH	FIRE-RETARDANT PAINT FINISH			CASE DECORATION
							LOG RETAINER
							ADJUSTABLE EXTENSION SLEEVE
							6 HEAT TUBES
							FAN

MANUFACTURER	LASSY TOOLS, INC.						
PRODUCT	"HEAT-CATCHER" STYLE E #E8	"HEAT-CATCHER" STYLE E # ER8	"HEAT-CATCHER" STYLE S #S18	"HEAT-CATCHER" STYLE B #36B	"HEAT-CATCHER" STYLE B #3681	"HEAT-CATCHER" STYLE B #42B	"HEAT-CATCHER" STYLE B #48B
OVERALL SIZE	25"H x 21 3/16"W(B) 23 3/8"W(F) x 19 1/2"D	25"H x 21 3/16"W(B) 23 3/8"W(F) x 19 1/2"D	28"H x 33"W(B) 35"W(F) x 21"D	25 1/2"H x 24"W(B) 30"W(F) x 14 1/2"D	21 1/2"H x 24"W(B) 30"W(F) x 14 1/2"D	25 1/2"H x 28"W(B) 34"W(F) x 14 1/2"D	25 1/2"H x 32"W(B) 38"W(F) x 14 1/2"D
BODY MATERIAL	16 GA. STEEL	16 GA. STEEL	16 GA. STEEL	16 GA. STEEL	16 GA. STEEL	16 GA. STEEL	16 GA. STEEL
ACCESSORIES	EMERGENCY BLOWER	EMERGENCY BLOWER	EMERGENCY BLOWER	EMERGENCY BLOWER	EMERGENCY BLOWER	EMERGENCY BLOWER	EMERGENCY BLOWER
	ADJUSTABLE EXTENSION SLEEVE		DUCTS FOR DEEP FIREPLACES				
	SPEED CONTROL						
	LOG RETAINER						
	1/2" RAILS (HEARTH)						
FEATURES	8 HEAT TUBES	8 HEAT TUBES	TWIN FORCED DRAFT CONTROL	9 HEAT TUBES	9 HEAT TUBES	9 HEAT TUBES	9 HEAT TUBES
	FAN	SPEED CONTROL	18 HEAT TUBES	FAN	FAN	FAN	FAN
		AIR FILTERS	STAGGERED CONFIGURATION				
		CASE DECORATION	FAN				
		LOG RETAINER					
		ADJUSTABLE EXTENSION SLEEVE					
		FAN					

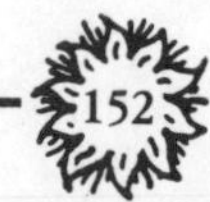

VII DIVISION — FIREPLACE ACCESSORIES SPECIFICATIONS

MANUFACTURER	LASSY TOOLS, INC.	MARKADE-WINNWOOD	RIDGWAY STEEL FABRICATOR, INC.				SHENANDOAH MFG., CO.
PRODUCT	"HEAT-CATCHER STYLE B #36BM	FIREPLACE ECONOMIZER #63	HYDROHEARTH #HA-17	HYDROHEARTH #HA-22	HYDROHEARTH #HA-28	HYDROGRILLE	FIRE GRATE FG-6
OVERALL SIZE	27"H X 22"W(B) 31"W(F) X 14½"D	30"H X 26"W 36"L	25"H X 15½"D 17"W	25"H X 15½"D 22"	25"H X 15½"D 28"W		22"H X 18½"D 28"W(F) X 20"W(B)
BODY MATERIAL	16 GA. STEEL		STEEL	STEEL	STEEL		
ACCESSORIES	EMERGENCY BLOWER	FAN					SPEED CONTROL
	DUCTS FOR DEEP FIREPLACES						ON/OFF SWITCH
FEATURES	9 HEAT TUBES	BI-FOLD DOOR SYSTEM	H_2O CIRCULATING GRATE	H_2O CIRCULATING GRATE	H_2O CIRCULATING GRATE	H_2O CIRCULATING BARBEQUE GRILL	6 TUBES
	FOR MAJESTIC PRE-FAB. FIRE.		CONNECTS EXIST. HEAT SYSTEM	CONNECTS EXIST. HEAT SYSTEM	CONNECTS EXIST. HEAT SYSTEM	H_2O CIRCULATING SWIM. POOL HEATER	50,000 BTU
	FAN						FAN

MANUFACTURER	SHENANDOAH MFG., CO.	STURGES HEAT RECOVERY, INC.			THERMOGRATE ENTERPRISES	VERMONT TECHNIQUES, INC.	
PRODUCT	FIRE GRATE FG-7	THRIFTCHANGER CHIMNEY FUMES	THRIFTCHANGER DIVERSION DAMPER	THRIFTCHANGER HEAT RECOVERY	THERMOGRATE	FIREPLACE STOVE	FIREPLACE STOVE
OVERALL SIZE	24"H X 20"D 31"W(F) X 23"W(B)	SIZE VARIES	SIZE VARIES		SIZE VARIES	31"H X 40"W	35"H X 44"W
BODY MATERIAL		METAL	METAL	REFRACTORY BRICK/GLASS	GA. STEEL	⅛" STEEL PLATE	⅛" STEEL PLATE
ACCESSORIES	SPEED CONTROL				FAN	MESH SCREEN	MESH SCREEN
	ON/OFF SWITCH				EXTENSION TUBES	HEATILATOR ADDITION	HEATILATOR ADDITION
					ELBOW KIT	WARMING OVEN	WARMING OVEN
						SPECIAL FITTINGS	SPECIAL FITTINGS
FEATURES	7 TUBES	WOOD BURNS HOT AND CLEAN	FUMES UP W/OPEN FIRE	WOOD BURNS SMOKELESS	AVAILABLE IN 28 SIZES		
	50,000 BTU	ELIMINATES SMOKE/CREOSOTE	DIVERTS FUMES THRU EXCHANGER		BURNS DIFFERENT FUEL		
	FAN	ELIMINATES AIR POLLUTION			CONVECTION PRINCIPLE		
		64 TUBES					

Division VIII Hot Water Heaters

EDITOR'S NOTE

Hot Water Heaters are defined as any product that can be added onto a combustion system to heat water.

VIII DIVISION — HOT WATER HEATERS SPECIFICATIONS

MANUFACTURER	BLAZING SHOWERS						
PRODUCT	STOVEPIPE HOT WATER HEATER	FIREBOX HOT WATER HEATER					
SIZE OF PIPE	15 FT.	12 FT.					
WEIGHT	11 LBS.	9 LBS.					
COIL MATERIAL	COPPER PIPE	COPPER PIPE					
GALS/HR	20 GAL/HR						
FEATURES	STORAGE TANK ADAPTER	DRUM TYPE STOVE DESIGN					
	BOOKLET	STORAGE TANK ADAPTER					
		FIREBOX HEATING UNIT					
		BOOKLET					

Division IX — Heat Reclaimers

EDITOR'S NOTE

Heat Reclaimers are defined as any product that can be added onto an existing combustion system to extract additional heat that would normally be lost in the discharge of flue gases.

*indicates optional at extra cost

IX DIVISION — SPECIFICATIONS / HEAT RECLAIMERS

MANUFACTURER	ISOTHERMICS	LANCE INTERNATIONAL	STURGES HEAT RECOVERY INC.	THERMAL RECLAMATION	TORRID MFG., CO., INC.	
PRODUCT	AIR-O-SPACE BTU-4K; BTU-5K	HEAT RECLAIMER	THRIFTCHANGER STACK FUMES	THERMOSAVER AIR-TO-AIR TR-1W	AIR HEATSAVER EA-60; RK-70	AIR HEAT-SAVER FA-9; FA-10
OVERALL SIZE		11"H X 17"W 18"L	10¾"H X 40"L	12"H X 10"W 12"L	18"H X 10"W 13"D	24"H X 11⅝"W 15½"D
WEIGHT	15 LBS.	32 S.W.		11 LBS.		
FLUE SIZE	FITS MOST FLUES	FITS MOST FLUES		5"-8"	6"-8"	9"-10"
BODY MATERIAL		22 GA. STEEL	METAL	26 GA. STEEL	20 GA. STEEL	20 GA. STEEL
THERMOSTAT	YES	YES			YES	YES
HEAT EXCHANGER	HEAT PIPES	3"D GA. STEEL 6 TUBES	64 STEEL TUBES	3½"D 4 TUBES	14 TUBES	14 TUBES
FAN (CFM)	100 CFM	150 CFM		300 CFM	210 CFM	210 CFM
BTU/HR	5,000-10,000				19,000	19,000
FEATURES	HORIZ./VERTICAL INSTALLATION		WORKS BY CONVECTION	HEAT RESISTANT PAINT	BUILT-IN TUBE CLEANER	BUILT-IN TUBE CLEANER
	STEAM, HOT AIR HYDRONIC SYSTEM		DIFF. MODELS AVAILABLE			DESIGN FOR FREE-ST. FIRE.
			ELIMINATES SMOKE-CREOSOTE			* CUSTOM COLOR MATCH.
			ELIMINATES AIR POLLUTION			

Division X — Barrel Stove Kits

EDITOR'S NOTE

Barrel Stove Kits are defined as any product that can convert a steel barrel to a heating appliance. The term kit refers to the obligation of the consumer to make the conversion.

*indicates drum not included

**indicates optional at extra cost

X DIVISION — BARREL-STOVE KITS / SPECIFICATIONS

MANUFACTURER	FISK STOVE	MARKADE-WINNWOOD					
PRODUCT	FISKSTOVE #3	BARREL STOVE KIT # H-74	BARREL STOVE KIT # H-30	BARREL STOVE KIT # H-55	BARREL STOVE KIT # V-55	BARREL STOVE KIT # V-30	BARREL STOVE KIT # H-31
DRUM SIZE		* 55 GAL.	* 30 GAL.	* 55 GAL.	* 55 GAL.	* 30 GAL.	* 30 GAL.
WEIGHT	DESIGN ONLY	25 S.W.	20 S.W.	25 S.W.	21 S.W.	20 S.W.	20 S.W.
FIREDOOR SIZE		12"x12"	9"x9½"	12"x12"	9"x9"	9"x9"	9"x9½"
PARTS MATERIAL		16 GA. STEEL	16 GA. STEEL	16 GA. STEEL	16 GA. STEEL	16 GA. STEEL	16 GA. STEEL
FLUE SIZE		6"	6"	6"	6"	6"	6"
FEATURES	MODULAR CONSTRUCTION	** DOUBLE-BARREL ADAPTER	** DOUBLE-BARREL ADAPTER	** DOUBLE-BARREL ADAPTER	VERTICAL	VERTICAL	** DOUBLE-BARREL ADAPTER
	AIR-TIGHT	DAMPER	DAMPER	DAMPER			HORIZONTAL
	DOWN/UPDRAFT DESIGN	HORIZONTAL	HORIZONTAL	HORIZONTAL			
	THERMOSTATIC CONTROL		DRAFT CONTROL	DRAFT CONTROL			
	FIREBRICK LINED						

MANUFACTURER	MARKADE-WINNWOOD	PORTLAND STOVE FOUNDRY	WASHINGTON STOVE WORKS	WHOLE EARTH ACCESS CO.
PRODUCT	BARREL STOVE KIT # H-52	DRUM STOVE	OIL DRUM CONVERSION	OIL DRUM CONVERSION KIT
DRUM SIZE	* 55 GAL.	* 55 GAL.	* 50 GAL.	* 55 GAL.
WEIGHT	25 S.W.	45 S.W.		
FIREDOOR SIZE	12"x12"		15"x15" OR 12"x12"	
PARTS MATERIAL	16 GA. STEEL	CAST IRON		CAST IRON
FLUE SIZE	6"	6"	6"	
FEATURES	** DOUBLE-BARREL ADAPTER			DAMPER
	HORIZONTAL			

EDITOR'S NOTE

Wood Splitters are defined as any product that can contribute to the more efficient splitting of fire wood.

XI DIVISION — WOOD SPLITTERS SPECIFICATIONS

MANUFACTURER	ALBRIGHT WELDING CO.	C+D DISTRIBUTOR	GARDEN WAY	HOUSEHOLD WOODSPLITTER	LA FONT CORP.		
PRODUCT	L.O. BALLS WOODSPLITTER	JIFFY WOODSPLITTERS	KNOTTY WOOD LOG SPLITTER	WOODSPLITTER	MICRO-MINI SPLITTER MMS-350	MINI-SPLITTER MS-400	LITTLE SPLITTER LS-500
WOOD LENGTH	26"-48"	30"	UP TO 8 FT.	18"	25"	20"	24"
OVERALL SIZE		36"H; 15"x15" BASEPLATE		33"H X 27"W 58"L			
WEIGHT		50 LBS.	210 S.W.				
BODY MATERIAL		STEEL		STEEL	STEEL/BRASS	STEEL/BRASS	
ENGINE			8 OR 12 H.P.	5 H.P./B+S	8 H.P.	8 H.P.	16 H.P.
POWER SOURCE	HYDRAULIC TAKE-OFF		3 POINT OR UNIVERSAL	HYDRAULIC	HYDRAULIC	HYDRAULIC	HYDRAULIC
RAM CYCLE			10 SEC.	12 SEC.	15 SEC.	12 SEC.	9 SEC.
FEATURES			DIFFERENT DESIGN	3 GAL. HYDRO. OIL RESERVE	5 GAL. HYDRO. OIL RESERVE	SPECIAL LENGTH * AVAILABLE; 7 GAL. HYDRO. OIL RESERVE	EXERTS 15 TONS HYDRAULIC FORCE; 15 GAL. HYDRO. OIL RESERVE; AUTOMATIC CYCLING

MANUFACTURER	LA FONT CORP.			NORTECH CORP.			
PRODUCT	SUPER SPLITTER SS-500	SAW-MATIC SM-500	3-PT. HITCH TRACTOR TS-400	"SCREW WEDGE" #28	"SCREW WEDGE" #39	"SCREW WEDGE" #PTO-1	"SCREW WEDGE" # GT-5
WOOD LENGTH			STANDARD 20" UP TO 8 FT.	UP TO 48"	UP TO 48"	UP TO 48"	UP TO 48"
OVERALL SIZE				24"H X 21"W 48"L	28"H X 26"W 58"L	21"H X 27"W 29"L	
WEIGHT		2,100 LBS.		210 S.W.	280 S.W.	260 S.W.	200 S.W.
BODY MATERIAL	RUGGED TUBULAR FRAME	RUGGED TUBULAR FRAME	STEEL FRAME BRASS SLIDE				
ENGINE	25 H.P. TWIN CYLINDER	25 H.P./2 CYLINDER ONAN		8 H.P./B+S	9 H.P./B+S		
POWER SOURCE	HYDRAULIC	HYDRAULIC	TRACTOR	ENGINE	ENGINE	FARM TRACTORS	GRAVELY TRACTORS
RAM CYCLE	5 SEC.		7-15 SEC.	12 SEC.	12 SEC.	12 SEC.	12 SEC.
FEATURES	EXERTS 18 TONS HYDRAULIC FORCE; AUTOMATIC CYCLING; CHANGEABLE WEDGE SYSTEM; * 16 H.P. ENGINE	SPLITS INTO HALVES/QUARTER; * CUSTOM MANUFACTURING; 15 GAL. HYDRO. OIL RESERVE; EXERTS 15 TONS HYDRAULIC FORCE		PORTABLE	PORTABLE	PORTABLE	PORTABLE

MANUFACTURER	TAOS EQUIPMENT MANUFACTURERS, INC.			
PRODUCT	THE STICKLER #1500	THE STICKLER #1000-U	THE STICKLER #1000-C5	THE STICKLER #1000-C6
WOOD LENGTH				
OVERALL SIZE				
WEIGHT	44 S.W.	41 S.W.	41 S.W.	41 S.W.
BODY MATERIAL	ALLOY STEEL TIP	ALLOY STEEL TIP	ALLOY STEEL TIP	ALLOY STEEL TIP
ENGINE				
POWER SOURCE	POWERED BY REAR AXILE OF CAR OR TRUCK			
RAM CYCLE				
FEATURES	ENGINE SHUT-OFF SWITCH; REPLACEABLE ALLOY TIP	ENGINE SHUT-OFF SWITCH; REPLACEABLE ALLOY TIP	ENGINE SHUT-OFF SWITCH; REPLACEABLE ALLOY TIP	ENGINE SHUT-OFF SWITCH; REPLACEABLE ALLOY TIP
	FOUR-BOLT PATTERN MODELS-FOR PICK-UPS, VANS, FOUR-WHEEL DRIVE, 90% AMERICAN CARS (TRACTORS AVAILABLE AT HIGHER PRICE)			

Reader Registration Card

I have purchased a copy of the *Woodburners Encyclopedia* from
Name ____________________________ City __________ State ____________
I would like to be notified when additional update (addendum) information is published.

Name. . . . .
Address . . . .
City. . . . . State. . . . . Zip. . . . .

Vermont Energy Resources
Box 1 Fiddlers Green
Waitsfield, VT 05673